HYDRAULIC ENGINEERING

PROCEEDINGS OF THE 2012 SREE CONFERENCE ON HYDRAULIC ENGINEERING AND 2ND SREE WORKSHOP ON ENVIRONMENT AND SAFETY ENGINEERING, HONG KONG, 21–22 DECEMBER 2012

Hydraulic Engineering

Editor

Liquan Xie

Department of Hydraulic Engineering, Tongji University, Shanghai, China

CRC Press
Taylor & Francis Group
Boca Raton London New York Leiden

CRC Press is an imprint of the
Taylor & Francis Group, an **informa** business

A BALKEMA BOOK

CRC Press/Balkema is an imprint of the Taylor & Francis Group, an informa business

© 2013 Taylor & Francis Group, London, UK

Typeset by V Publishing Solutions Pvt Ltd., Chennai, India
Printed and bound in Great Britain by CPI Group (UK) Ltd, Croydon, CR0 4YY.

Published by: CRC Press/Balkema
             P.O. Box 11320, 2301 EH Leiden, The Netherlands
             e-mail: Pub.NL@taylorandfrancis.com
             www.crcpress.com – www.taylorandfrancis.com

ISBN: 978-1-138-00043-8 (Hbk)
ISBN: 978-0-203-74448-2 (eBook)

Hydraulic Engineering – Xie (Ed.)
© 2013 Taylor & Francis Group, London, ISBN 978-1-138-00043-8

Table of contents

The 2nd SREE workshop on environment and safety engineering

Hydraulic Engineering – Xie (Ed.)
© 2013 Taylor & Francis Group, London, ISBN 978-1-138-00043-8

Preface

The 2012 SREE Conference on Hydraulic Engineering (CHE 2012) showcases the exciting and challenging developments occurring in the area of hydraulic engineering today, and serve as a major forum for researchers, engineers and manufacturers to share recent advances, discuss problems, and identify challenges associated with engineering applications in the hydraulic engineering. The Second SREE Workshop on Environment and Safety Engineering (WESE 2012) is held in CHE 2012 and hopes to offer researchers an occasion to exchange their experiences about all aspects of environment and safety engineering.

CHE 2012 and WESE 2012 have received more than 140 papers, and 56 technical papers are published in the proceedings. Each of the papers has been peer reviewed by recognized specialists and revised prior to acceptance for publication. This proceedings reviews recent advances in several areas that are important for hydraulic engineering and environmental engineering development. The papers related to hydraulic engineering mainly focus on flood prediction and control, hydropower design and construction technology, water & environment, comprehensive water treatment, urban water supply and drainage. The papers related to environmental issues address on environmental prediction and control techniques in environmental geoscience, environmental ecology, atmospheric sciences, ocean engineering, safety engineering and environmental pollution control.

Last but not least, we would like to express our deep gratitude to all authors, reviewers for their excellent work, and Léon Bijnsdorp, Lukas Goosen and other editors from Taylor & Francis Group for their wonderful work.

Hydraulic Engineering – Xie (Ed.)
© 2013 Taylor & Francis Group, London, ISBN 978-1-138-00043-8

Sponsor

Sponsored by *Society for Resources, Environment and Engineering*

2012 SREE conference on hydraulic engineering

Hydraulic Engineering – Xie (Ed.)
© 2013 Taylor & Francis Group, London, ISBN 978-1-138-00043-8

Evaluation and selection on multi-dimensional regulation schemes for water cycle system: A case study of *Haihe River* basin

Kai Xu, Lin Wang, Zhiguo Gan & Haiping Ji
State Key Laboratory of Simulation and Regulation of Water Cycle in River Basin,
China Institute of Water Resources and Hydropower Research, Beijing, China

ABSTRACT: Multi-dimensional regulation is an effective means to achieve sustainable development of water resources in water-deficiency basins. Due to the complicacy of evaluation and selection on multi-dimensional regulation schemes, it is necessary to establish a practical method to evaluate multi-dimensional regulation schemes and to identify more effectual scheme. In this study, based on entropy theory and synergetic theory, control variables (order parameters) are selected as the evaluation indices for each subsystem. The functions of coordination degree for single schemes and integrated distance for combined schemes are set to quantitatively evaluate the effect of single and combined multi-dimensional regulation schemes. *Haihe River* Basin is taken as a case to demonstrate the applicability of the method.

1 INTRODUCTION

With the economic-social development, water cycle system shows dualistic (natural-social) characteristics (Liu *et al.* 2010). The high-intensity human activities cause changes to water cycle process, deriving society, economy and environment subsystems (social characteristic), and lead to resource subsystem weakened and ecological subsystems (natural characteristic) degenerated (Cao *et al.* 2012). To guarantee economic-social sustainable development, it is required to explore the efficient mode of water resources development (Wang *et al.* 2010). The multi-dimensional regulation of water cycle system is an effective means to promote the synergetic development of five-dimension (resource, economy, society, ecology, and environment) subsystems (Wang *et al.* 2003). To optimize water regulation, evaluation and selection on regulation schemes are required.

There are a large number of studies about evaluation and selection methods on regulation schemes. The fuzzy theory was applied to water resources redistribution scheme evaluation and decision-making in Tseng-Wen and Kao- Ping River Basins by Chen and Chang (2010). Srdjevic *et al.* (2004) combined basin simulation and evaluation technology of regulation schemes as an improved TOPSIS method based on entropy for evaluation on Jacuipe River Basin regulation schemes. Locus *et al.* (2007) incorporated hydrological model, reservoir operation model and basin water demand prediction model, and took the deficiency amount of basin water resources as an index to evaluate water resource management measures in Pinios River and Lake Karla Basin. *GA* was used to guide the restoration measures of water pollution, so as to complete the optimal selection on restoration measures (Mohammed and Larry 1993). Multiple decision analysis theory based on Analytic Hierarchy Process (AHP) was used to evaluate water supply schemes in Kwara State, Nigeria by Okeola and Sule (2012). Based on entropy theory, order degree entropy function was set to evaluate and select multi-dimensional regulation schemes in *Haihe River Basin* (Gan *et al.* 2010).

The methods mentioned aimed at the evaluation and selection of schemes in one level year series. Recommended scheme in each level year according to time series are combined as the best combination regulation scheme. Studies have shown that the result of such a simple combination may not be optimal. This study is aimed at developing a new method based on entropy theory and synergetic theory, which use coordination degree of water cycle system and integrated distance of order degree as indices for the schemes in one level year and combination schemes in a series of level years respectively. The evaluation and selection method was extended to combination regulation schemes. *Haihe River Basin* was taken as an example to demonstrate the applicability of the method.

2 THEORIES

2.1 *Entropy theory*

With the growing water use in a basin, the competitive relationship among five dimensionals becomes acute. If a regulation scheme leads to one certain subsystem orderly, another may out-of-order, and the change of the water cycle system is unclear. With entropy theory, the relationship between order degree of subsystems and that of water cycle system is established for multi-dimensional regulation schemes evaluation and selection.

2.2 *Synergetic theory*

The objective of multi-dimensional regulation is to optimize water allocation among five subsystems and achieve the orderly evolution of the water cycle system. It is feasible to apply the synergetic theory to multi-dimensional regulation. The coordination degree is introduced to quantify the coordination among subsystems. The integrated distance of order degree was used to reflect the windage after the combined regulation.

3 METHODOLOGIES

Figure 1 shows the progress for evaluation and selection on schemes.

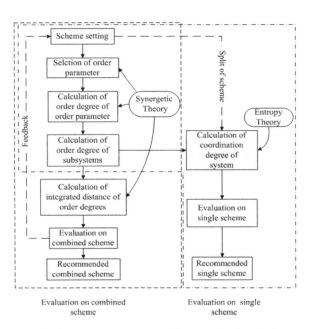

Figure 1. Progress of evaluation and selection on multi-dimensional regulation schemes.

3.1 Order degree function

Due to the limits of water resources, the order parameter of each subsystem in a certain period should be within a threshold. The order degree can be express as:

$$U_{ji}(e_{ji}) = 1 - \left| \frac{e_{ji} - c}{a_{ji} - b_{ji}} \right| \qquad U_j = \sum_{i=1}^{n} \lambda_i U_{ji}(e_{ji}) \quad j = 1,2,3,4,5 \tag{1}$$

where, $U_{ji}(e_{ji})$ is the order degree of e_{ji}; e_{ji} is the i-th order parameter of j-th subsystem; a_{ji} and b_{ji} are the maximum and minimum values of e_{ji} respectively; c is the desired value within the threshold, i.e. $c \in [b_{ji}, a_{ji}]$; U_j is the order degree of j-th subsystem; λ_i is the weight of order parameter for i-th order parameter.

If e_{ji} is within the threshold of the i-th order degree, $U_{ji}(e_{ji}) \in [0,1]$; the more approaching to c the e_{ji} value is, the closer to 1 the $U_{ji}(e_{ji})$ will be.

3.2 Coordination degree function

The criterion of regulation schemes refers to the coordination of water cycle system. The coordination degree function of water cycle system is established based on the geometric means method:

$$H(t) = \theta^* \sqrt[5]{\left| \prod_{j=1}^{5} U_j \right|} \qquad \theta = \frac{MIN(U_j)}{\left| MIN(U_j) \right|} \quad j = 1,2,3,4,5 \tag{2}$$

where, $H(t)$ is the coordination degree of water cycle system; θ makes the coordination degree of water cycle system positive only if the order degree of each subsystem was positive.

Coordination degree $H(t) \in [0,1]$. The greater the $H(t)$ is, the more coordinated the water cycle system will be in one level year, and the better the regulation scheme will be.

3.3 Integrated distance function

When a regulation scheme persists multiple level years, there are no one comprehensive indice to reflect the effect of combination regulation schemes. The integrated distance of order degrees is introduced to overcome the problem.

Assuming that a combination regulation scheme persists N level years, the integrated distance of order degrees function is shown as follow:

$$\rho(t_1, t_2 \ldots t_N) = \left(\prod_{n=1}^{N} \sqrt{\sum_{j=1}^{5} (1 - U_{n,j})^2} \right)^{\frac{1}{N}} \quad j = 1,2,3,4,5 \tag{3}$$

where, $\rho(t_1, t_2 \ldots t_n)$ represents the integrated distance of order degrees for a combination regulation scheme; $U_{n,j}$ represents the order degree for j-th subsystem in n-th regulation period.

The smaller the $\rho(t_1, t_2 \ldots t_n)$ is, the more approaching to desired state the water cycle system will be, and the more effective the combination regulation scheme will be.

4 APPLICATION

4.1 Study area

Haihe River Basin is located in North China (112°E–120°E, 35°N–43°N). The total area is about 318,000 km². The basin lies in the semi-humid, semi-arid continental monsoon climate zone, which has the main characteristics of less water resources and uneven special and temporal distribution of precipitation. The total water resource is about 37.2 billion m³, and per-capita water resource is 305 m³, only 1/7 of nationally and 1/28 of globally.

4.2 *Schemes setting and evaluation indices*

Multi-dimensional regulation schemes are set according to certain hydrological conditions. Based on the hydrological conditions in *Haihe River Basin* from 1956 to 2000 and comprehensive water resources programming of Haihe River Basin, boundary conditions in different level years were combined to form several regulation schemes with Year 2007 as the base year. Among the schemes, five combination schemes are selected to demonstrate the applicability of the method.

Considering the characteristics of water resources utilization in *Haihe River Basin*, ten evaluation indices of regulation schemes (order parameter) are selected (Table 2). The trade-off method (Gan *et al.* 2010) was applied to determine the ideal point and threshold of evaluation indices.

Table 1. Evaluation indices of regulation schemes.

Subsystem	Evaluation indices (order parameter)	Represent	Ideal point	Bottom	Upper	Weights
Resource	Rate of surface water exploited [%]	surface water	50	45	67	0.4
	Groundwater over pumping [billion m³]	Ground water	0	0	3.6	0.6
Economy	GDP per capita [thousand yuan]	GDP	107.6	60	107.6	0.4
	Average water use of 104 GDP [m³]	efficiency	30	30	55	0.6
Society	Rate of rural-urban water consumption per capita	rural/urban	0.78	0.6	0.8	0.3
	Food production per capita [kg]	food	375	350	375	0.7
Ecology	Outflow to sea [billion m³]	outflow	7.5	5.5	7.5	0.4
	ecological water use [billion m³]	ecological	5.5	2.8	5.5	0.6
Environment	COD to river [thousand t]	COD	300	300	800	0.5
	Qualification rate of water [%]	rate	100	75	100	0.5

DAMOS (Gan *et al.* 2007) was applied to simulate and calculate the regulation results of schemes, and the results of evaluation indices (order parameter) are shown in Table 2.

Table 2. Value of evaluation indices (order parameters).

Scheme	Level Year	Resources		Economy		Society		Ecology		Environment	
		surface water	ground water	GDP	efficency	rural/urban	food	out flow	ecological	COD	rate
F0	2020	56	3.6	68	48.4	0.74	356	6.4	3.75	580.2	100
	2030	48.6	0	107	30.5	0.79	352	6.8	4.23	328.1	100
F1	2020	59.4	1.6	68	48.7	0.74	356	6.4	3.75	582.4	100
	2030	51.1	0	106	30.5	0.79	365	6.8	4.23	326.9	100
F2	2020	56.3	1.6	67	48.1	0.74	356	6.4	3.75	564.1	100
	2030	60.8	0	104	30.1	0.79	366	6.8	4.23	316.4	100
F3	2020	55.1	1.6	68	48.4	0.74	356	6.4	3.75	565.1	100
	2030	43.2	0	105	30.1	0.79	365	9.3	4.23	316.6	100
F4	2020	55.1	1.6	68	48.1	0.74	356	6.4	3.75	568.1	100
	2030	61.6	0	105	30.1	0.79	365	6.8	4.23	316.6	100

Table 3. Order degree and coordination degree of water cycle system.

Scheme	Level Year	Order degree					Coordination degree	Integrated distance of order degrees
		Resource	Economy	Society	Ecology	Environment		
F0	2020	0.291	0.226	0.408	0.391	0.72	0.376	1.04
	2030	0.975	0.983	0.341	0.578	0.972	0.712	
F1	2020	0.562	0.218	0.408	0.391	0.718	0.426	0.81
	2030	0.98	0.975	0.705	0.578	0.973	0.823	
F2	2020	0.619	0.224	0.408	0.391	0.736	0.439	0.82
	2030	0.804	0.967	0.733	0.578	0.984	0.798	
F3	2020	0.641	0.226	0.408	0.391	0.735	0.442	0.94
	2030	0.876	0.976	0.705	0.358	0.983	0.733	
F4	2020	0.641	0.233	0.408	0.391	0.732	0.445	0.83
	2030	0.789	0.976	0.705	0.578	0.983	0.79	

5 RESULTS AND DICUSSION

Table 3 shows the result of equations for the evaluation and selection of regulation schemes.

5.1 *Evaluation and selection on single schemes*

The coordinate degree of water cycle system in *Haihe River Basin* increased after schemes regulation. In 2020 level year, coordination degree of water cycle system reached the maximum with F4 scheme. And in 2030 year, the maximum occurred to F1 scheme. The result indicates that as considering 2020 and 2030 level year, the regulation measures of F4 and F1 schemes were the most effective respectively.

5.2 *Evaluation and selection on combination schemes*

F4 scheme in 2020 and F1 scheme in 2030 with maximum coordination degree in a single level year were combined which participated in schemes evaluation and selection. According to equation (3), the integrated distance of order degrees for each selected scheme was calculated.

The smaller the integrated distance is, the better combination regulation scheme will be. The results show that considering multi-dimensional regulation effect in 2020–2030 level year series, F1 is the optimal solution among six regulation schemes, and the water cycle system is closest to the desired state. Although F41 is the combination scheme with maximum coordination degree of water cycle in each level year, its integrated distance (0.814) is greater than that of F1 (0.809), indicating that coordination of water cycle system of F41 is less than that of F1.

6 CONCLUSIONS

With the increased influence of human activities on water cycle system, the contradiction between water supply and demand is becoming more prominent. To guarantee the support of water resources for economic and social sustainable development, it is crucial to carry out evaluation and selection on multi-dimensional regulation schemes.

Based on the dualistic characteristic of water cycle system under the influence of high-intensity human activities, entropy theory and synergetic theory were introduced to the evaluation and selection method on multi-dimensional regulation schemes of water cycle system based on system theories. Firstly, order parameters were selected as evaluation indices for five-dimensional subsystems of water cycle system and used to establish coordination degree function of various dimentional subsystems and order degree function of water cycle system. Subsequently, the coordination degree was used for evaluation and selection on a single regulation schemes, and the integrated distance of order degree was applied to quantify the continuing effect on the water cycle system of combination regulation schemes in cross-level year series. The evaluation method was proven reasonable and feasible with the practical application of evaluation and selection on multi-dimensional regulation schemes in *Haihe River Basin*, providing a reference for evaluation and selection on multi-dimensional regulation schemes in other basins.

ACKNOWLEDGEMENTS

The study is financially supported by the Special Fund Program for Scientific Research on Public Causes of Ministry of Water Resources (NO.201001018, NO.201101016) and 973 Program (No. 2006CB403408).

REFERENCES

Cao, Y.B, Gan H. & Wang L., *et al.*, 2012. *Multi-dimensional Overall Regulatory Threshold Values and Modes for Water Cycle of Haihe River Basin.* Beijing: Science Press

Chen, H.-W. & Chang, N.-B. 2010. Using fuzzy operators to address the complexity in decision making of water resources redistribution in two neighboring river basins. *Advances in Water Resources.* 33, 652–666.

Gan H., Zhu, Q.L. & You J.J. *et al.* 2010. Alternative avaluation and selection based on order degree entropy: A case study of the Haihe River basin in China. *Journal of Food, Agriculture & Environment.* 8(2), 1062–1066.

Gan, Z.G., Jiang Y.Z. & Shen Y.Y. 2007. Water resources allocation model with ET as a core concept. *In: Youth Science and Technology Forum of CHES, ed. Proc of the 3rd Youth Science and Technology Forum of CHES.* Chengdu: Yellow River Hydraulic, 118–125.

Gan, Z.G., Gan H. & Wang L, *et al.*, 2010. Tradeoff approach of multiple objective analysis in the Haihe River Basin. *Journal of Food, Agriculture & Environment.* 8(3&4), 991–995.

Liu, J.H., Qin D.Y. & Wang H., *et al.* 2010. Dualistic water cycle pattern and its evolution in Haihe River Basin. *Chinese Science Bulletin.* 55(16), 1688–1697.

Loukas. A., Mylopoulos.N & Vasiliades. L. 2007. A Modeling System of the Evaluation of Water Resources Management Strategies in Thessaly, Greece. *Water Resources Management.* 21, 1673–1702.

Mohammed, L. & Larry, W.C. 1993. Alternatives Evaluation and Selection in Development and Environmental. *ENVIRON IMPACT ASSESS REV.* 13, 37–61.

Okeola, O.G. & Sule, B.F. 2012. Evaluation of management alternatives for urban water supply using Multicriteria Decision Analysis. *Journal of King Saud University-Engineering Sciences.* 24, 19–24.

Srdjevic, B. Medeiros. Y & Faria .A. 2004. An Objective Multi-Criteria Evaluation of Water Management Scenarios. *Water Resources Management.* 18, 35–54.

Wang, H., Qin D.Y. & Wang J.H. *et al.* 2003. State identification and multiple regulation of regional water resources shortage. *Resources Science.* 25(6), 2–7.

Wang, H., Yan D.H. & Jia Y.W. *et al.* 2010. Subject system of modern hydrology and water resources and research frontiers and hot issues. *Advances in Water Science.* 21 (4), 479–489.

Hydraulic Engineering – Xie (Ed.)
© 2013 Taylor & Francis Group, London, ISBN 978-1-138-00043-8

Characteristics of uniform turbulent flow with steep slopes

Jun Wan & Huiling Duan
State Key Laboratory for Turbulence and Complex System, Department of Mechanics and Aerospace Engineering, College of Engineering, Peking University, Peking, China

Jianjing He
Environmental Science and Engineering, Hohai University, Jiangsu, China

ABSTRACT: Based on experimental data measured by the LDV system, vertical velocity distributions are well compared for steep and small slopes in different hydraulic conditions in an open channel. New Equation of wall-normal velocity has been proposed and their physical interpretations are explained. Different from previous studies, the variation of water surface can generate the mean-vertical velocity, resulting in studying the open channel flow more convenient. It is found that The Karman value for steep slopes is smaller than that for small slopes. This study also explains the significant influence of the near-wall velocity on the turbulence structures, such as the re-distributions of turbulence intensities. The equations for defect-law in steep flow have been developed by taking the effect of non-uniformity (the wake parameter Π) into account, and the relationship between the maximum velocity and wake intensity has been established. Good agreement between the measured and predicted turbulence intensities has been achieved. Based on this investigation, the following conclusions can be reached: The velocity profiles of both small and steep slopes well fitted the data. The Reynolds number does not obviously affect the Karman value. Therefore, the logarithmic law adequately approximates the data throughout the entire channel depth. Compared log-law with defect-law, Due to its limitation, the defect-law is not better than the log-law in describing the velocity profile. Valuable conclusions are drawn and references are provided for further studies on the hydraulic characteristics of steep slopes.

1 INTRODUCTION

In natural streams, open channel flow is a ubiquitous phenomenon that is not fully understood even for the simplest case—varied slopes flow and high Froude numbers in open channel (Yamamoto Y, 2011; Li Xingyu et al, 1995). Turbulence structures are crucial for predicting sediment and contaminant motions in rivers, lakes, and coastal waters where the effect of Froude number is significant. Over the half-century, the turbulence characteristics of uniform flow have been studied extensively by many researchers, such as Coles D, 1956; Cardoso AH, et al, 1989; Graf, 1984; Ram B, 2005; Yang SQ, 2008; Hubert C, 2012 etc., but few researchers have studied the effect of channel slopes on the turbulence characteristics. The turbulent characteristics in small slope are different from that in steep slope. The accelerating flow generally hampers the turbulence whereas the decelerating flow strengthens the turbulence (Huai WX, 2009). Unfortunately, the underlying mechanism for the phenomenon has not been well revealed, and the quantitative description for the turbulent characteristics is not available in the literature.

The small slope flow has been widely investigated theoretically and experimentally (Coles D, 1956; Prinos P, 1995; He, 2003; Nezu, 2004; Yang SQ, 2006). Coles (1956) has suggested a wake function (Π) for further explaining turbulence boundary layers. Nezu (2004) have experimentally found that Π is a function of the Reynolds number (Re). The wake strength Π approaches a constant of 0.2 for $Re > 10^5$, whereas $\Pi \approx 0$ in a small Fr flow. They have also verified that the log law has a Karman constant (κ) of 0.412 and an integration constant of 5.29, where it fits in the inner region. In the outer region, the wake function can be applied. Cardoso and Graf (1989) have measured the velocity, turbulence intensity, and friction velocity. They have found that the law of the wall fits the experimental data over the entire channel depth. They have also found that the constants κ and B had values of 0.40 and 5.1, respectively. However, such studies on turbulent flow in steep slopes with high Fr are very few. Similar studies in open channels with steep slopes are limited due to difficulties in measurement arising from small flow depths and high velocity fluctuations. These studies are important in establishing the resistance law of mountain rivers and solving sediment erosion and transport problems under such conditions (Pcggi D, 2002). Whether conclusions for turbulent flow in small slopes needs to be reexamined and further studied in more experiment (Adaramola MS, et al, 2010; 2012). As a continuous effort to investigate the effect of steep slope, this study concentrates on the turbulence characteristics in steep slope flow. The objectives of this paper include (1) to investigate the influence of the wall-normal velocity on turbulent structures; (2) to establish new relations between the wall-normal velocity and turbulent structures; and (3) to compare the measured wall-normal velocity with the developed model.

2 EXPERIMENTAL PROCEDURE AND MEASUREMENTS

The experiments were conducted in one of the flumes in the Hydraulics Lab of Hohai University. The flume was 8 m long with a cross-section 0.3 m × 0.4 m. The sidewalls of the flume were made of glass, whereas the bed was composed of steel plates with anti-rust paint. Water depth (H) was measured by steel probes. The LDV system was developed at the Hydraulic Laboratory of Tsinghua University. It had the advantages of high resolution and high signal–noise ratio. The experimental data were processed by a high-precision real-time digital analyzer JSP-1, which provided the mean velocity, square, triple, and quadruple of velocity fluctuation.

In these experiments, the slope was varied from 1/400 to 1/200. Uniform flows became available in small slopes by adjusting the switch and in steep slopes by adjusting the entrance switch. Considering the influence of the entrance flow, velocity measurements were conducted at a station 2.95 m from the channel entrance. Table 1 shows the main flow characteristics of all experiments.

As shown in Figure 1a and b. For the experimental data, as $\log y^+ > 1.4$ ($y^+ > 25$), the flow obeyed the universal law of the wall, whether in small or steep slopes. Therefore, the law of the wall can be applied throughout the channel depth except in the viscous sublayer and buffer layer. The flow in the small and steep slopes both had a linear region (Fig. 1c), that is, $u^+ = y^+$. The average viscous sublayer scale was $0 < y^+ < 15$. Between the viscous sublayer scale and the turbulence layer was a buffer layer, whose scale was $15 < y^+ < 25$. In the small slopes, the viscous sublayer scale had a bigger variety than that in the steep slopes. For example, in Flow 1, $y^+ < 15$, and in Flow 2, $y^+ < 8$.

Table 1. Flow parameters of hydraulic conditions.

Experiments	Bed slope (i)	Flow depth h (mm)	Flow discharge Q/(L/s)	B/h	Re	Fr
1	0.0025	46.8	7.91	6.41	20432	0.83
2	0.0025	51.1	9.68	5.87	21432	0.89
3	0.0025	56.2	11.18	5.34	20983	0.89
4	0.0025	65.7	13.69	4.57	19667	0.87
5	0.005	35.5	7.91	8.45	33471	1.26
6	0.005	38.9	9.68	7.71	34739	1.34
7	0.005	42.6	11.18	7.04	34110	1.35
8	0.005	49.9	13.69	6.01	31595	1.31

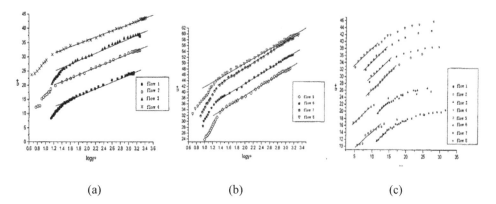

(a)	(b)	(c)

Figure 1.   Velocity distribution in small slopes (a), steep slopes (b) and in near wall region.

3   EXPERIMENTAL ANALYSES AND RESULTS

3.1   *Turbulence flow structures*

3.2   *Velocity distribution*

Turbulent flow in smooth open channels can be described by the log law, written as:

$$u^+ = \frac{2.3}{\kappa}\log\left(y^+\right) + B \tag{1}$$

where $u^+ = u/u_*$, $y^+ = u_* y/v$, and $u_* = \sqrt{\tau_0/\rho}$. In a smooth open channel, turbulent flow has a viscous layer in the near region. u^+ linearly varies with wall distance y^+. Thus, τ_0 can be calculated by the velocity gradient. We have found that flow in an open channel is very complicated, and that κ, B of the log wall are not always constant values, but have a variety of ranges. Table 2 shows that in small slopes, the mean value and standard deviation of Karman's constant are: $\kappa = 0.384 \pm 0.025$ and $B = 6.26 \pm 0.87$ whereas in steep slopes, $\kappa = 0.317 \pm 0.023$ and $B = 5.69 \pm 0.95$. The κ value is smaller in steep slopes. In a small slope, Cardoso have obtained $\kappa = 0.401 \pm 0.016$ and $B = 5.75 \pm 0.241$ in the inner region. Nezu have obtained $\kappa = 0.412 \pm 0.011$ and $B = 5.10 \pm 0.96$ in the inner region. Therefore, when the κ value is small, the velocity distribution is described by a single log wall.

Figure 2 show that the experimental data agreed with the findings by Nezu as well as Graf in that $10^4 < \mathrm{Re} < 10^6$ in a small slope. The Re does not obviously affect the Karman value. B cannot also be analyzed in the present data. According to Table 2, u^+ in small slopes can be expressed as:

$$u^+ = 5.99\log\left(y^+\right) + 6.26 \tag{2}$$

In steep slopes, u^+ can be expressed as:

$$u^+ = 7.26\log\left(y^+\right) + 5.69 \tag{3}$$

We suggested a new equation of velocity well applied in our experiment data

$$\frac{u}{\bar{u}} = 0.2736\log\frac{1000\,y}{h} + 0.31 \tag{4}$$

11

Table 2.   Velocity parameters and friction velocity.

Experiments	u_* (m/s)	κ	B
1	0.025597	0.372	6.46
2	0.028226	0.383	6.41
3	0.027895	0.365	7.13
4	0.032499	0.414	5.05
Average		0.384	6.26
5	0.03425	0.315	5.89
6	0.03633	0.31	4.74
7	0.03508	0.303	5.66
8	0.03857	0.341	6.45
Average		0.317	5.69

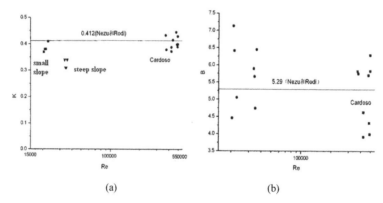

(a)   (b)

Figure 2.   Variation of κ and Variations in the flume width (B) with Re.

here, $\bar{u}$ is the vertical mean velocity. The wall-normal velocity can be induced by steep slopes. Accelerating flows produce downward velocity whereas decelerating flows generates upward velocity. The influence of wall-normal velocity on turbulent structures should not be underestimated. The fluctuating velocities are proportional to the vertical mean velocity, and this equation can be extended into the universal. Moreover, it does not have a parameter u_* and can be more convenient in studying the open channel flow. The present data agreed with the findings, as shown in Figure 3. The velocity profiles of both small and steep slopes well fitted the data.

3.3   *Discussion on the velocity-defect law*

The present data were well explained by the log law, and the wake region was subsequently investigated. Turbulent flow was divided into two parts, namely, the inner ($y/h<0.2$) and outer ($y/h>0.2$) regions. The Maximum of velocity could approximate the present data to the water surface used by Cardoso. The intercept of the regression lines at the water surface is $u_{max} - u/u_*$, which is equal to $2\Pi/\kappa$. The defect-law can then be approximated by:

$$\frac{u_{max} - u}{u_*} = -\frac{2.3}{\kappa}\log\frac{y}{h'} + \frac{2\Pi}{\kappa} \tag{5}$$

The parameters of the defect-law are shown in Table 3. In small slopes, $\kappa_1 = 0.377 \pm 0.034$, as obtained from Eq. 5. The mean value of the wake parameter $\Pi = -0.073$. In steep slopes, $\kappa_1 = 0.317 \pm 0.024$ and $\Pi = -0.074$, which showed that the wake parameter was too weak to

12

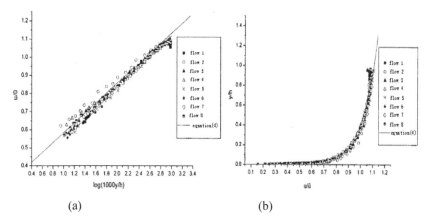

(a)   (b)

Figure 3.   Relative velocity distribution compared experimental data with the new formula.

Table 3.   Parameters of the defect-law.

Experiments	H (cm)	U_{max} (cm/s)	h'	κ_1 (y/h < 0.2)	Π
1	46.8	68.5	44.928	0.367	−0.059
2	51.5	73.8	49.44	0.372	−0.095
3	56.2	72.85	53.952	0.359	−0.144
4	65.7	80.1	63.072	0.411	0.008
Average				0.377	−0.073
5	35.5	94.11	34.08	0.316	−0.046
6	38.9	102.47	37.344	0.312	−0.078
7	42.6	103.42	40.896	0.300	−0.061
8	49.9	105.94	47.904	0.341	−0.11
Average				0.317	−0.074

be regarded. Graf has obtained: $\kappa_1 = 0.413 \pm 0.028$ and $\Pi = 0.08 \pm 0.093$. The present data was observed to have some deviations. Graf has already remarked that in the inner region ($y/h < 0.2$), the velocity-defect law is also formally valid. In the outer region ($0.2 < y/h < 0.7$), a limited strength ($\Pi = 0.08 \pm 0.093$) does exist, but the retarding flow in the near-face-zone ($0.7 < y/h < 1$) is possibly due to weak secondary currents. Thus, it tends to compensate for the wake effect.

The good results in the present study confirmed that the velocity-defect law was a good fit for the data, as shown in Figure 4 (dash lines are the best fit curves for the measurements in the inner region). A wake does exist both in small slopes and steep slopes because the log-wake law does not better describe the small slopes and steep slopes because the log-wake law does not better describe the velocity profile than the logarithmic law. The formula was also complicated and cannot be applied very conveniently.

3.4   Mixing length

Prandtl's mixing length hypothesis of $l = \kappa y$, with $\kappa = 0.4$, best fits the inner region ($y/h < 0.2$). Schlichting (1968) provides the following equation:

$$u_*^2 = \frac{\tau_0}{\rho} = l^2 \left| \frac{du}{dy} \right| \frac{du}{dy}$$

(6)

13

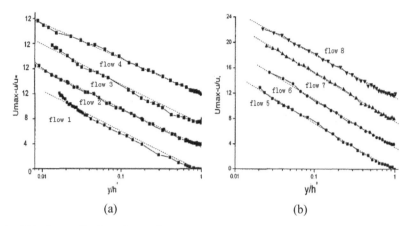

(a)   (b)

Figure 4.   Velocity profiles in wall coordinates compared experimental data with the defect formula.

and

$$\varepsilon = l^2 \left| \frac{du}{dy} \right|$$

where ε is the eddy viscosity. These equations can be transformed into

$$\varepsilon = \frac{u_*^2 (1 - y/h)}{du/dy}$$

and

$$l = \frac{u_* \sqrt{1 - y/h}}{du/dy} \tag{7}$$

τ is the shear stress at a distance y from the bottom, and τ_0 is the bottom shear stress, satisfying $\tau = \tau_0(1 - y/h)$. Figure 5 shows the distributions of the dimensionless mixing length (l/h). The present data seemed to be reasonably good up to the inner region ($y/h < 0.2$), in agreement with Prandtl's mixing length hypothesis. However, important differences and scatters were identified in the outer region ($y/h > 0.2$). Flows 5, 6, 7, and 8 showed bigger deviations, which indicated that the mixing length in steep slopes was smaller than that in small slopes in the outer region ($y/h > 0.2$).

3.5   Turbulence intensity and shear stress

The corresponding velocity fluctuation u' may reflect the turbulence intensity. Here, turbulence intensity is defined as the root-mean-square value of u': $\sqrt{\overline{u'^2}}$. The following function is proposed for the turbulence intensity distribution:

$$N = \frac{\sqrt{\overline{u'^2}}}{\overline{u}} \tag{8}$$

where N stands for the turbulence intensity and $\overline{u}$ stands for the vertical mean velocity. The turbulence intensity distributions are shown in Figure 6. The biggest N value scale was $1 < \log y^+ < 1.25$, as shown in Figure 6. In small and steep slopes, the biggest N values were 0.138 and 0.145, respectively. At the core of the outer region, ($0.2 < y/h < 0.7$), the turbulence intensity declined with increased relative water depth. The intensity distributions in small and steep slopes had similarities. The biggest N value in steep slopes was bigger than that in small slopes, probably due to the higher flow rate.

14

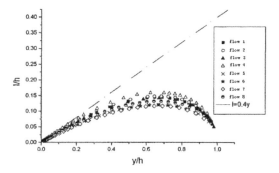

Figure 5.   Distribution of the mixing length in small and steep slope flow.

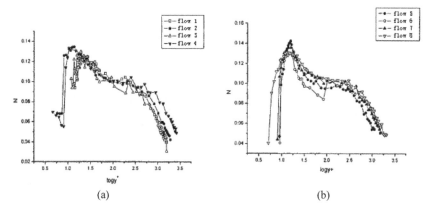

(a)   (b)

Figure 6.   Turbulence intensity in small and steep slope flow.

Table 4.   Calculated τ_0 value.

Q(l/s)	τ_0 (small slope)	τ_0 (steep slope)
7.91	0.42	0.86
9.68	0.80	0.98
11.18	0.52	0.98
13.69	1.06	1.13

   The determination of τ_0 is important in turbulent flow. Thus far, there are four methods for measuring τ_0. The viscous sublayer can be precisely measured by the LDV; hence, τ_0 was calculated according to the velocity distribution in the viscous sublayer, as shown in Table 4. The τ_0 value in steep slopes is bigger than that in small slopes.

4   CONCLUSIONS

Based on the presentation and discussion of the present data set, the following conclusions are drawn:

1. Small and steep slopes have the same layers, namely, viscous sublayer, Buffer layer, and log-law layer. These different layers take the scale values of $0 < y^+ < 15, 15 < y^+ < 25$, and $y^+ > 25$, respectively. The logarithmic law adequately approximates data throughout an entire channel depth in both small and steep slopes. In small slopes, $\kappa = 0.384 \pm 0.025$ and

15

$B = 6.26 \pm 0.87$. In steep slopes, $\kappa = 0.317 \pm 0.023$ and $B = 5.69 \pm 0.95$. The κ value in steep slopes is smaller than that in small slopes. The Re does not obviously affect the Karman value. In small slopes, u^+ can be expressed as: $u^+ = 5.99\log(y^+) + 6.26$. In steep slopes, u^+ can be expressed as: $u^+ = 7.26\log(y^+) + 5.69$.

2. In agreement with our findings: $\dfrac{u}{u} = 0.2736\log\dfrac{1000 y}{h} + 0.31$. The velocity profiles in both small and steep slopes well fit the data.

3. A wake does exist. The mean value of the wake parameter ($\Pi = -0.0735$) is very small. Being limited, the log-wake law cannot better describe the velocity profile than the logarithmic law. The formula is also complicated and cannot be applied very conveniently.

4. The mixing length in steep slopes is smaller than that in small slopes in the outer region $(y/h > 0.2)$.

5. The biggest N values of steep and small slopes are approximately 0.138 and 0.145, respectively. The biggest N value in steep slopes is larger, probably due to the larger velocity. Thus, the τ_0 value in steep slopes is bigger.

ACKNOWLEDGEMENTS

Support by the following agencies and program is acknowledged: National Basic Research Program of China under Grant No. 2008CB418203; The National Science and Technology Specific Project of China under Grant No. 20080ZX07422.

REFERENCES

Adaramola M.S, Bergstrom D.J, David S, 2012. Characteristics of turbulent flow in the near wake of a stack. *Experimental Thermal and Fluid Sciences,* **40**:64–73.

Adaramola M.S, Sumner D, Bergstrom D.J, 2010. Effect of velocity ratio on the stream wise vortex structures in the wake of a stack. *Journal of Fluids and Structures*, **26**:1–18.

Cardoso A.H, W.H. G., G. Gust, 1989. Uniform flow in a smooth open channel. *Journal of Hydraulic Research*, **27**(5): 603–615.

Coles. D, 1956. The law of the wake in the turbulent boundary layer. *Journal of Fluid Mechanics*, **1**:191–226.

Graf, W.H., 1984. The disscusion of "Velocity distribution in smooth rectangular open channel. *Journal of Hydraulic Engineering, ASCE,* **110**:268–273.

He Jianjing, H.-W., 2003. Turbulence characteric of non-uniform flow in a smooth open channel. *Journal of Hohai University (Natural Sciences)*, **31**(5): 513–517.

Huai W.X, Y.H.Z., Z.G. Xu, Z.H. Yang, 2009. Three-layer model for vertical velocity distribution in open channel flow with submerged rigid vegetation. *Advances in Water Resources*, **32**(4): 487–492.

Hubert C, Nicholas D.J, 2012. Turbulent velocity measurements in open channel bores. *European Journal of Mechanics B-fluids*, **32**:52–58.

Li Xingyu D.Z., Chen Changzhi, 1995. Turbulent flows in smooth-wall open channels with different slope. *Journal of Hydraulic Research,* **33**:333–347.

Nezu I, Azuma R., 2004. Turbulence characteristics and interaction between particles and fluid in particle-laden open channel flows. *Journal of Hydraul Engineering*, **130**(10): 988–1001.

Pcggi D, A.P., L. Ridolfi, 2002. An experimental contribution to near-wall measurements by means of a special LDA technique. *Experiment Fluids*, **32**(3): 366–375.

Prinos, P, A.Z., 1995. Uniform flow in open channel flow. *Journal of Hydraulic Research*, **33**:705–719.

Ram, B. 2005. Velocity measurements in a developed open channel flow in the presence of upstream perturbation. *Journal of Hydraulic Research*, **43**(3): 258–266.

Schlichting H., 1968. Boundary-Layer Theory. *Sixth edition, Mc Graw-Hill Company.*

Yang SQ, Alex T. Chow, 2008. Turbulence structures in non-uniform flows. *Advances in Water Resources*, **31**:1344–1351.

Yang SQ, Lim SY, 2006. Discussion of "shear stress in smooth rectangular openchannel flows". *Journal of Hydraulic Engineering*, **131**(1): 30–37.

Yamamoto Y, Kunugi Tomoaki, 2011. Direct numerical simulation of a high-Froude-number turbulent open-channel flow. *Physics of Fluid*, **23**(12): 30–7.125108-125119.

Hydraulic Engineering – Xie (Ed.)
© 2013 Taylor & Francis Group, London, ISBN 978-1-138-00043-8

Multi-objective function based on sensitivity for calibration Xinanjiang model

Li Qian
Zhejiang institute of Hydraulics and Estuary, Hangzhou, China
College of Water Resources and Hydrology, Zhejiang University, Hangzhou, China

Liang Guoqian & Qian Jinglin
Zhejiang institute of Hydraulics and Estuary, Hangzhou, China

ABSTRACT: An automatic calibration scheme for Xinanjiang model (CSS) was developed. A new weighted objective functions based on sensitivity analysis was introduced. The sensitivity analysis can performed the specific role that model parameters have on the model processes. The proposed CSS procedure combining the new objective functions mimic the course of the manual calibration. This is done by minimizing different objective functions for different parameters. The steps were in a fixed order in the CSS according to the XAJ model structure. Two tests are produce in three testing watersheds with different hydrological characteristic. Results indicated that the optimal values of parameters were stable, with different initial values varying in consideration ranges. The CSS procedure gave the better model performance in three testing basins.

1 INTRODUCTION

The accuracy of a model output is dependent on the quality of the input data, the model structure and the calibration. In Xinanjiang (XAJ) model, as in all hydrological models, a number of parameters are not directly measurable and have hence to be calibrated. Calibration can be formulated as: to obtain a unique and conceptually realistic parameter set so that the model becomes specific to the system it simulates and performs well. Manual calibration is often a tedious trial and error procedure (Boyle et al, 2000; Cheng et al, 2002), whereby the parameters are adjusted by matching the input/output behavior of the watershed to that of the model. The disadvantage of manual calibration and the increases in computing power have prompt the development of automatic calibration procedure at least three and a half decades. (Yapo et al, 1998; Vrugt et al, 2003; Fenicia et al, 2007). However, if the objective function is computed for the whole period, the parameters interaction would create noise on the objective function, with respect to the studied parameter (Harlin, 1991).

Therefore, this paper proposes a calibration scheme developed for the XAJ model, which utilizes the physical representation of the model components and the experience from manual calibrations. This is done by splitting the whole calibration period into sub periods, within which one specific process dominates the runoff.

2 MODEL STRUCTURE AND CALIBRATION PARAMETERS

The model used in this paper is a semi distributed conceptual rainfall-runoff model—Xinanjiang model developed in 1973 and published in 1980 (Zhao et al, 1980). The flow chart of Xinanjiang model is shown in Figure 1. All symbols inside the blocks were variables including inputs, outputs, state variables and internal variables while those outside the blocks were parameters.

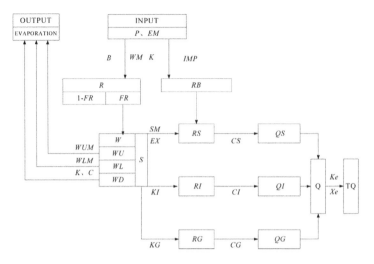

Figure 1.   Schematic description and structure of the Xinanjiang model.

The model had 16 parameters. Evapotranspiration parameters K, UM, LM, C; Runoff production parameters WM, B, IM; Parameters of runoff separation SM, Ex, KG, KI; Runoff concentration parameters CG, CI, CS; Muskingum parameters Ke, Xe. The output is more sensitive to 7 parameters. They are K, SM, CS, CI, CG, KE, and XE.

3   WEIGHTED OBJECTIVE FUNCTIONS BASED ON SENSITIVE ANALYSIS

The sub periods where the effects of individual parameter are relatively dominant can be identified by the weighted function based on the sensitivity analysis; it can be expressed as:

$$\omega(t) = \left(\sum_{i=0}^{n-1} \frac{TQC(t)_{i+1} - TQC(t)_i}{P_{i+1} - P_i} \middle/ (n-1) \right) \tag{1}$$

where $\omega(t)$ is the weight for each t; TQC stands for the calculated discharge; P is the parameter.

3.1   *Criteria agreement and iteration loop*

The calibrated parameters described in section 2 all have clear physical meanings, according to the layered structure of XAJ model; these parameters can be divided into three parts: evapotranspiration parameter K, parameters of runoff separation SM, and runoff concentration parameters CS, CI, CG, XE, and KE.

The actual evapotranspiration of the basin is related to potential evapotranspiration and to soil moisture condition. An empirical coefficient K is used to transferring the pan evaporation to potential evapotranspiration, called 'the ratio of potential evapotranspiration to pan evaporation'. The output is particularly sensitive to this parameter which controls the water balance. The water balance error, being the difference between the total observed and the total computed runoff:

$$FY = \sum_{i=1}^{Y} \left(\sum_{j=1}^{N} TQ_{ij} - \sum_{j=1}^{N} TQC_{ij} \right)^2 \tag{2}$$

18

where, Y and N are the number of year and that of day in a year, respectively, in the period of calibration. TQ and TQC represent total observed and computed runoff.

The area mean of the free water capacity of the surface soil layer. SM represents the maximum possible deficit of free water storage. Surface runoff is sensitive to the values of this parameter. The objective function for SM is:

$$FSM = \sum_{t=1}^{n} \omega_{SM}(t)(TQ(t) - TQC(t))^2 \qquad (3)$$

CS stands for the recession constant in the "lag and route" method for routing through the channel system within each sub-basin. CI is the recession constant of the lower interflow storage. CG is the recession constant of the groundwater storage. If the recession constant is bigger, the concentration time of runoff is longer. The sensitivity analysis indicates that CS is sensitive to peak of outflow; CG is relative sensitive to recession stage, CI fall in between. Considering the flow concentration parameters mainly affect the sharp of the hydrograph, the adopted function given as:

$$FCi = \sum_{t=1}^{n} \omega_{Ci}(t)(TQ(t)/TQ(t+1) - TQC(t)/TQC(t+1))^2 \qquad (4)$$

where, i can be expressed as CS, or CI or CG.

KE may be approximately taken the travel time over a sub-reach. The time—peak error is considered as the criteria for KE:

$$FKE = \sum_{t=1}^{n} \omega_{KE}(t)(TP - TPC) \qquad (5)$$

where, TP the time of observed peak storage occurrence, TPC the time of calculated peak storage occurrence.

XE is the parameter of the Muskingum method, representing the river way and hydraulic characteristic. It also has influence on the sharp of hydrograph. The evaluation function is showed:

$$FXE = \sum_{t=1}^{n} \omega_{XE}(t)(Q(t)/Q(t+1) - QC(t)/QC(t+1))^2 \qquad (6)$$

Table 1. Calibration order and objective function used in the calibration loop.

Order	Parameters	Objective functions
1	*Evapotranspiration and runoff production*	
	K	Eq. (4)
2	*Separation of runoff component*	
	SM	Eq. (5)
3	*Runoff routing*	
	CS	Eq. (6)
	CI	Eq. (6)
	CG	Eq. (6)
	KE	Eq. (7)
	XE	Eq. (8)
4	*All parameters*	
		$MSE = \dfrac{1}{n}\sum_{t=1}^{n}(TQ(t) - TQC(t))^2$

19

Different criterions used for different parameters are set, and then an iteration loop is performance over the whole model. The parameters are calibrated one at a time in a set order starting with evapotranspiration and runoff production, over the runoff separation, and finally the flow routing phase. When a loop over all parameters has been finished, the mean square error (RMSE) for whole periods is evaluated using the values of parameters calibrated in the loop. The iteration loop continues until the parameters stabilized, i.e, when the RMSE criterion for the whole calibration period stopper changing. Calibration order, objective function for each parameter is given in Table 1.

4 APPLICATION EXAMPLE

4.1 Study area and data

The Xinanjiang model linked the proposed calibration procedure is applied at three test basins: HJ, TX and QFL. They will stand for 1, 2, and 3. These basins represent different watershed characteristic. Hydrological data used for calibration and validation of the hydrological model's parameters, including daily and hourly rainfall and runoff and daily evaporation rate. The rainfall data is obtained from rain gauges contained in the catchment. The daily evaporation data is obtained by using daily evaporation pan data from evaporation station in the catchment.

4.2 Results

It is of practical interest to extensively test the proposed calibration procedure to see whether its optimization features are sufficient to obtain robust and reasonable parameter set. This study produce two tests: 1) 100 sets of initial parameter values including some extreme and unrealistic values are produced. Then the uncertainty and robustness of the values of parameters optimized by CSS are analyzed, comparing with those optimized by conventional single objective function RMSE based on the simplex method (called RMSE for short in following section).

It is well known that the robustness of data series can be measured by mean square deviation, which is expressed as:

$$\sigma = \sqrt{\frac{1}{n}\sum_{i=1}^{n}(x_i - \bar{x})^2} \qquad (7)$$

The mean square deviation obtained by the proposed method and traditional RMSE method are compared in Table 2.

It is evidence, seen from Table 2, that the mean square deviation about all parameters is decreased by CSS, though the improvement about K is slight. K is the parameter represent

Table 2.   Mean square deviation obtained by CSS compared with RMSE.

	Hengjing		Tunxi		Qingfenlin	
	RMSE	CSS	RMSE	CSS	RMSE	CSS
K	4.35E–02	4.56E–02	6.78E–02	5.44E–02	4.18E–02	3.42E–02
SM	7.68E+00	1.95E+00	1.51E+01	1.71E+00	6.47E+00	3.09E+00
CS	5.55E–02	4.40E–02	1.38E–01	5.35E–02	1.01E–01	7.98E–02
CI	9.54E–02	4.03E–02	1.39E–01	3.54E–02	9.23E–02	4.06E–02
CG	3.11E–03	2.68E–03	5.67E–03	3.42E–03	4.55E–03	3.92E–03
KE	3.62E–01	1.52E–01	8.75E–01	1.97E–01	4.44E–01	2.21E–01
XE	8.42E–01	1.16E–01	4.26E–01	1.36E–01	9.96E–01	9.93E–02

Table 3. The performance of CSS in model simulation in terms of three evaluations.

Watersheds	PBAIS (%)	RPSE (%)	TPE	NS
1	−0.35%	3.23%	1	0.785
2	1.19%	1.72%	0	0.788
3	8.40%*	7.65%*	0	0.805

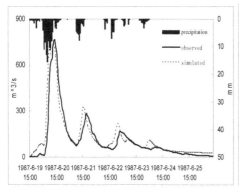 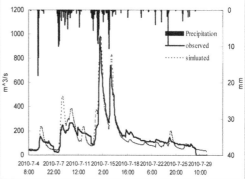

Figure 2. a) the hydrographs for flood event b) The hydrographs for flood event 31100704 in 31870619 in HJ TX.

water balance, therefore, it is calibrated by RMSE is effective and efficient. However, for other parameters, such as CS, CI, CG, XE representing the sharp of hydrograph and KE simulating the lag time of system, the traditional objective function RMSE is helpless. The comparison results show that the robustness about calibration results are improved by the weighted objective function based on the sensitivity analysis. It is indicated that the proposed weighted objective functions has the ability for utilizing the information about parameters with different physical response in calibration process.

2) Second test: the model performance of parameters calibrated by CSS is compared with performance produced by manual calibration. Four evaluation criterions are expressed as PBIAS, NS, RPSE, TPE. Table 3 reveals that the model performance of CSS for calibration flows in three testing watersheds.

Then in Figure 2 the simulation hydrograph by using CSS are illustrated and compared with observed hydrograph. These figures also illustrate the better results by using CSS for calibration. The weighted functions combining in CSS do not emphasize a special flow stage but are developed according to the physical meaning and role on model processes of parameters. From figures, the better fit to each flow stage (high-flow, peak, and low-flow) is indicated that the weighted functions can utilize the physical information contained in parameters effectively and efficiency, and the CSS procedure, which implies a stepwise calibration of sets of parameters that are associated with specific aspects of hydrology response, can do better in calibrating XAJ model.

5 CONCLUSION

In this study, a procedure for XAJ model is developed. A new weighted objective functions based on sensitivity analysis which perform based on the specific role that model parameters have on model processes, are introduced. The proposed procedure combining the strength of manual calibration use different objective functions for different parameters. The weighted objective functions and the fixed order in CSS are important design in order to avoid the effects of parameter interaction and reduce the number of parameters being optimized simultaneous.

The CSS procedure is straightforward, simple and consistently performed well. It is preferred to the traditional direct search method using only one objective function (RMSE), because it takes advantage of our understanding of the physical system, the model structure and the manual calibration experience. The calibration procedure is, however, not dependent on the XAJ model, and it should be applicable to other models as well.

ACKNOWLEDGEMENT

This study is supported by the Key Program of Science and Technology Department of Zhejiang Province (No. 2009C13011).

REFERENCES

Boyle, D.P., Gupta, H.V., Sorooshian, S. (2000). Toward improved calibration of hydrologic models: combining the strengths of manual and automatic methods. *Water Resour. Res.* 36(12):3663–3674.

Cheng, C.T., Ou, C.P., Chau, K.W. (2002). Combining fuzzy optimal model with a genetic algorithm to solve multi-objective rainfall-runoff model calibration. *Journal of Hydrology.* 268:72–86.

Fenicia, F., Savenije, H.H.G., Matgen, P., Pfister, L. (2007). A comparison of alternative multi-objective calibration strategies for hydrological modeling. *Water Resour. Res.* 43:W03434.

Harlin, J. (1991). Development of a process oriented calibration scheme for the HBV hydrological model. *Nordic Hydrology.* 22:15–36.

Vrugt, J.A., Gupta, H.V., Bastidas, L.A., Bouten, W., Sorooshian, S. (2003). Effective and efficient algorithm for multiobjective optimization of hydrologic models. *Water Resour. Res.* 39:1–19.

Yapo, P.O., Gupta, H.V., Sorooshian, S. (1996). Automatic calibration of conceptual rainfall-runoff models: sensitivity to calibration data. *Journal of Hydrology.* 181:23–48.

Zhao, R.J. (1992). The Xinanjiang model applied in China. *Journal of Hydrology.* 135:371–381.

Hydraulic Engineering – Xie (Ed.)
© 2013 Taylor & Francis Group, London, ISBN 978-1-138-00043-8

Study on longitudinal dispersion caused by vertical shear in ice-covered rivers

Zhigang Wang, Yongcan Chen, Dejun Zhu & Zhaowei Liu
Department of Hydraulic Engineering, Tsinghua University, Beijing, China

ABSTRACT: The presence of ice cover alters the velocity and longitudinal dispersion as well. Based on the logarithmic law of velocity distribution in ice-covered rivers, the longitudinal dispersion produced by vertical shear in ice-covered rivers are analyzed, and the results show that: the hydraulic slope, the flow depth and the ratio of Manning roughness coefficients are the main factors which influence the longitudinal dispersion. When the hydraulic slope or the flow depth increases, the longitudinal dispersion coefficient will increase as well. However, the influence of the Manning roughness coefficients behaves more complicated. When the Manning roughness coefficient of the ice cover is far different from that of the river bed, the presence of the ice cover would enlarge the longitudinal dispersion capacity. And if the both Manning roughness coefficients are almost of the same magnitude, the longitudinal dispersion coefficient would be decreased, until achieving the smallest value.

1 INTRODUCTION

Longitudinal dispersion, embodying the influences of shear velocity on the transfer and mixing of the substances dissolved or suspended in water, has been one of the highlights in the environmental hydraulic research since the longitudinal dispersion concept was proposed by Taylor (1954). Elder (1959) extended Taylor's concept into wide rectangular channels and derived a triple integral as

$$D_L = -h^2 \int_0^1 \hat{u}(\eta) d\eta \left[\int_0^\eta \frac{1}{D_t} \left(\int_0^\eta \hat{u}(\eta) d\eta \right) d\eta \right] \tag{1}$$

where D_L is the longitudinal dispersion coefficient produced by vertical shear; D_t is the vertical turbulent diffusivity; h is the flow depth; η is a dimensionless coordinate equal to z/h, and z is the vertical coordinate upwards; $\hat{u}(y)$ is the deviation of local velocity at η from the average velocity. According to Eq. 1, longitudinal dispersion is in good relationship with velocity distribution and vertical turbulent diffusivity.

Afterwards, the longitudinal dispersions caused by vertical shear in various open waters are analyzed using Eq. 1 (Elder 1959; Xu and He 1990; Zhou 1992). However, after the river is closed by ice covers, the vertical velocity distribution would be changed to be different from that of ice-free rivers, with the maximum velocity line shifting down towards the river bed and the velocity contours closed (Ke et al. 2004). Thus the longitudinal dispersion caused by vertical shear is expected to be different from that in open waters.

In the paper, we established the logarithmic law of vertical velocity distribution on Manning roughness coefficients and derived the theoretical calculation of the average vertical turbulent diffusivity first. And based on this, the characteristics of longitudinal dispersion produced by vertical shear are analyzed using Eq. 1.

2 VERTICAL DISTRIBUTION OF VELOCITY

As is mentioned previously, the presence of ice cover can alter the vertical distribution of longitudinal velocity greatly, and the average velocity would decrease. In order to describe it in mathematics, a lot of investigations were proposed (Teal et al. 1994; Meyer 2009), of which the logarithmic law has been drawing the most interests owing to its simpler frame, better theoretical foundation and better agreement with experimental data. Thus the logarithmic law is also used in this research.

2.1 Vertical velocity distribution

According to the logarithmic law, it is easy to know that: (1) the cross section is divided into two zones by the maximum velocity line: the ice-covered zone and the riverbed zone. In the ice-covered zone, the flow is controlled mainly by the ice cover rather than the river bed, while in the riverbed zone, the flow is controlled mainly by the river bed instead of the ice cover; (2) the vertical velocity distribution obeys the logarithmic law in both the ice-covered zone and riverbed zone separately.

$$u(z) = \begin{cases} \dfrac{u_{*i}}{\kappa} \ln\left[\dfrac{30h - 30z}{K_i} \right] & h_b < z \le h \\[3mm] \dfrac{u_{*b}}{\kappa} \ln\left[\dfrac{30z}{K_b} \right] & 0 < z \le h_b \end{cases} \tag{2}$$

where u is the streamwise velocity; κ is the von Karman's constant, commonly taking the value of 0.41; u_{*i} and u_{*b} are shear velocities of the ice-covered zone and the riverbed zone separately; K_i and K_b are the equivalent roughness heights of the bottom of the ice cover and the surface of the river bed, respectively; h is the flow depth of the river and h_b is the flow depth of the riverbed zone.

2.2 Equivalent roughness heights

As is described in Eq. 2, the equivalent roughness heights K_i and K_b are key parameters to get the velocity distribution. However, in most cases, they are not well ready and need to be transformed by other information, of which the Manning roughness coefficients are always preferred. Hereafter, the relationship between the equivalent roughness heights and the Manning roughness coefficients are established.

Based on Eq. 2, the average velocities of ice-covered zone and riverbed zone can be calculated by integration, with the result as

$$U_i = \frac{u_{*i}}{\kappa} \ln\left[\frac{30h - 30h_b}{K_i} \right] - \frac{u_{*i}}{\kappa}, \quad U_b = \frac{u_{*b}}{\kappa} \ln\left[\frac{30h_b}{K_b} \right] - \frac{u_{*b}}{\kappa}, \tag{3}$$

where U_i and U_b are the average velocities of ice-covered zone and riverbed zone, respectively. At the same time, U_i and U_b can also be obtained by Manning Formula, shown as

$$U_i = \frac{1}{n_i} [h - h_b]^{2/3} S^{1/2}, \quad U_b = \frac{1}{n_b} h_b^{2/3} S^{1/2} \tag{4}$$

where S is the hydraulic slope; n_i and n_b are the Manning roughness coefficients of ice cover and river bed separately. Combining Eqs. 3 and 4, the relationship of equivalent roughness heights and Manning roughness coefficients is obtained as below

$$n_i = \frac{\kappa}{\sqrt{g}} \cdot \frac{(h - h_b)^{1/6}}{\ln\left[30(h - h_b)/K_i\right] - 1} \tag{5}$$

$$n_b = \frac{\kappa}{\sqrt{g}} \cdot \frac{h_b^{1/6}}{\ln\left[30 h_b/K_b\right] - 1} \tag{6}$$

3 VERTICAL TURBULENT DIFFUSIVITY

Vertical turbulent diffusivity, representing the effects of the flow turbulence and vertical advection, is one of the most important parameters to analyze the longitudinal dispersion produced by the vertical shear. Up to the very present, a theoretical expression of D_t is not available and we are right in a position to evaluate it. Reynolds analogy states

$$D_t = \left| \frac{\tau/\rho}{du/dz} \right| \tag{7}$$

where, τ is the viscous shear which can be regarded as linear in either the ice-covered zone or the riverbed zone (Lau and Krishnappan 1980; Meyer 2009); ρ is the density of water. Based on the linear assumption, substituting Eq. 2 into Eq. 7 can get

$$D_{ti} = \frac{u_{*i}\kappa[z - h_b][h - z]}{h - h_b} \quad \text{in ice-covered zone} \tag{8}$$

$$D_{tb} = \frac{u_{*b}\kappa z[h_b - z]}{h_b} \quad \text{in riverbed zone} \tag{9}$$

where D_{ti} and D_{tb} are the vertical turbulent diffusivities of the ice-covered zone and riverbed zone, respectively. Based on Eqs. 8 and 9, the depth-averaged vertical turbulent diffusivity can be easily obtained by integration with the result as

$$D_t = \frac{\kappa u_{*i}[h - h_b]^2 + \kappa u_{*b} h_b^2}{6h} \tag{10}$$

4 CHARACTERISTICS OF LONGITUDINAL DISPERSION

Due to the influences of the ice cover, the characteristics of the longitudinal dispersion coefficient in ice-covered rivers are different from that of ice-free rivers. Through further analysis, it is found that the longitudinal dispersion caused by vertical shear is influenced mainly by three types of factors: hydraulic slope S, flow depth h and ratio of Manning roughness coefficients r defined as

$$r = n_i / n_b \tag{11}$$

In order to make the analysis much easier, define relative influence factor RIF as

$$RIF = \frac{D_L - D_{LO}}{D_{LO}} \tag{12}$$

where D_{LO} is the longitudinal dispersion coefficient of the corresponding ice-free rivers. Apparently, $RIF > 0$ means the presence of ice cover enlarges the longitudinal dispersion coefficient, while $RIF < 0$ means the dispersion capacity of the ice-covered rivers is smaller than that of the corresponding ice-free rivers.

4.1 Hydraulic slope

Hydraulic slope is one of the most important hydraulic factors, and it would influence the velocity distribution by changing the acceleration exerted by gravity. Furthermore, the characteristics of the longitudinal dispersion by vertical shear would be altered as well. Imaginary cases of $H = 5$ m, $n_b = 0.03$; $S = 0.0009/0.0012/0.0015$ were designed with the results as in Figure 1.

From the calculated results shown in Figure 1, it is known that as the hydraulic slope S increases, the longitudinal dispersion coefficient produced by vertical shear D_L increases as well. However, the relative influence factor RIF doesn't change with S.

4.2 Flow depth

Flow depth can also exert influences on the velocity, and longitudinal dispersion as well. The cases with $S = 0.0009$, $n_b = 0.03$; $H = 3$ m/5 m/10 m were designed and the calculated results were shown in Figure 2. Accordingly, it's concluded that when the flow depth is large, the longitudinal dispersion capacity would be large, and vice versa. However, the change rates of D_L with flow depth H are different, depending on the ratio of Manning roughness coefficients r.

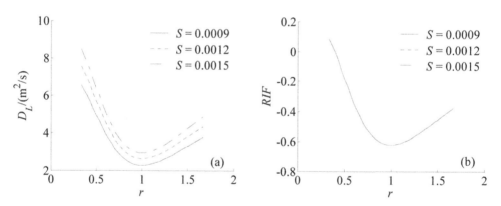

Figure 1.   Longitudinal dispersion coefficients V.S. hydraulic slope ($H = 5$ m; $n_b = 0.03$).

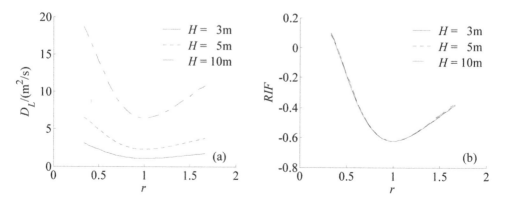

Figure 2.   Longitudinal dispersion coefficients V.S. flow depth ($S = 0.0009$; $n_b = 0.03$).

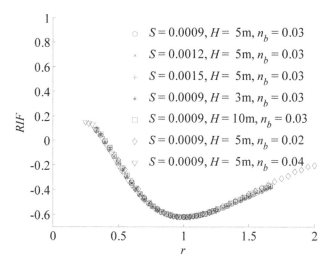

Figure 3. The longitudinal dispersion V.S. ratio of Manning roughness coefficients.

When r approaches 1, the change rate will decrease, until achieving the smallest. In addition, it is verified that flow depth H does nothing to the relative influence factor RIF.

4.3 Manning roughness coefficients

Manning roughness coefficients are the most important external factors that can influence the velocity distribution. They exert their influence by changing the wall conditions of the river. Here two cases of $H = 5$ m, $S = 0.0009$, $n_b = 0.02/0.04$ were designed, and the results were showed in Figure 3. The calculated results of previous cases are also present in Figure 3.

Through Figure 3, it is obvious that no matter what value H or S takes, RIF is in good relationship with r. For most designed cases, r is minus, which means the presence of the ice cover has reduced the longitudinal dispersion capacity. When r approaches 1.0, RIF achieves its smallest and the influence of the ice cover behaves the largest.

5 CONCLUSIONS

Ice-covering is a common phenomenon in middle and high latitudes on the earth. When it happens, the velocity distribution of natural rivers would be changed, and the longitudinal dispersion coefficients would be changed as well, especially that produced by vertical shear. In the paper, the characteristics of the longitudinal dispersion coefficients produced by vertical shear are analyzed in theory, and the conclusions are as follows:

1. The presence of the ice cover reduces the longitudinal dispersion capacity in most cases;
2. The longitudinal dispersion coefficients are in great relationship with hydraulic slope, flow depth and Manning roughness coefficients; but the relative influence factor only depends on the ratio of the Manning roughness coefficients.

ACKNOWLEDGEMENTS

The authors gratefully acknowledge the financial support of the National Natural Science Foundation of China (No. 51039002 and No. 51279078) and the National Water Pollution Control and Management S&T Special Projects of China (No. 2012ZX07201002).

REFERENCES

Elder, J.W. 1959. The dispersion of a marked fluid in turbulent shear flows. *Journal of Fluid Mechanics*, 5(4): 544–560.

Ke, S.J., Wang, M., Rao, S.Q. 2002. Ice research on the Yellow River. Zhengzhou: Yellow River Conservancy Press. (in Chinese).

Lau, Y.L., Krishnappan, B.G. 1980. Ice cover effects on stream flows and mixing. *Journal of Hydraulics Division*, (107): 1225–1242.

Meyer, Z. 2009. An analysis of the mechanism of flow in ice-covered rivers. *Acta Geophysica*, 58(2): 337–355.

Taylor, G.I. 1954. Dispersion of soluble matter in solvent flowing slowly through a tube. *Proceedings of Royal Society of London, Series A*, 219: 186–203.

Teal, M.J., Ettema, R., Walker, JF. 1994. Estimation of mean flow velocity in ice-covered channels. *Journal of Hydraulic Engineering*, 120(12): 1385–1400.

Xu, X.P., He, L. 1990. Analytical solution of longitudinal dispersion in double falling film on a wall. *Journal of Hydrodynamics, A series*, 5(4): 129–134. (in Chinese).

Zhou, X.D. 1992. Study of longitudinal dispersion coefficient of turbulence density current. *Journal of Hydrodynamics, A series*, 7(2): 185–191. (in Chinese).

Hydraulic Engineering – Xie (Ed.)
© 2013 Taylor & Francis Group, London, ISBN 978-1-138-00043-8

Distribution scheme study of the Guhai pump-water project from Yellow River in salt chemical industry park, Guyuan, Ningxia

Li Jin-Yan & Zhang Wei-Jiang
School of Civil and Hydraulic Engineering, Ningxia University, Ningxia, China
Engineering Technique Research Center for Water Saving Irrigation and Water Resources in Ningxia, Ningxia, China

ABSTRACT: Through the analysis of Guhai Pump-water system, it shows that, deducting planning using water project, three different schemes can be considered about rich water right of Yellow River mainstream in Yuanzhou rigion. The rich quantity is 10.36 million m^3/a, 24.60 million m^3/a and 20.00 million m^3/a respectively. We will review rich flow of each pump station in Guhai Pump-water system and water supply capacity of Nanping reservoir which take on water supply task of the salt chemical industry park further more. Through the analysis above, we have two schemes about the pump water supply capacity, respectively is 27.22 million m^3/a, 24.60 million m^3/a. At the same time, the analysis of the dam stability shows that it needs to be heighten 8 m and 10 m respectively for Naping reservoir to ensure dam's safety and reliability of water supplying.

1 INTRODUCTION

Guyuan is one of the poorest minority cities in Ningxia Hui Autonomous Region and even in the whole country. The industrial backwardness has seriously confined the socio-economic development of the city. Based on newly discovered rocking salt resources, coal, limestone and regional mineral resources, the government of Guyuan is planning to construct industrial chain based on salt chemical source. It was called as Guyuan Salt Chemical Industry Park. The construction of the industry park is an inevitable choice for the industry development of Guyuan city, and it can also break the bottlenecks of industrial development.

The planning 'Salt Chemical Industry Park' locates between Shen Zhuang and Da Geda village in Pengpu, Yuanzhou district, Guyuan. The project demands large quantity and high quality of water, while the water resource is very lack and ecological environment is really bad in the project area. So a single water source can not meet the water requirement. Therefore, local water source, Guhai Pump-water system and other water source must be taken into account to supply water for the industry park together. Guhai Pump-water project is the main water resource among them. So it is greatly necessary to research the pump-water supply scheme because of it's particularity for the project (Yang, 2010. Zhang, 2002).

2 OVERVIEW OF GUHAI PUMP-WATER SYSTEM FROM YELLOW RIVER

Guhai pump-water project was the first largely public poverty alleviation project in Yellow River irrigation area of Ningxia. It has made great contribution to change people's poverty, agriculture and economy in southern Ningxia. It also contributed to the improvement of ecological environment, the national unity and social development. Water in taking of Guhai pump-water project is from High Canal at 13+071 m in Zhongning County and ends in Yuanzhou district of Guyuan. The total length of main canal is 169.6 km. There are 12 pump stations in whole system whose maximum net head is 470.2 m, design flow is 12.7 m^3/s.

After the adjustment of Yuanzhou district, the irrigation area involved in Guhai Pump-water system is 122,000 acres.

3 ANALYSIS OF THE INFLOW WATER OF GUHAI PUMP-WATER SYSTEM IN YUANZHOU DISTRICT

The design flow of 11th pump station in Yuanzhou district of Guhai Pump-water project is 5.86 m³/s, and design flow of 11th channel is 5.86 m³/s too. There are six pumps in the station, four big and two small. After the implementation of Haiyuan water supply project, the 11th pump station supply water for it. There are two water supplying pumps for Haiyuan water supply project, single pump flow is 1.38 m³/s and double pump flow is 2.38 m³/s. We plan supply water for the industry park by the Haiyuan water supply project.

4 ANALYSIS OF AVAILABLE WATER SUPPLY QUANTITY OF GUHAI PUMP-WATER SYSTEM IN YUANZHOU DISTRICT

4.1 *Planning water requirement from Guhai pump-water system in Yuanzhou district*

It is necessary to make the totally planning water requirement of Yuanzhou district from Guhai Pump-water system clearer firstly in this study (deducting water demand of salt chemical industry park). According to survey and statistical analysis, the planning water demand from Guhai Pump-water system in Yuanzhou district is as Table 1.

4.2 *Rich water right index analysis of Yellow River in Yuanzhou district*

According to 'Initial Water Right Distribution Scheme of the Yellow River in Ningxia', it shows that the initial water quantity from main stream of the Yellow River distributing to Yuanzhou district is 48 million m³. It belongs to agriculture using and ecology using only. Through the table 1, it plans to take water from Guhai Pump-water system 23.39 million m³ in Yuanzhou district. Considering water demand for irrigation, drinking water and the different hydrological year in Yellow River, the water rights in main stream of Yellow River can be included as three demonstrations in Yuanzhou district below.

Scheme one: *Considering the water distribution principle of the Yellow River as 'the abundant increasing and the dry cutting'*

Design irrigation assurance rate of agriculture is 75% in the pump water irrigation area, design assurance rate of drinking water is 95% in Ningxia. This project belongs to the industrial water supplying project, assurance rate should be considered as 95%.

Table 1. The planning water demand from Guhai Pump-water system in Yuanzhou district Unit: Wan Mu, Wan Ren, Wan m³.

Water consumption category	Water saving irrigation project after 11th pump station	Irrigation area of Qi Jiazhuang ecological immigrant	Drinking project after 11th pump station	Drinking water projects of Qi Jiazhuang ecological immigrant
Present irrigation area	8.9	3.6		
Planning settled people			6.27	0.865
Gross consumption water	1890.9	264.6	161.4	22.8

30

In accordance with the water distribution principle of 'the abundant increasing and the dry cutting', if the assurance rate increases, irrigation water quantity should be cut with same proportion. The reduction factor based on the volume of surface runoff in 95% assurance rate, 75% assurance rate and average estimate of Yellow River for many years. Through the above analysis, the rich water right of the Yellow River in Yuanzhou district is as below in 95% assurance rate.

We Can calculate Surplus water quantity in assurance rate of 95% as below.

$$4800 \times 0.6 - (1890.9 + 264.6) \times 0.77 - (161.4 + 22.8) = 10.36 \text{ (million m}^3) \text{ (Liu, 2003. Pei, 2006)}$$

If there are no new project added to use water of Yellow River in pump-water irrigation of Yuanzhou district, redundant water index is 10.36 million m^3 in accordance with the distribution principle of 'the abundant increasing and the dry cutting' in $P = 95\%$ assurance rate.

Scheme two: *Regardless of the dispatch principle 'the abundant increasing and the dry cutting' in the Yellow River*

The flow of Yellow River decreased seriously in 2003, the Office of water resources in Ningxia draw up 'Emergency Water Dispatching Plan of Yellow River in Irrigation Area from Month 4 to Month 6 in 2003' (Called as the Emergency Plan below). The 'Emergency Plan' was approved by the people's government of the autonomous region to implement. Water allocation principle according to the 'Emergency Plan' is: water supplying quantity according with monthly average water in latest five years, distributed as 'the abundant increasing and the dry cutting' according with the same ratio. But the supplying quantity of Guhai pump-water, Yanhuanding pump-water, Hongsipu pump-water system will not decrease. Therefore, the department of water resources of Ningxia will not consider the effect of hydrological year on Guhai pump irrigation when dispatching water in the main stream of the Yellow River.

Based on the above analysis, $P = 95\%$ guaranteed rate of the Yellow River, the rich water right index is as below:

$$4800 - (1890.9 + 264.6) - (161.4 + 22.8) = 24.60 \text{ (0 million m}^3)$$

If there are no new project added to use water of Yellow River in pump-water irrigation of Yuanzhou district, redundant water index is 24.60 million m^3 not in accordance with the distribution principle of 'the abundant increasing and the dry cutting' in $P = 95\%$ case.

Scheme three: *According to the "Government Conference" principles*

According to the Government Conference of Ningxia Hui Autonomous Region on April 23, 2010, Guhai pump-water system should supply water of 20.00 million m^3 for project of phase 1 in Salt Chemical Industry Park every year, precisely ensure drinking water and agricultural irrigation water.

4.3 *Analysis of available water supply of Guhai pump-water system*

Through analysis in 4.2, we can see that it still has a certain amount of surplus water right index after meeting planning water requirements in Yuanzhou district on 'Water Right Distribution of Yellow River'. It still needs to analyze water supplying capacity of the whole Guhai pump-water system further more, to make sure the water supply capacity is rich or not and whether it can meet the water need of the project. The study select some typical and key pump stations of Guhai pump-water system such as 11th pump station, 8th pump station, 1st pump station to review it's water supply capacity.

1. Water supplying capacity review of the 11th pump station
   Two main water using categories was planned for 11th pump station. They are irrigation water after 11th canal and water put into Nanping reservoir. Nanping reservoir supplies water for Haiyuan County and drinking water of rural people in Haiyuan. Based on 'the

Plan Report of Water Supply System in the New Area of Haiyuan County', there will be 20.08 million m³ water to be put into Nanping reservoir in 2020 (16.15 million m³ for Haiyuan County, 3.93 million m³ for drinking). Average inflow into reservoir is 1.43 m³/s (1.15 m³/s for Haiyuan County, 0.28 m³/s for drinking).

Based on the above analysis, planning flow of 11th pump station (regardless of salt chemical industry project of Guyuan) should be accumulative water of 11th canal and 12th canal for agriculture and water used for the Haiyuan county in planning year. Flow process is shown in Figure 1.

2. Water supplying capacity review of the 1st pump station

According to 'Design Materials of Pump Water from Yellow River Project of Ningxia', design water flow of 1st pump station is 12.7 m³/s. Reviewing water supplying capacity is according to the water supply process of 1st pump station in 2011. Flow process is shown in Figure 2.

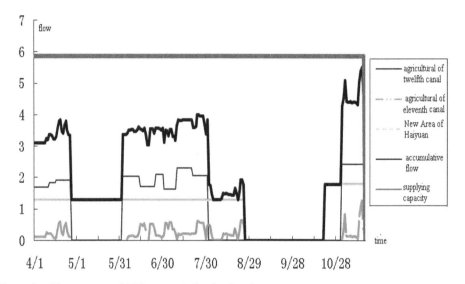

Figure 1.   Flow process of 11th pump station in planning year.

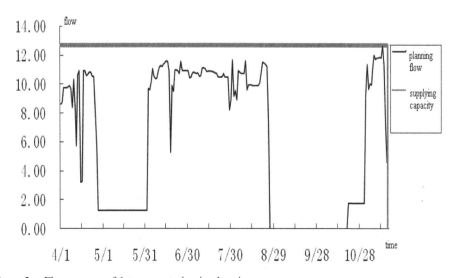

Figure 2.   Flow process of 1st pump station in planning year.

3. Comprehensive rich flow of Guhai pump-water system

Basing on rich flow of 1st pump station, 8th pump station and 11th pump station in planning year, we can calculate the integrated rich flow process of the whole pump-water system. It is shown in Figure 3.

4.4 *Water supplying capacity review of the Napping reservoir*

By the above analysis, we can conclude that there is rich water index of Guhai pump-water system not only in the distribution index but also in water supply capacity. Water from Yellow River for salt chemical industry park will be planned to take from Nanping reservoir combined with water supply project of Haiyuan County. So it is necessary to analyze and review water supply capacity of Nanping reservoir further more.

Based on 'Master Plan Report of Water Supply Project in the New Area of Haiyuan County (2008.8.29)', water consumption from pump-water project is 17.07 million m³ (49100 m³/d) in the Haiyuan county in planning year of 2020. Irrigation inflow is 1.29 m³/s in autumn and summer, and 1.79 m³/s in winter. From year 2009 to 2011, the actually average water consumption volume is only 0.66 million m³. It is more less than the planning quantity. Accordingly, the water supplying company of Xinhai, Ningxia forecast that water supplying quantity is only 20,000 m³/d. in the planning year.

Scheme one: *Considering the rich flow of Guhai pump-water system, water supplying capacity of Nanping reservoir and planning water supply scheme of 20,000 m³/d in Haiyuan county*

Nanping reservoir is an injection reservoir and water is from 11th pump station. Based on 'Master Plan Report of Water Supply Project in the New Area of Haiyuan County (2008.8.29)', it has ascertained water supplying time of each main canal. Guhai Pump-water system works for 177d in a year, total suspension time is 188d. Water supply relies on Nanping reservoir during suspension period. The maximum suspension time is from November 23 to April 4 annually, totally 133d. But now the 11th pump station supplies water from April 1 to August 25 each year which is regarded as summer supplying period, from November 1 to November 16 is the winter supplying period. Therefore, we recommend winter water supply period adjusted from October 16 to November 16 in Option one.

From Figure 4, we can see the integrated rich flow of Guhai pump-water system is 0.85~4.57 m³/s during the entire irrigation period in planning year of 2016. The status design flow of Nanping reservoir is 1.38 m³/s, larger flow is 2.38 m³/s. Water supplying flow of the reservoir based on the larger flow of 2.38 m³/s in the whole water supplying period.

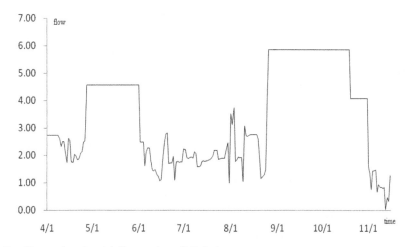

Figure 3.   Comprehensive rich flow review of Guhai pump-water system.

Table 2. Nanping reservoir scheme regulating calculation result.

Periods	Days (d)	The flow into reservoir (m³/s)	The water quantity into reservoir (wan m³)	The flow out of reservoir (wan m³/d)	The water quantity out of reservoir (wan m³)	Evaporation and leaking loss (wan m³)	Entering water–consuming water (wan m³)
4.1–4.28	28	2.38	575.77	9.46	264.88	14.83	296.06
4.29–5.10	12	2.38	246.76	9.46	113.52	7.42	125.82
5.11–5.15	5	2.38	102.82	9.46	47.3	3.08	52.43
5.16–5.26	11	2.38	226.20	9.46	104.06	6.80	115.34
5.27–6.4	9	2.38	185.07	9.46	85.14	5.56	94.37
6.5–6.17	13	2.38	267.32	9.46	122.98	8.04	136.31
6.18–6.30	13	2.38	267.32	9.46	122.98	8.04	136.31
7.1–8.25	56	2.38	1151.54	9.46	529.76	33.97	587.81
8.26–10.15	51		0.00	9.46	482.46	33.97	−516.43
10.16–11.16	32	2.38	658.02	9.46	302.72	20.95	334.35
11.17–12.31	46		0.00	9.46	435.16	27.98	−463.14
1.1–3.31	89		0.00	9.46	841.94	57.29	−899.23
Total	365		3680.81		3452.9	227.91	0.00

Table 3. Option four Nanping reservoir regulation computation results.

Periods	Days (d)	The flow into reservoir (m³/s)	The water quantity into reservoir (wan m³)	The flow out of reservoir (wan m³/d)	The water quantity out of reservoir (wan m³)	Evaporation and leaking loss (wan m³)	Entering water–consuming water (wan m³)
4.1–4.28	28	2.95	713.66	11.65	326.2	16.32	371.15
4.29–5.10	12	2.95	305.86	11.65	139.8	8.16	157.90
5.11–5.15	5	2.95	127.44	11.65	58.25	3.39	65.80
5.16–5.26	11	2.95	280.37	11.65	128.15	7.48	144.74
5.27–6.4	9	2.95	229.39	11.65	104.85	6.11	118.43
6.5–6.17	13	2.95	331.34	11.65	151.45	8.84	171.06
6.18–6.30	13	2.95	331.34	11.65	151.45	8.84	171.06
7.1–8.25	56	2.95	1427.33	11.65	652.4	41.47	733.46
8.26–10.15	51		0.00	11.65	594.15	37.37	−631.52
10.16–11.16	32	2.75	760.32	11.65	372.8	23.05	364.47
11.17–12.31	46		0.00	11.65	535.9	30.78	−566.68
1.1–3.31	89		0.00	11.65	1036.85	63.02	−1099.87
Total	365		4507.06		4252.25	254.81	0.00

Pump stations and canal of Guhai pump-water system can be controlled within its supplying capacity.

Based on adjustment calculation, water quantity into Nanping reservoir is 36.81 million m³ yearly, available quantity is 34.53 million m³. In accordance with water supply capacity of pump station and rich water quantity of Nanping reservoir is still 27.23 million m³, deducting planning water supplying quantity of 20,000 m³/d to the Haiyuan county in Scheme two. Total capacity of the reservoir needs to be 19.50 million m³, dam of reservoir needs to be heightened 8 m on the basis of the existing height. It is not need to be modified from 1st pump station to 11th pump station, the same as all of the main canal.

Scheme two: *Considering the rich flow of Guhai pump-water system, water rich right of Yuanzhou district and improvement of water supplying system*

Through the rich flow analysis of 1st pump station, 8th pump station and 11th pump station, we can see integrated rich flow of Guhai pump-water system is 0.85~4.57 m³/s during the entire irrigation period in planning year of 2016. In accordance with this rich flow, combined with the remaining water right indicator of 24.60 million m³/a in Yuanzhou district, the reservoir can supply water of 42.52 million m³ by regulating calculation of reservoir. Surplus water supply capacity of of Nanping reservoir is still 24.60 million m³ deducting preliminary water supplying quantity of 49100 m³/d. Regulating calculation result is shown in Table 3. Total capacity need to be 23.26 million m³ and dam of Nanping reservoir need to be heightened 10 m on the basis of the existing height. The 11th pump station of Guhai pump-water system and water pipeline of Nanping reservoir is required to be reformed by the calculation.

5 CONCLUSIONS

1. It is showed that rich water right in main stream of the Yellow River in Yuanzhou district can be in accordance with three different Scheme deducting planning projects taking water from Guhai pump-water system. They are 10.36 million m³/a, 24.60 million m³/a, 20.00 million m³/a respectively.
2. In accordance with the rich flow of pump station in Guhai system and water supplying capability of Nanping reservoir, there are two water supplying options of Guhai system. They are 27.23 million m³/a and 24.60 million m³/a respectively.
3. Through the analysis, we can see that Guhai pump-water system has capacity to distribute water for Guyuan Salt Chemical Industry Park, but the supplying quantity should combined with local water resources.

REFERENCES

Liu Sai-guang & Liu Zhi-rong. 2003. Strategic Countermeasures of Sustainable Utilization on Water Resources of Irrigation Area in Yellow River Area in Ningxia. *Ningxia Agriculture and Forestry Science and Technology* (6): 97–99.
Pei Yuan-sheng & Zhao Yong. 2006. Quantitative Study of Water Circulation in Response to Water Resources Allocation—Taking Ningxia as an Example. *Resource science* (28): 89–91.
Yang Zhi, Wang Xiu-nei. 2010. Yanhuanding Pump-water from Yellow River Project Distribution of Water Resources Engineering in Ningxia. *Technology* (9): 130–132.
Zhang Feng & Yang Shao-jun. 2002. Present Situation of Management and Suggestions on Yanhuanding Pump-water from Yellow River Project in Shanxi-Gansu-Ningxia Province. *People Yellow River* (24): 38–41.

Hydraulic Engineering – Xie (Ed.)
© 2013 Taylor & Francis Group, London, ISBN 978-1-138-00043-8

Burst detection in water distribution systems based on artificial immune system

Hai-Dong Huang & Tao Tao

College of Environmental Science and Engineering, Tongji University, Shanghai, China

ABSTRACT: A novel burst detection method is presented in this work based on Artificial Immune System (AIS). Data corresponding to normal operation are trained through application of clonal immune algorithm (CSA) to generate a detector set which is used for monitoring the system. The state of a water distribution network is then determined in terms of Euclidean distance between a new data sample and the corresponding detector. Monte Carlo Simulation (MCS) is here applied to create artificial burst data. Then the artificial burst data are used to generate another detector sets which are used for burst identification. According to Euclidean distance between a new data sample and the corresponding detector, possible burst locations are identified. A case study with three real burst events is illustrated to evaluate the proposed method. The results obtained from the case study show that the method has the potential to be useful tool for burst detection.

1 INTRODUCTION

Leakage reduction has now been given more and more attentions around the world as a result of water scarcity caused by rapid population growth, urbanization and climate change. Water loss mainly results from pipe bursts and leaks in water distribution networks. Hence, to minimize water loss caused by pipe breaks, a fast and efficient burst detection method is required through which the time prior to burst awareness can be minimized and eventually reduce bursts duration.

Leakage/bursts detection methods can generally be classified into two types: equipment-based methods and software-based methods. However, this work focuses on the latter methods. Over the past decades, considerable effort has been spent on two major groups of these methods. They are transient based methods and data driven approaches respectively. Transient based methods usually detect leakage/burst through investigating and modeling the behavior of a water distribution network under transient events generally initiated by any change (e.g. often very large) in pressure and/or flow in the system (e.g. open/close hydrant) (Liggett & Chen, 1994, Vítkovský et al. 2007). As for data driven approaches, they detect leakage/bursts often followed by application of a series of data mining or Artificial Intelligence (AI) techniques through which sensor data obtained from a distribution system can be analyzed efficiently and eventually detect leakage/bursts based on different principles (Mounce et al. 2010, Ye et al. 2011). In this work, an alternative method based on the Artificial Immune System (AIS) is proposed to detect bursts and identify bursts hotspots including a small number of nodes. Compared to the methods mentioned above, the immune inspired method has some advantages: computational efficiency, less data required for training and providing more accurate information on bursts (e.g. the precise area where a burst occurred). The proposed method is described in more detail below, followed by a case study.

2 METHODOLOGY

The AIS as a new artificial intelligence methodology is increasingly attracting more and more attentions in different fields of engineering, especially in monitor engineering system due to its powerful ability of pattern recognition, learning and memory. Numerous applications of the AIS in the area of engineering processes (e.g. fault detection and diagnosis) have been reported (Ghosh & Srinivasan, 2011, Silva et al. 2012). In practice, a pipe burst can be also regarded as a fault in water distribution systems, however, burst detection applications using AIS have yet to be reported. Generally, fault detection and diagnosis has four main steps: 1) data acquisition; 2) data pre-process; 3) feature abstraction and selection; 4) pattern recognition. The pipe burst detection procedure is similar to the steps mentioned above. However, considering the particularity of burst detection, some specific details should be further presented to better understand the proposed method. The concrete procedure of the proposed method is listed below.

2.1 Data acquisition

Data can be grouped into two classes: normal data obtained from normal operation of networks and fault data derived from burst events. Normal data can be easily obtained from records (e.g. pressure and flow rate) of a Supervisory Control and Data Acquisition (SCADA) system, for online monitoring technology has long been employed by water utilities due to the rapid advancement of modern communication technology. It is reported that flow measurement data are more sensitive to a burst or leak than the pressure measurement data (Ye et al. 2011). However, flow data combined with pressure data may be more effective to identify anomaly of water distribution systems. Note that flow and pressure under normal operation between two adjacent weeks have little change for most of water distribution networks. Thus, in this work, the last seven days (a week) normal flow and pressure data are required to predict next week condition in water distribution systems. However, it should be pointed out that the above finding is case specific. Compared to normal data, available burst data are usually limited. Consequently, some artificial data should be derived from simulation of pipe burst based on computer platform in order to evaluate the proposed method. The process of establishing a database that contains node pressure, pipe flow, and other information about bursts is detailed below.

Step 1: All pipe Burst Events (BEs) should be specified during the day (24 h) according to time step Δt(minutes) prior to generate artificial burst data. Considering the characteristic of pipe bursts, a BE can be stated as BE $= (p, t, f)$ for a system including P pipes, where $p \in (1, ..., P)$, $t \in (0, ..., 1440)$ with an given increment Δt, $f \in (minf, ..., maxf)$ also with an given increment Δf, $minf$ and $maxf$ represents the minimal and maximal burst flow respectively. In this study, the increment Δt was equal to the signal sampling frequency of sensors (e.g. usually 15 minutes). The increment Δf can be determined through field tests. In the field tests, hydrants can be opened with different flow to test the sensitivity of sensors. If sufficient tests are available, the feasible value of Δf can be estimated. It is recommended that the difference of pressure between two adjacent BE should be slightly significant. The determination of $minf$ and $maxf$ will be discussed in case study. To make the analysis simple, it is generally assumed that bursts occur at nodes. Thus, given a system consists of N nodes, a BE will be expressed as BE $= (n, t, f)$, where $n \in (1, ..., N)$. Besides, it must be indicated that a BE here is caused by the burst of a single pipe without any simultaneous pipe bursts due to their very low probability of occurrence. An extra demand, equal to the burst flow f, was imposed on the node where the burst occurred to simulate burst effects during a BE.

Step 2: The extended period simulation hydraulic analysis for each specific BE was performed based on EPANET hydraulic network solver, in turn. Considering pipe bursts may result in pressure-deficient in water distribution networks and the standard EPANET is not suitable for analysis of water distribution networks with low operating pressures, original EPANET should be modified before burst events are simulated. An extension of EPANET based on a Modified Pressure-Deficient Network Algorithm (M-PDNA) proposed by Babu and Mohan (2012) is applied to simulate bursts in water distribution networks. Pipe roughness and water consumption at a node are usually considered to be the most uncertain variables in a water distribution system. To achieve more accurate results, Monte Carlo Simulation (MCS) is applied in this study to estimate the uncertainties of pressure and demand at nodes resulted from the uncertain input variables in a hydraulic simulation model. However, only the uncertainty of water consumption at nodes is taken into account here, for pipe roughness is usually time invariant or varying very slowly.

Step 3: The simulation information of each BE was recorded. The information includes the following: a) pressure value at each pressure monitoring point, b) flow value at each flow monitoring point, c) the time when the BE occurs, d) burst flow, and e) BE location.

2.2 Data pre-process

Data pre-process is a very important step for the following procedures. As for the normal data, 96 past values at a 15-min time step at each sensor are mapped to a two-dimensional matrix $X \in R^{SxJ}$, where S designates the total sampling number of each sensor during a day (here, equals to 96), J denotes the total number of sensors (flow and pressure). The burst data is also mapped to a two-dimensional matrix $D \in B^{SxJ}$. The actual values of these variables in matrix X and matrix D are scaled or normalized in the range [0, 1] by the following formula:

$$x(i,j) = \frac{\hat{x}(i,j) - \min_j}{\max_j - \min_j} \tag{1}$$

where, $\hat{x}(i,j)$ is the element in the ith row and jth column of matrix X or matrix D; $x(i, j)$ is the normalized value corresponding to $\hat{x}(i,j)$; $\min_j$ and $\max_j$ are the minimal value and the maximal value respectively in jth column of matrix X or matrix D.

2.3 Burst detection and location

The proposed approach for bursts detection and location also comprises the following: 1) training phase; 2) bursts detection and location as described below.

In the training phase, the AIS consisting of detectors for normal operation and detectors for burst conditions is constructed. Various immune algorithms based on different principles or purposes are available to construct the AIS. Of these algorithms, two major algorithms have been widely used in AIS: Clonal Selection Algorithm (CSA) and Negative Selection Algorithm (NSA). In this study, the CSA algorithm is used to create the AIS network. The procedure of CSA algorithm generally consists of eight steps: 1) define antigen; 2) generate random antibody; 3) compute affinity value (antigen and antibody); 4) select; 5) clone; 6) hypermutation; 7) reselect; 8) if specified criterion is satisfied, the whole procedure is end, or else return to step (2). More details can be found in the relative literature (De Castro et al. 2002). After the normal data has been processed, the CSA algorithm is applied to generate a detector set M_0, corresponding to normal condition. This detector set M_0 is used to estimate the state in water distribution system. Similar process can be applied to the burst data to generate another detector sets $M_{i,T}$ (i denotes the number of BEs at time T, T represents the time when a BE occurs). These detector sets are used for burst location.

In the burst detection and location phase, the detector set M_0, generated from the training data corresponding to normal process operation, is first used to detect the state in water distribution networks (normal or abnormal). If the system is considered to be in an abnormal state, then, the other detector sets $M_{i,T}$ are used for burst location. For a new sample $s(t)$ (flow value), obtained from SCADA record at time t, it is first normalized by Eq. (1). Then, $s(t)$ is projected on detector set M_0. If $s(t)$ matches the detector $M_{0,t}$ which is the detector corresponding to the normal data obtained at time t, then the water distribution system is considered to be in a normal state; or else, the system is under an abnormal condition. $s(t)$ is said

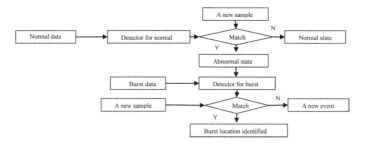

Figure 1. Schematic of the proposed method.

to match the detector $M_{0,t}$ if the Euclidean distance between them is less than or equals to a given threshold σ_0. Consequently, the system condition is detected based on $s(t)$ as follows:

$$\text{Condition}(s(t)) = \begin{cases} \text{Normal, if } \| s(t) - M_{0,t} \| \leq \sigma_0 \\ \text{Abnormal, if } \| s(t) - M_{0,t} \| > \sigma_0 \end{cases} \quad (2)$$

In this study, it is assumed that system operators can distinguish the pipe burst and other abnormal situations like fire flow and temporal large industrial usage. Consequently, once the abnormal state is detected, a new data set $q(t)$, including flow and pressure value of all sensors at time t, is projected on the detector sets $M_{i,T}$ generated from the training of bust data. The burst location can be conclusively identified if the data set $q(t)$ matches a detector $M(i,t)$ generated from burst data obtained at time t. The Euclidean distance between them is also used to determine if they match each other. However, if no detector in the detector sets $M_{i,T}$ matches the sample $q(t)$, the burst event corresponding to the data set $q(t)$ is considered to be a new burst event. Given a specific threshold σ_b, similar representation can be given by:

$$\text{State}(q(t)) = \begin{cases} \text{Identified, if } \| q(t) - M(i,t) \| \leq \sigma_b \\ \text{New event, if } \| q(t) - M(i,t) \| > \sigma_b \end{cases} \quad (3)$$

The schematic presentation of the proposed method is described in Figure 1.

3 CASE STUDY

The effectiveness of the proposed method for burst detection and location is illustrated using a DMA case study. The DMA examined with only one inlet includes approximately 5,000 properties. One flow meter is installed at the inlet, while 10 pressure meters are deployed on the network. Consequently, the data used consists of one flow value and 10 pressure value. The network consists of 100 nodes and 135 pipes varying from 100 and 500 mm in diameter. A simplified network of the DMA is presented in Figure 2.

The data used to test and validate the proposed method derives from the historical records about real burst events. Both flow and pressure data were measured at intervals of 15 min and were recorded using the SCADA system. A specific burst event occurred at about 16:35 on July 28th is illustrated to show how the proposed approach performs real data.

To evaluate the proposed method, the normal data obtained are first used to map to a two-dimensional matrix X. Each row in matrix X is denoted by a vector given by:

$$x(i) = (f_{1,t}, p_{1,t}, \dots, p_{j,t}) \quad (4)$$

where, $i \in (1, \dots, 96)$, $f_{1,t}$ represents the flow value at time t; $p_{j,t}$ denotes the pressure value at jth sensor at time t. Then, the matrix X is normalized by Eq. (2). After the normal data has been normalized, corresponding detector set DN are generated by performing the CSA algorithm.

The burst data used to generate the burst detectors are then obtained from the MCS procedure. Before the MCS procedure is implemented, all the BEs should be specified.

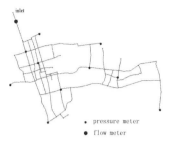 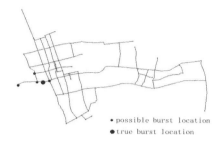

<table>
<tr><td></td><td>• possible burst location</td></tr>
<tr><td>• pressure meter</td><td></td></tr>
<tr><td>• flow meter</td><td>• true burst location</td></tr>
</table>

Figure 2.   Schematic diagram of examined network.   Figure 3.   Possible burst locations.

As stated above, a BE is represented as BE = (n, t, f). In general, burst flow is recognized as a pressure dependent flow that flows out of the water distribution network. The greater the pressure at the burst location, the greater the burst flow is. Although burst flow is usually unknown underground, it can frequently be represented as an orifice flow. Its hydraulic behavior can be depicted through an orifice flow equation. On the whole, an orifice flow equation is defined as follows:

$$q = \mu A \sqrt{2gH} \tag{5}$$

where, q is the orifice discharge, μ is the orifice discharge coefficient, A is the orifice cross-sectional area, H is the free water pressure at the orifice, and g is the gravitational acceleration.

The minimal burst flow can be determined by field test or simulation based on a hydraulic model, while the maximal burst flow can be estimated based on Eq. (5). In this work, the increment Δf is set to be 2 L/s, while the minimal burst flow and the maximal burst flow is set to be 10 L/s and 50 L/s.

To perform the MCS simulations, range of variation for nodal demand should be estimated from historical data by applied some statistical tests. In this study, variation for nodal demand is in the range [−0.15, 0.15]. On the basis of range of variation, some noise is imposed on each node to simulate the uncertainty of nodal demand. Then, a total of 1000 MCS are performed for each BE based on an extension of EPANET. Once 1000 MCS simulations have been evaluated, the mean of results (including flow and pressure) obtained from the total 1000 MSC simulations is used as the burst data for each BE. After the burst data have been processed like the normal data, the CSA algorithm is used to generate burst detector sets DB corresponding to burst conditions.

Considering flow and pressure data were measured at intervals of 15-min, real burst data should be obtained at 16:45. Once the AIS is constructed, the real burst data is then projected on the normal detector set generated from the normal data obtained at 16:45. Later, the real burst data is projected on the burst detector sets generated from the burst data obtained at 16:45 to identify the location of the burst event. An abnormal state is efficiently confirmed based on Eq. (2). 4 possible locations including the true burst area are successfully identified in terms of Eq. (3). The results are showed in Figure 3.

Figure 3 shows that all of the other locations seem to be in the neighborhood of the true burst location. Considering these locations are connected with each other, only three pipes are identified from a total of 135 pipes, and one of them is the true burst location. The burst area is narrowed down to a great degree through application of the method.

In order to further evaluate burst detection method, the proposed method is applied to another two burst events. Based on the same principles and similar procedure, these burst events can be successfully detected and located just like the aforementioned event. 5 and 4 nodes are identified as possible burst locations respectively. The results are described in Figures 4 and 5.

As can be seen from Figures 3–5, these case studies share many commonalities. For example, A few possible locations including true burst area can be identified as possible burst areas. All of the other locations are in the vicinity of real event area. Note that these locations usually

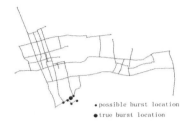

Figure 4.   Possible burst location for 2nd event.   Figure 5.   Possible burst location for 3rd event.

connect with each other. Just a few pipes are distinguished from a total of 135 pipes. In this sense, it can be drawn a conclusion that the proposed method is effective for burst detection. However, it is clear that like most of methods, the proposed method can only locate the burst to a few of pipes or a small amount of nodes and is unable to pinpoint the exact burst position.

4   CONCLUSIONS

Burst detection is very important in water distribution systems to reduce water losses and improve service quality of water utilities. Here, burst detection is regarded as the problem of pattern recognition. In this work, a novel burst detection methodology based on the AIS network has been presented. Unlike ANNs based methods, the current method does not rely on density model to predict the further data. As a result, less data are required to perform the method here. The data corresponding to normal operation are trained to generate a detector set which is used to monitor the water distribution system. Data corresponding to each BE are used to generate a separate detector set. These detector sets corresponding to BEs are used for pinpoint the potential burst hotspots. The method has been validated by applying to a case study with 3 real burst events. Results obtained from the case study illustrate that the method is effective for burst detection. Despite initial success, a series of issues need to be investigated further in order to improve the proposed method. For example, how to create artificial burst data more accurately and effectively is still needed to further study. Besides, how to set various parameters properly also remains to be solved. However, the proposed method has a potential to be a useful tool for burst detection in water distribution networks.

REFERENCES

Badu, K.S. & Mohan, S. 2012. Extended Period Simulation for Pressure-Deficient Water Distribution Network. *J. Comput. Civ. Eng,* 26: 498–505.

De Castro, L.N. & Von Zuben, F.J. 2002. Learning and Optimization Using the Clonal Selection Principle (PDF). *IEEE Transactions on Evolutionary Computation, Special Issue on Artificial Immune Systems* (IEEE), 6(3): 239–251.

Ghosh, K. & Srinivasan, R. 2011. Immune-System-Inspired Approach to Process Monitoring and Fault Diagnosis. *Industrial & Engineering Chemistry Research*, 50(3): 1637–1651.

Liggett, J.A. & Chen, L. 1994. Inverse Transient Analysis in Pipe Networks. *Journal of Hydraulic Engineering*, 120(8): 934–955.

Mounce, S.R., Boxall, J., & Machell, J. 2010. Development and verification of an online artificial intelligence system for detection of bursts and other abnormal flows. *J. Water Resour. Plng. and Mgmt.*, 136(3): 309–318.

Silva, G.C., Palhares, R.M., & Caminhas, W.M. 2012. Immune inspired Fault Detection and Diagnosis: A fuzzy-based approach of the negative selection algorithm and participatory clustering. *Expert Systems with Applications*, 39: 12474–12486.

Vítkovský, J.P., Lambert, M.F., Simpson, A.R., & Liggett J.A. 2007. Experimental Observation and Analysis of Inverse Transients for Pipeline Leak Detection. *Journal of Water Resources Planning and Management*, 133(6): 519–530.

Ye, G., & Fenner, R.A. 2011. Kalman filtering of hydraulic measurements for burst detection in water distribution systems. *J. of Pipeline Systems Engineering and Practice*, 2(14): 14–22.

Hydraulic Engineering – Xie (Ed.)
© *2013 Taylor & Francis Group, London, ISBN 978-1-138-00043-8*

The application of SWMM model based on GIS in the piedmont rain control: A case study in Beiwucun gravel pit of Beijing, China

Wu-Qing Li
Xi'an University of Technology, Xi'an, China
Beijing Municipal Commission of Development and Reform, Beijing, China

Gang Kong
Xi'an University of Technology, Xi'an, China
Beijing Hydraulic Research Institute, Beijing, China

Qiang Huang
Xi'an University of Technology, Xi'an, China

Jian-Gang Chen
Beijing Municipal Commission of Development and Reform, Beijing, China

ABSTRACT: SWMM model based on GIS was used to simulate piedmont rain-runoff in Beiwucun gravel pit, Beijing. The result showed that, with the increase in rainfall return period, the peak time significantly advances and the peak flow increases significantly. The Gravel pit in Beiwucun not only evidently mitigates piedmont flood, but also improves the urban river drainage standard and increase the amount of groundwater resources, due to the storage, detention, infiltration of the gravel pit.

1 INTRODUCTION

With the development of urbanization, the city's water resources and rain-flood disasters become more and more serious problems. On the one hand, urbanization leads to the increase in the demand of the city's water resources and brings about urban water crisis; the other hand, urban rain-flood drainage system pressure is increasing due to the change in the characterization of urban hydrology along with the increase in rain-flood disasters. Therefore, how to effectively make use of rainwater resources and choose a model that reflects the characteristics of urbanized area rainwater has become the urgent topic to be studied. General methods are only applicable to the simple urban drainage system with the following characteristics: (1) drainage area is less than 3 km²; (2) dendritic pipe network system; (3) simple outlet. With the increasing levels of urbanization, the complexity of underlying conditions is increasing. Urban flood control and environmental protection problems become more prominent. Urban hydrology study has not only limited to urbanization area, it should be able to analyze runoff yield and confluence characteristics of the large urban area, simulate complex hydraulic phenomena (pressure flow, backwater, overloading, backflow) and annular pipe network system to provide the basis for the layout, design, installation of the city sewer and hydraulic construction. So there is an urgent need of a more advanced and comprehensive method.

This research uses SWMM model based on GIS technology to set up piedmont rainwater utilization model in the center of Beiwucun gravel pit, Haidian, Beijing and simulate the piedmont rainwater utilization.

2 SWMM MODEL

SWMM (Storm Water Management Model, SWMM) is a comprehensive computer model of rainwater quantity and quality analysis for the urban area. Since first version developed in 1971, through the generations of upgrades, it is widely used in the planning, analysis, design and management of rainstorm runoff, combined pipelines, sewers and other pipelines in urban areas. In addition, there are a lot of applications in non-urban areas. The latest version, SWMM 5.0 is produced in October, 2004 by the Water Supply and Water Resources Division of the U.S. Environmental Protection Agency's National Risk Management Research Laboratory with assistance from the consulting firm of CDM Inc. The version with the Windows operating platform, friendly visual interface environment and improved processing capabilities can edit input data, simulate hydrologic and hydraulic process, water quality as well as provide the result of the rich expression.

SWMM can simulate both of a single rainfall event and the continuous rainfall events (independent of the calculation of the limit on the number of time step). The SWMM typical applications include: (1) to design the size of the facility of the drainage system to control flood; (2) to design the size of detention storage facilities, ancillary facilities and the facilities of water quality protection; (3) to draw flood plain of natural channels; (4) develop control strategies that reduce combined pipeline overflow; (5) to assess the impact of the inflow and infiltration of sewer overflow; (6) to study of non-point source pollution; (7) to assess practices effect of best management that reduce the pollution load during the rainy season.

3 STUDY AREA

Beiwucun gravel pit located on the west of Beiwucun Road, on the south of the Gold River, Haidian. The south is a large green area and it is abandoned artificial sand mining pit in history. Currently it covers an area of approximately 40,000 m², the bottom of the pit area of about 20,000 m². Its volume is about 160,000 m³ around with planted shrubs and the entire slope and the bottom of the pit is almost no vegetation. Due to the lack of a drainage system in west side of the pumping station compound 1, rain water and sewage mixed inflow into Gold River. The Gold River starting point at the intersection of Nanhan River and Wan'an East, end import Kunyu River, flows through the village of Zhongwu, Beiwu, Houyao, Chuanying Etc. The total length is 5.02 km and the drainage area is 5.2 km². It is a major flood-relief channel in the northeast of Nanhan River. Gold River is a seasonal flood relief river. Gold River is a seasonal flood-drainage river, this place throughout the year from June to September is the rainy season, according to field survey on the primary river management units, the river throughout the day rainfall runoff to reach more than about 20 mm will form. The river will form runoff when rainfall exceeds 20 mm and it need to open the downstream Kunyu connection at the check gate, in order to avoid downstream diffuse embankment when rainfall exceeds 25 mm. If rainfall exceeds 45 mm it need to shut the check gate between Nanhan River and Gold River to avoid too mush water from Nanhan River inflowing into Gold River, causing the river diffuse embankment. In the region, the annual average rainfall is about 600 mm, precipitation not only changes very uneven inter-annual but also within the year, mostly concentrated in the June to September. Average annual water surface evaporation is about 1200 mm (E601) with the annual average wind speed of 2.6 m/s. The gravel pit is located in the south side of Gold River, connected by two 1 m diameter pipe jacking with Gold River the main north–south dry river drainage. Gravel pit pool of rainwater mainly comes from the Xiangshan, Xiangquan Roundabout region and nearby areas' rainwater. Firstly, the rainwater feeds into Nanhan river and Beihan river, and then joins Gold River. Finally, the rainwater enters the Gravel pit through two pipes to store and infiltrate to achieve the purpose of elimination of flood disaster and groundwater complement.

4 GENERALIZATION OF THE STUDY AREA

The study area is the catchment of the Beiwucun gravel pit. The water enters the nearest drainage ditches, roadside drains and other drainage facilities firstly, after the runoff generation of the rainfall in the area. And then it flows into the nearby river. Eventually it reaches the gravel pit to be infiltrated and stored. After investigation, the upstream catchment area of gravel pit mainly includes five parts: (1) River area; (2) the Xiangquan Roundabout south of the Fifth Ring Road, Min Zhuang Road and the arid South River surrounded area; (3) Xiangshan area in the west of Xiangshan Roundabout; (4) Botanical Garden area in the north of Xiangquan Roundabout; (5) Xiangshan South Road and an area surrounded Min Zhuang Road and Fifth Ring Road. We use ArcGIS to match Remote sensing image with Beijing 1:1000000 basic geographic data. The DEM of the study area is cut from the 100 m * 100 m DEM of Beijing. In order to truly reflect the flow of water, we need to expand the scope of study. Based on the P-code with DEM, a large study area is divided into 174 catchments: For the mountain, P coding carved out of the ridge, and reasonable coding can be used as the basis for the determination of the model boundary; for most of the plain area of the city, P encoded by sub-basin is not adaptable. In SWMM, on the basis of the P-code division sub-basin, according to the data of the pipe network, the study area catchment area are redistricted, eventually study area is divided into 43 catchment areas. The results are shown in Figure 1.

5 THE ESTIMATESION OF MODEL PARAMETERS

The rainwater simulation parameters are the typical values of the reference the SWMM model user manual and learn from the gravel pit entrance flow, river flow, determination of geological parameters and infiltration rate measurement research. Horton infiltration model is used to simulate rainfall infiltration of the study area in the process of deducing each sub-basin runoff, Models need to enter the maximum infiltration rate, the smallest infiltration rate and the attenuation coefficient taken as 76.2 mm/h, 3.81 mm/h and 0.0006. Nonlinear reservoir model is used to simulate confluence calculation. The main parameters include surface slope, permeable surfaces, impervious surface, Manning coefficient of pipeline, depression storage capacity of permeable surface and impervious surface. Based on actual observations, depression storage capacity of permeable surface and impervious surface were taken for the 14 mm and 3 mm, the average surface slope of 1/5000. Referring the rainfall runoff water quantity

Figure 1.   Catchment area division of SWMM model.

and quality simulation results of three typical cities in South Florida by Vassilios A. Tsihrintzis et al, and in accordance with the underlying surface of the gravel pit area, permeable surface, impervious surface and pipeline Manning coefficient were taken 0.03, 0.015, and 0.013.

6 APPLICATION OF THE MODEL

6.1 *Calculation of design rainfall*

Storm frequency statistical period chooses 10 min, 30 min, 1 h, 6 h, 12 h, 24 h, 3 d, 7 d. According to the layout of the gravel pit nearby rainfall stations and the scope of its control, the gravel pit is controlled by Yihedong Water Gate (the rainfall observation stations have a long series of rainfall observation data). We can get period the rainfall statistical parameters, mean rainfall in each period, the coefficient of variation, and skewness by referring Beijing rainstorm Atlas, and thus calculate the return period of 1 year, 5 years, 10 years, 20 years, 50 years period design rainfall results in Table 1.

6.2 *Distribution of design storm process*

We use design rainfall values of different reproduction period to calculate 1 year, 5 years, 10 years, 20 years and 50 years of design storm process. As shown in Figure 2.

6.3 *Results and analysis*

Using the model to calculate return period in the outlet section of the study area that were 1 year, 5 years, 10 years, 20 years and 50 years of peak flow and total runoff. The simulation process and simulation of the outlet section is similar to the output, so we will only focus on Gravel pit inlet. Gravel pit in the different frequencies under design conditions inlet rainstorm simulation results contrast in Figure 3, Gravel pit inlet flow comparison shown in Figure 4.

Figure 3 is a simulation comparison of the results of four frequencies rainstorm. With the return period increase runoff volume gradually increase, this is mainly because of infiltration by rainfall intensity greater impact. When the rain powerful stable ground infiltration rate stability into the leachate infiltration, infiltration is no longer influenced by the rainfall intensity. As shown in Figure 4, peak time significantly advance, the duration of the flood peak and peak flow increase significantly with return period increasing significantly. This means that the catchment time will obviously increase, watershed within the low-lying areas

Table 1. Analysis results of time rainfall frequency in gravel pit.

Statistical parameter	Maximum							
	10 min	30 min	1 h	6 h	12 h	24 h	3 d	7 d
Mean	16	31.2	40.3	62.1	74.4	87.9	104.6	143
Cv	0.32	0.38	0.47	0.43	0.45	0.45	0.39	0.49
Cs/Cv	3.5	3.5	3.5	3.5	3.5	3.5	3.5	3.5
Year frequency	8.2	13.3	17.2	26.7	31.9	37.7	44.7	61.1
5 years frequency	19.8	39.7	53	80.5	97.3	114.9	133.7	189.1
10 years frequency	22.9	47.1	65.4	97.6	119	140.5	159	235.5
20 years frequency	25.7	54.1	77.5	114.2	140	165.3	183.5	281.1
50 years frequency	29.3	63	93.3	135.3	167.2	197.5	214.6	341

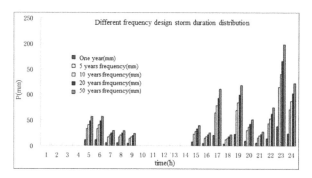

Figure 2. Different frequency design storm duration distribution.

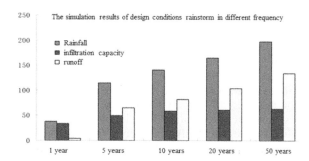

Figure 3. The simulation results of design conditions rainstorm in different frequency.

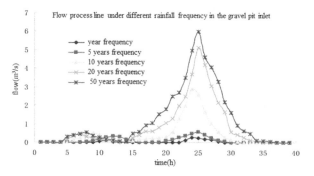

Figure 4. Flow process line under different rainfall frequency in the gravel pit inlet.

submerged time increases, the people and property are suffered a greater threat. However, on the one hand the presence of a gravel pit improves the urban river drainage standards, on the other hand increases the amount of ground water resources, enhance the effects of environmental, which benefits for the sustainable development of city.

6.4 *The analysis of flood detention and abatement flood peak*

It can be obtained in certain rainfall conditions of the peak and of the gravel pit and detention effect with infiltration analysis on the volume of sand and gravel pit. Gravel pit is divided into 18 layers when measuring, the actual measurement of the area of each layer and each layer corresponding elevation can calculated water storage volume of each layer and the total water storage volume, storage volume elevation curve that shown in Figure 5.

47

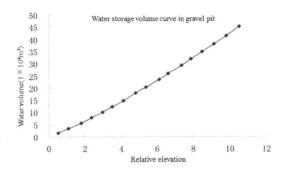

Figure 5.   Water storage volume curve in gravel pit.

Based on the above results can be drawn, Relative to a one-year return period rainfall gravel pit assimilate rainwater was 39.7%, 5-year return period rainfall consumptive rainwater was 12.7%, 9.8% of the 10 year return period rainfall assimilate rainwater was once in 20 years rainfall to dissolve rainwater was 8.6%, in 50 years of rainfall elimination satisfied rainwater was 7.5%. From the above results can be obtained: under the conditions of 945.3 hectares of land in the catchment area, the impervious area accounted for 21%, if the year is a case of row needed to equal the rainfall no runoff the infiltration conditions gravel pit water storage capacity of 1.5 million m³, five years case of row needed to equal the rainfall no runoff the infiltration conditions gravel pit water storage capacity of 4.64 million m³ year return period rainfall no runoff efflux requires equal infiltration conditions under gravel pit water storage capacity of 5.93 m³, two decades once in rainfall is no runoff efflux requires equal infiltration under the conditions of the gravel pit water storage capacity of 6.82 million m³, 50 years in case of rainfall no runoff efflux infiltration conditions under equal gravel pit the water storage capacity of 8.12 million m³; However, due to an area of investment, water balance, and other factors limit the size of the gravel pit can not be unlimited expand, and therefore able to control the general year return period rainfall runoff can not produce outflow.

7   CONCLUSION

1. The SWMM used in storm flood simulation of Beiwucun gravel pit, Beijing, is able to reflect the processes of runoff generation and routing, which provides technical support for the city's flood control and drainage.
2. The analysis focused on "the different rainfall frequency conditions" based on SWMM model. The results showed that, the simulated daily runoff increases significantly with the increase of the rainfall intensity, under the condition of unchanged land surface.
3. The utilization of storage volume in the city to store and infiltrate monsoon floods, not only improves urban river drainage standards, but also increases the amount of ground-water resources, which enhances the effects of environmental benefits for the sustainable development of cities.

REFERENCES

Cong Xing-yu. 2006. The rain flood simulation analysis based on SWMM in Beijing. Water Resources and Hydropower Engineering. Volume 41. 64–67.
Liu Jun, Xu Xiang-yang. 2001. The application of urban rainwater drainage analysis calculation model in Tianjin. Haihe Water Resources. Volume 1. 9–11.
Liu Lin-lin, He Jun-shi. 2006. The impact of urbanization on urban rainwater resources. Journal of Anhui Agricultural Sciences. 16 rainstorm Atlas. 1999. Beijing hydrological Manual. Beijing Water Authority. Volume I SWMM model user manual.
Zhao Shu, Qi Jin, Cun Tian. 2009. The application of SWMM in Beijing. Water supply and Water Drainage. 448–451.

Hydraulic Engineering – Xie (Ed.)
© 2013 Taylor & Francis Group, London, ISBN 978-1-138-00043-8

Study on applications of neural network to flood forecasting in Yiluo River

Yue Yanbing
College of Hydrology and Water Resources, Hohai University, Nanjing, China
Shanxi Conservancy Technical College, Yuncheng, Shanxi, China

Li Zhijia
College of Hydrology and Water Resources, Hohai University, Nanjing, China

ABSTRACT: The applications of BP neural network to flood forecasting in Yiluo River based on the characteristics of channel flood propagation are studied in the paper. By training of network according to the historic flood data of White Horse Temple, Luohe and Longmen town, Yi River at the upstream of Yiluo River, and the results of simulating Heishiguan flood showed that BP neural network is useful on Yiluo River Flood Forecasting.

1 INTRODUCTION

There are many artificial neural network forms, of which the feedforward neural network technology based on BP algorithm is much more active in the hydrological forecasting (Gu and Wang. 2004; Yuan and Zhang. 2004; Wang. 2010). It is a special nonlinear mapping method, according to unary function's multiple conform to approximate the multivariate function, and it has good mathematical basis. Generally, neural network consists of three-layers, namely: input layer, hidden layer, and output layer (Chen and Xie. 2006). Input layer neurons usually represents certain physical significance. For instance, input layer neurons are often regarded as the main factor affecting the prediction precision in hydrological forecasting. The number of output layer and input layer neurons is related to the problems studied, and is easy to determine, while hidden layer need to be tested and often does not represent any physical significance. At present, artificial neural network has already had successfully application in many nonlinear timing modeling and there are also many literatures studying flood evolution problems by using neural network (Kang and Jiang. 2005).

Channel flood propagation has strong nonlinear effect. As neural network has strong ability to deal with large-scale complex nonlinear dynamics system, neural network technology has been applied in channel flood prediction. However, the result of channel flood prediction by using neural network model is not ideal (Li and Kong. 1997). Neutral network model is used to channel forecasting in Yiluohe channel, and some results have been obtained with certain accuuacy.

2 NEURAL NETWORK MODEL TEST

2.1 *Overview of the channel*

We select Heishiguan station, Yiluo River as downstream, and White Horse Temple in Luohe and Longmen town station in Yihe study channel as upstream. See Figure 1.

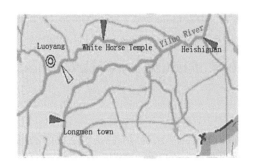

Figure 1.   Diagram of river.

Table 1.   Flood forecasting features of 4 floods at Heishiguan.

No.	Start time	End time	Forecasted peak flow (m³/S)	Measured peak flow (m³/S)	Peak absolute error (m³/S)	Average absolute error (m³/S)	Determined coefficient
1984092108	1984-9-21 8:00	1984-10-8 20:00	2386.4	2400	13.6	211	0.8470
1985091410	1985-9-14 10:00	1985-9-20 23:00	1508.4	1470	38.4	140	0.8239
1988080810	1988-8-8 10:00	1988-8-14 7:00	1522.8	1430	92.8	147.2	0.7340
1989070910	1989-7-9 10:00	1989-7-16 4:00	1271.6	1230	41.6	129.9	0.8266

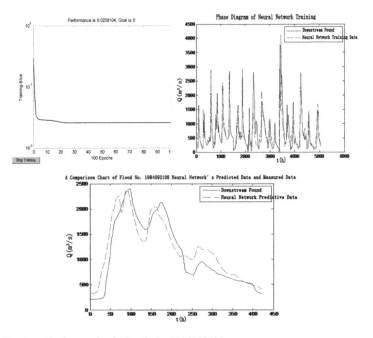

Figure 2.   Simulated hydrograph of Flood No. 1984092108.

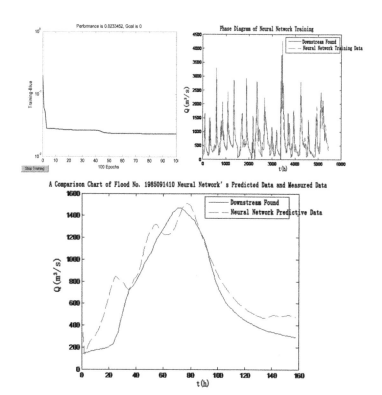

Figure 3.   Simulated hydrograph of Flood No. 1985091410.

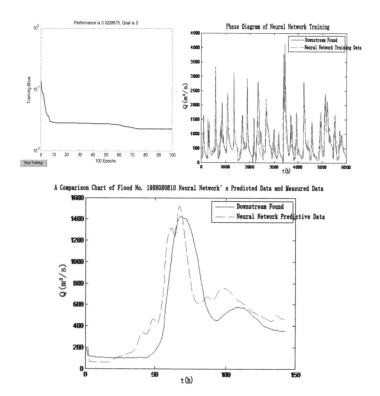

Figure 4.   Simulated hydrograph of Flood No. 1988080810.

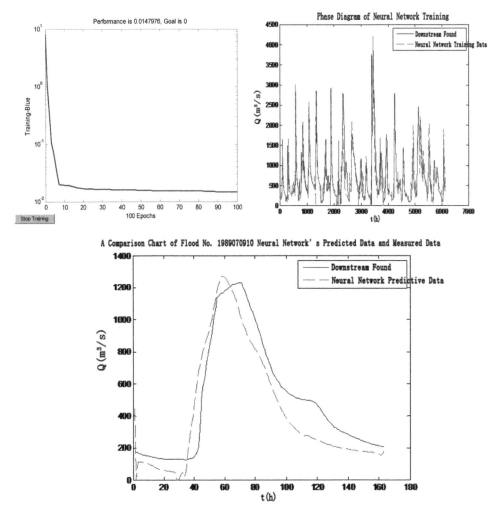

Figure 5. Simulated hydrograph of Flood No. 1989070910.

2.2 *Neural network model test*

Network is been trained based on the historical flood data of White Horse Temple and Longmen town, and the results of simulating flood forecast at four periods in Heishiguan is in Table 1. Two input nodes, five implied nodes, and an output node are used in the neural network structure. The results shows in Table 1 and Figures 2–5.

3 CONCLUSION

Neutral network technology could be used to improve accuracy of flood forecast in channel flow as shown in the results of the test of Heishiguan. However, there are certain choices in the neural network technology forecasting process, and the accuracy could be improved. Observed hydrological date is not enough in the training network. Later on we will gradually enrich measured dates, and other neutral network as well as method of data fusion so as to improve forecast accuracy.

REFERENCES

Chen Tianqing, Xie Jiancang. 2006. Dynamic Parameter Estimation for Muskingum Routing Model Based on BP Artificial Neural Network. *Journal of Hydroelectric Engineering.* 3(6):31–38.

Gu Xiao-ping, Wang Chang-yao. 2004. Research on the BP Neural Network for Hydrologic Forecast. *Ecology and Environmental Sciences.* 13(4):524–527.

Kang Ling, Jiang Tiebing. 2005. Study on Artificial Neural Network to Identify the Farameters of Flood Wave Equation. *Journal of Hydroelectric Engineering.* 1(2):98–101.

Li Zhijia, Kong Xiangguang. 1997. Channel Flood Routing Model of Artificial Neural Network. *Journal of Hohai University.* 5(9):7–12.

Wang Qi-hu. 2010. Application of BP Neural Network Model in Forecasting Runoff of Liuxihe Reservoir. *Environmental Protection Science.* 3(6):19–21.

Yuan Jing, Zhang Xiao-fen. 2004. Real-time Hydrological For ecasting Method of BP Neural Network Based on Forgetting Factor. *Advances in Water Science.* 11(6):787–792.

Hydraulic Engineering – Xie (Ed.)
© 2013 Taylor & Francis Group, London, ISBN 978-1-138-00043-8

Application of Ensemble Kalman filter in parameter calibration of Muskingum model

Yue Yanbing
College of Hydrology and Water Resources, Hohai University, Nanjing, China
Shanxi Conservancy Technical College, Yuncheng, Shanxi, China

Li Zhijia
College of Hydrology and Water Resources, Hohai University, Nanjing, China

ABSTRACT: Flood forecasting procedure and its characteristics of Ensemble Kalman Filter technology have been discussed. Parameters for Muskingum model has been calibrated dynamically by adopting the Ensemble Kalman Filter technology. The observed flood data of Heishiguan of Yellow River has been tested for compared test and the results are good. Such kind of dynamic calibration of parameters has a high accuracy and good effect.

1 INTRODUCTION

Muskingum flow routing algorithm proposed by G.T. MaCarthy in 1938 was widely applied in rivers of the world. A lot of further discussions and research had been carried out and improved gradually by many scholars (Rui. 2002). The channel storage equation and flow algorithm equation of Muskingum model are shown as follows (Zhao. 1983):

$$W_t = K\left[X_t I_t + (1 - X_t) O_t\right]$$ (1)

$$O_{t+1} = C_0 I_{t+1} + C_1 I_t + C_2 O_t$$ (2)

In the above equation, I is the upstream input flow, O is the downstream output flow, K is the slope of the relation curve for storage and flow, X is the proportion coefficient of the flow, C is the flow algorithm equation coefficient.

One of the important problems in the practical application of Muskingum model is the calibration of model parameters X, K or C_0, C_1 and C_2. Study on the dynamic calibration of Muskingum model parameters has been conducted with Ensemble Kalman Filter technology in this paper.

2 ENSEMBLE KALMAN FILTER TECHNOLOGY

In Kalman Filter, the background error was not supposed as a constant, and instead, it was calculated according to the forecasting equation of covariance matrix (Gao. 2005). Aiming at the problems such as the calculation instability in the covariance matrix forecasting model of the Kalman Filter, Ensemble Kalman Filter was proposed, whose main idea was to discard the covariance matrix forecasting model and take direct advantage of multi-integral model of Monte-Carol method. Thus the covariance matrix for the background error could be obtained with smaller computation storage than that of the Kalman Filter (Jurgen and Gunter. 2008).

Ensemble Kalman (EnKF) was a pure statistical Monte-Carol method, whose assembly of model state was changed in the state space with time. The average value of assembly was the optimal estimation for the model forecasting value, and the divergent distribution of the assembly would represent error variance. In the measurement, each measuring value would be replaced by the other assembly, and the estimated value would be the optimal estimation of the measured value, the variance would reflect the measurement error. EnKF assimilation method consists of forecast and analysis (Burgers and van. 1998; Reichle and McLaughlin. 2002; Evensen. 2003).

Forecast: for a hydrological calculation model, the analysis assembly obtained in analysis phase would be regarded as the initial field which will proceed to the next data observing period (namely the data assimilation moment t):

$$X_{bt}^n = MX_{at}^{n-1} \tag{3}$$

In equation (3), X is the state variable, and M is forecasting mode.

Analysis: the model forecasting assembly was obtained in the forecasting part, and actual value was replaced by the average assembly value, then the average solution to the assembly could be analyzed by assimilating the observing data. The flow measurement will be uncertain since the flood discharge in the river was a stochastic dynamic process. In the parameter estimation of Ensemble Kalman Filter, the uncertain on-spot observed value could be simulated by exerting observation noise on the observed value. For Ensemble Kalman Filter, addition of certain noise could also improve the stability of the filter, so that the estimated standard deviation would not decrease dramatically[5]. Simulate white Gaussian noise with Monte Carlo method (average value is 0, and standard deviation could be obtained by multiply the observed value with noise scale factor), and flow observation vector would become a group of undisturbed observation assembly, in which the ith sample could be represented as:

$$X_i' = HX_i + \varepsilon_i \tag{4}$$

In equation (4), H is observation operator and ε is observation error. According to the actual value and test value, the model parameters can be updated directly by combining the forecasting error matrix.

$$X_{at}^n = X_{bt}^n + K(X_t^n - HX_{bt}^n) \tag{5}$$

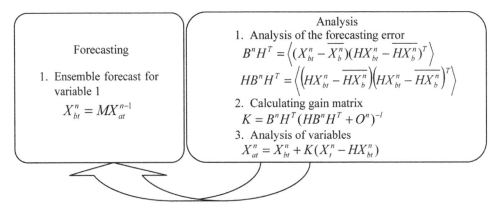

Figure 1.   Calculation procedure of Ensemble Kalman Filter.

$$K = B^n H^T (HB^n H^T + O^n)^{-l} \tag{6}$$

$$B^n H^T = \left\langle (X_{bt}^n - \overline{X_b^n})(HX_{bt}^n - \overline{HX_b^n})^T \right\rangle \tag{7}$$

$$HB^n H^T = \left\langle \left(HX_{bt}^n - \overline{HX_b^n}\right)\left(HX_{bt}^n - \overline{HX_b^n}\right)^T \right\rangle \tag{8}$$

In the above equation, K is Kalman gain matrix, O is the observation error covariance matrix and B is the background error covariance matrix. The calculation process of data forecasting for Ensemble Kalman Filter is shown in Figure 1.

3 CALIBRATION AND TEST OF MUSKINGUM MODEL PARAMETER WITH ENSEMBLE KALMAN FILTER

3.1 Overview of river channel for study

The downstream Heishiguan station of Yiluo River, the upstream White Horse Temple station of Luo River and upstream Longmen town station of Yi River have been selected for study, as shown in Figure 2.

3.2 Dynamic calibration and test for model parameters

The flood data for White Horse Temple and Longmen town has been adopted for the simulation of flood forecasting of 4 floods at Heishiguan during the forecasting period, and the results have been shown in Table 1 and Figure 3.

Filter divergent analysis: although the filter gain method has been adopted for error compensation, it cannot be estimated that the forecast value is related to the actual value due to the finite test assembly, so that the assembly supposed to be zero is not zero, and the correlation between the forecast and actual value is over-estimated. With the delay in forecasting period, evident filter divergent problems may occur, which will drive the analysis field close to the background field, and finally the forecasting value will deviate from the actual value. In the river flood forecasting of Ensemble Kalman Filter, the number of assembly and proper extended forecasting period should be added to conduct tests which will prevent the filter from diverging. Further studies on the application of Ensemble Kalman Filter data assimilation technology in the hydrological forecasting domain should be carried out.

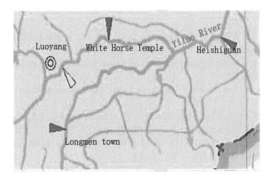

Figure 2. Scheme of river.

Table 1. Flood forecasting features of 4 floods at Heishiguan.

No.	Start time	End time	Forecasted peak flow (m³/S)	Measured peak flow (m³/S)	Peak absolute error (m³/S)	Average absolute error (m³/S)	Determined coefficient
1984092108	1984-9-21 8:00	1984-10-8 20:00	2399	2400	1	23	0.9970
1985091410	1985-9-14 10:00	1985-9-20 23:00	1466.2	1470	3.8	25.4	0.9942
1988080810	1988-8-8 10:00	1988-8-14 7:00	1419.6	1430	10.4	27.7	0.9820
1989070910	1989-7-9 10:00	1989-7-16 4:00	1229.6	1230	0.4	22.4	0.9872

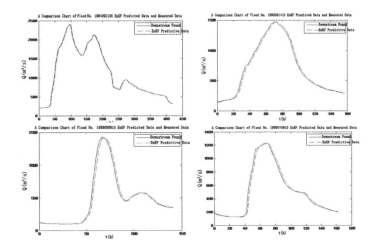

Figure 3.   Comparison of measured and calculated hydrograph of 4 floods at Heishiguan.

4   CONCLUSION

Ensemble Kalman Filter technology can integrate information from multiple sources, which can improve the accuracy of river flood forecasting effectively. Good results have been achieved from 4 floods at Heishiguan, as well as the dynamic calibration of Muskingum Model Parameter with Ensemble Kalman. The hydrological observation data has been adopted for testifying the algorithm in this paper. And in the future, indirect observation information with satellite or radar can be employed in information integration, which can continuously revise the forecasting track of the model, improve the forecasting accuracy and extend the forecasting ahead period.

REFERENCES

Burgers G, van Leeuwen PJ, Evensen G. 1998. Analysis scheme in the Ensemble Kalman Filter. *Mon Weather Rev*. 126:1719–1724.
Evensen G. 2003. The Ensemble Kalman Filter: Theoretical formulation and practical implementation. *Ocean Dynamics*. 53:343–367.
Gao Shuanzhu. 2005. Review on Ensemble Kalman Filter Data Assimilation. *Meteorological Monthly*. 6:3–8.
Jurgen Komma, Gunter Bloschl, Christian Reszler. 2008. Soil moisture updating by Ensemble Kalman Filtering in real-time flood forecasting. *Journal of Hydrology*. 2008;357:228–242.
Reichle RH, McLaughlin D, Entekhabi D. 2002. Hydrologic data assimilation with the Ensemble Kalman Filter. *Monthly Weather Rev*. 130:103–14.
Rui Xiao-fang. 2002. Some theoretical studies on the Muskingum method and its successive routing in subreaches. *Advances in Water Science*. 13(6):682–688.
Zhao Renjun. 1983. Catchment Hydrological Modelling—Xinanjiang and Shanbei Modelling. China Water Power Press: Beijing.

Hydraulic Engineering – Xie (Ed.)

Discussion on the selection of influent water quality of municipal wastewater treatment plant

Gao Ting
China Three Gorges University, Yichang, China

Zhang Kai
North China Municipal Engineering Design and Research Institute, Tianjin, China

Jin Tiantian
China Institute of Water Resources and Hydropower Research, Beijing, China

ABSTRACT: The design and actual influent water quality from 27 municipal wastewater treatment plants was compared in this paper. The results showed that: the design range was concentrated and could not reflect the real influent water quality received by individual plant. Most of the design value is close to the average influent water quality, but could not represent the worst water quality conditions. In most cases, the actual influent water quality deviated from the design water quality fluctuated largely. Vague concept of the design influent quality, indefinite calculation method, and inconsistent reference data are the major causes of "favor run, slight design" which is not conducive to the operating and managing of water plant. Specifying the connotation of the design water quality, giving emphasis on statistical analysis of actual data, and introducing scenarios with different water quality and quantity in design can make the design scientific and reliable.

1   INTRODUCTION

Municipal wastewater treatment plant is a critical infrastructure which be one of the fields that the country makes great efforts on for cutting pollutant emissions, reducing the environmental burden, improving the water environment. By the end of late 2010, the daily processing capacity of 102.62 million cubic meters and municipal wastewater treatment rate of 76.9% in the 1214 municipal wastewater treatment plants played an important role in the protection of the ecological environment, and the improvement of the urban living environment.

The reasonable design and stable operation of municipal wastewater treatment plants guarantee the river basin water environment quality. The influent quality and quantity is the key basis for determining the size and process of the municipal wastewater treatment plants, however, the water quality and quantity forecasts (especially water quality forecast) are weak links existing commonly in the engineering design of current municipal wastewater treatment. In the project, the actual influent quality deviated from the design, often affects the target rate of effluent quality and the economical operation of water plants. According to the representative drain outlet water quality data measured regularly, and determining the design value based on frequency of influent pollutants concentration (Ju and Peng, 2007; Zhou and Zhou, 2006), etc. can reduce deviations caused by subjective factors in the design. Inevitably, the discrepancy exists between the actual influent quality and the design quality because of the absoluteness of water quality fluctuation. This article by the analysis of the data of the design and operation of the wastewater treatment plants, quantified the differences, discussed the connotation and positioning of the design quality and proposed the possible ways to reduce the adverse effects caused by the subjective and objective differences in the design for the researchers to discuss and reference.

2   DATA AND PROCESSING

2.1   *Data source*

Accessing to the literatures of the past 10 years on the design and operation of municipal wastewater treatment water plant, from which the 27 instances with a comprehensive design on influent quality and actual influent quality data are available for reference. The actual influent quality is the maximum or minimum value and the mean of the average monthly value of conventional water quality such as COD, BOD_5 (BOD), SS, TN, TP, etc. If there is no special version in this paper, "minimum" refers to the minimum of average monthly value, "average value" refers to the average value of average monthly value that is similar to the annual average value, and "maximum value" is the maximum value of average monthly value.

2.2   *Data processing*

Due to the varying original data sources, there are missing values in some of the data. Apart from the individual missing average value of COD, BOD, SS that has been completed with the average value, there is no additional handling missing values.

In order to facilitate the discussion of the differences between the actual influent quality and the design influent quality, two other values are defined:

1. R refers to the ratio that is defined as the ratio of the actual influent quality and the design influent quality;
2. Δ refers to the degree of difference that is defined as $\Delta = |R - 1|$.

The values of R and Δ corresponding to various indicators were calculated by definition. RCOD refers to the ratio of the actual influent COD mean and design influent COD; RCOD1 refers to the ratio of actual influent water COD maximum value and the design influent water COD, ΔCOD is the degree of difference of COD mean value, ΔCOD1 is the degree of difference of actual influent water COD maximum value, and so on.

In addition, among the original data, there are NH_3-N data in some water plants, TN data in others. They are not strictly separated in the ratio analysis, but labeled TN in the discussion.

3   STATISTICS AND ANALYSIS

Statistical Product and Service Solutions (SPSS) is applied in statistics for the frequency descriptive statistics analysis. The statistics include the variable group maxima or minima and range, standard deviation, mean, median, skewness, kurtosis, and percentile values that include quartile and tens percentile. The output includes frequency statistics scale, frequency tables and histograms.

3.1   *Water quality data statistics and analysis*

Without regarding to the corresponding relationship among the data in each case, this paper only makes statistical analysis of the data characteristics of various water quality indicators available, compares to the distribution characteristics of each index design value, the actual influent water average value and maximum value.

As *N* is not unified, there is no analysis and discussion on it. The main statistical results of water quality data are shown in Table 1. The frequency distribution of design, average and maximum values of COD, BOD, SS, and TP is reflected orderly in Figure 1a~d. The abscissa

Table 1. Statistics of water quality data.

	COD design value	BOD design value	SS design value	TP design value	COD mean	BOD mean	SS mean	TP mean	COD1	BOD1	SS1	TP1
N												
Effective value	25	27	27	22	27	26	24	21	21	21	21	16
Missing value	2	0	0	5	0	1	3	6	6	6	6	11
Mean	365.2	178.2	235.7	5.3	327.6	154.8	225.9	4.8	482.4	230.3	339.4	7.8
Median	350.0	165.0	220.0	4.0	252.5	122.3	173.0	3.6	367.0	182.0	280.0	4.8
Standard deviation	133.795	68.432	98.339	3.747	258.331	135.971	205.63	5.865	334.931	192.073	267.248	11.838
Skewness	2.812	2.824	2.915	3.191	2.861	3.022	3.599	3.995	2.097	2.441	3.195	3.702
Kurtosis	10.699	9.761	12.217	11.714	9.785	10.666	15.392	17.201	5.255	7.380	12.386	14.262
Range	700	350	550	18	1270	671	1066	28	1488	869	1262	50
Minimum value	200	100	100	2	93	31	37	1	100	36	122	1
Maximum value	900	450	650	20	1362	702	1103	29	1588	905	1384	51

Note: "COD design value" refers to the COD design value data group, "COD mean" refers to COD actual average value data group, "COD1" refers COD actual maximum data group, and so on.

Table 2. Statistics of R.

	RCOD	RBOD	RSS	RTP	RTN	RCOD1	RBOD1	RSS1	RTN1	RTP1
N										
Effective value	25	26	24	21	23	19	21	21	17	16
Missing value	2	1	3	6	4	8	6	6	10	11
Mean	0.843	0.825	0.895	0.945	0.987	1.191	1.175	1.325	1.245	1.427
Median	0.757	0.718	0.848	0.846	0.864	1.118	1.200	1.160	1.096	1.229
Standard deviation	0.399	0.454	0.478	0.579	0.626	0.558	0.639	0.602	0.648	1.139
Skewness	1.743	1.795	1.062	2.155	3.002	1.245	1.339	1.023	0.921	2.520
Kurtosis	4.205	5.034	1.365	6.609	11.374	2.594	3.137	2.010	-0.349	7.685
Range	1.81	2.18	2.02	2.80	3.03	2.39	2.84	2.63	2.04	4.93
Minimum value	0.37	0.24	0.18	0.13	0.44	0.40	0.28	0.41	0.48	0.19
Maximum value	2.18	2.41	2.21	2.93	3.46	2.79	3.12	3.04	2.52	5.12

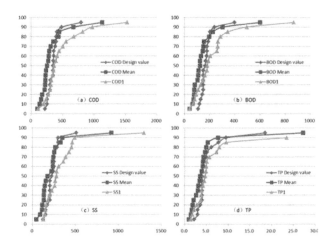

Figure 1.   Frequency distribution of design, average, max values of each index.

indicates pollution concentration (mg·L⁻¹), and the vertical coordinate is the percentage that is not more than certain concentration.

In Table 1, there is no clearly difference of mean and median in the design value data group, and the skewness is at about 3, which reflects the design value in line with the normal distribution, slightly left-skewed, and the relative concentrated value range (kurtosis 9.8~12.2). Figure 1 shows that 80% or more of the design water value of COD, BOD, SS, TP concentrated in the range 250~400, 120~200, 150~300, 2~7 (mg·L⁻¹), which is substantially the conventional, typical water quality in the Reference Design manual and design examples in the related courses.

There are some differences between the actual quality data and the design values. First, observing the distribution of the average value and maximum value of the actual influent within the aforementioned common design value range, and the ratio of the 4 indexes can be calculated according to the frequency table, that are 55% and 43%, 34% and 32%, 41% and 42%, 70% and 77% respectively. It is clearly that conventional design value interval cannot reflect exactly the actual situation except TP data. Second, the range of the actual influent quality data is 1.6~1.9 times as that of the design value, its range of maximum value is more than 2 times (2.1 to 2.8) as that of design value, and the standard deviation is also relatively large, all of which reflect the difference between the actual influent quality and design water quality, but also illustrate great differences (larger than expected) of the influent quality among the different water plants. If designing as the typical water quality without the specific analysis, the design may be not match with the actual situations. Third, the variable mean shows that the average value of actual influent quality is close to or slightly lower than that of design water quality. The values of COD, BOD, SS, and TP are lower to 11%, 13%, 4%, 9% respectively, while the maximum values are significantly greater than the design values, 32%, 29%, 44%, and 47% higher respectively. The differences will be much greater, when applied in daily water quality maximum value, and hourly water quality maximum value.

3.2   The statistics and analysis of differences between the actual influent quality and design quality in each case

3.2.1   The ratio statistics and analysis

Table 2. The great difference between the minimum and maximum values of the ratio (0.13~5.12) reflects the big differences of the running load in different water plants. The value of R that is calculated by the average value of influent quality reflects the long-running state, from which, we can see that reserved too much and expected shortfall are both in the design.

As shown in Figure 2, the ratio distribution can be found out combined with the Table 2 and the histogram output from SPSS. In addition to RBOD1, the mean value of R is slightly larger than its median, and slightly left-skewed distribution; the kurtosis values are not high reflecting that the distribution concentration ratio is not high; the abnormal values are in the upper part of the box plot, and the abnormal values of COD and BOD of the organic pollution, SS of the insoluble pollutants, and TN, TP of the nutritional elements are all from different cases, which suggests that the high value are not necessarily appeared synchronously in each index, and some cases including the actual value of single indicator far larger than the design value. RTN kurtosis is relatively large and the mean is 0.99, while RTN1 is the left-skewed bimodal form, whose data are relatively scattered. This can reflect the design value of the indicator in line with the average concentration of influent water; therefore, it cannot reflect clearly the peak concentration.

R distributed on both sides of "1" shows that the actual influent water concentration is lower or higher than the proportion of the design value. By the frequency table, the average values of the actual influent water COD, BOD, SS, TN, TP are lower than the proportion of design values, (i.e. R < 1), which are 72%, 69%, 71%, 70% and 71%; the maximum values are higher than the design values (R > 1), the proportions of which are 63%, 57%, 76%, 53% and 69%, respectively. These results indicate that the working conditions of the design water quality are usually worse than that of the average water quality, but can be not sufficient to characterize the adverse working conditions of water quality, as the most unfavorable monthly average, daily average, even hourly average.

3.2.2 Statistics and analysis of the degree of differences

The design theory and working practice both recognize and permit slight discrepancies between actual values and design values. Figure 3 directly reflects the degree of difference of the actual water quality compared to the design water quality.

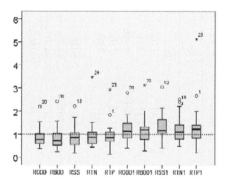

Figure 2. Box-plot of R.

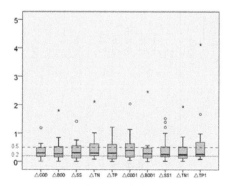

Figure 3. Box-plot of Δ.

Table 3. Statistics of water quality fluctuation.

	Average monthly MAX/average value					Average monthly MAX/average monthly MIN				
	COD	BOD	SS	TN	TP	COD	BOD	SS	TN	TP
Effective value N	21	21	20	18	16	21	21	20	18	16
Mean	1.40	1.36	1.41	1.32	1.36	2.23	2.16	2.51	1.95	2.13
Skewness	1.05	1.67	1.31	2.86	0.35	2.06	1.96	2.80	2.37	1.26
Kurtosis	−0.036	2.642	0.421	9.466	−1.062	4.658	3.604	9.634	5.691	1.458
Minimum value	1.07	1.09	1.08	1.02	1.10	1.15	1.20	1.17	1.04	1.21
Maximum value	2.10	2.18	2.17	2.64	1.75	6.15	5.50	8.60	5.98	4.32
Concentration factor										
1~1.2	28.6%	33.3%	30.0%	50.0%	31.3%	14.3%	4.8%	5.0%	22.2%	0.0%
1~1.5	71.4%	81.0%	80.0%	88.9%	68.8%	23.8%	28.6%	25.0%	55.6%	31.3%
1~2.0	95.2%	95.2%	85.0%	94.4%	100.0%	71.4%	61.9%	50.0%	77.8%	56.3%
1~3.0	100.0%	100.0%	100.0%	100.0%	100.0%	81.0%	81.0%	75.0%	88.9%	87.5%
1~4.0	100.0%	100.0%	100.0%	100.0%	100.0%	90.5%	90.5%	90.0%	88.9%	93.8%

Note: Excluding an abnormal value, the minimum value of SS in Case 3.

The percentage of Δ value that is not greater than 0.2 and 0.5 can reflect respectively the proportions (within 20%~50%) of the actual water quality relative to the design values. Considering each water quality index, there are about 2/5 of the difference (≤20%), 3/4 of the difference (≤50%) between the average water quality and the design value; less than 1/3 of the difference (≤20%), and 3/5 of the difference (50%) between the average monthly maximum and the design values. Therefore, most of the working conditions deviate far from the design values in most water plants. This is bound to virtually increase the economic running difficulty of the water plants, and improve the requirements on the level of the management technology of water plants.

3.3 Water quality fluctuations statistics and analysis

There's no equalizing reservoir generally in municipal water treatment plants because of the smaller fluctuation of the water quality and quantity of sanitary wastewater compared with industrial wastewater. In order to observe the fluctuation of the actual influent quality, calculate the ratios of monthly average maximum value and the average value and monthly average minimum value of the actual influent quality, the main statistics results are listed in Table 3.

By statistical data, the mean of R of monthly average maximum and the average value is at about 1.4, up to 2.64; the mean of R of monthly average maximum and monthly average minimum is greater than 2, up to 8.60, reflecting large fluctuation of the actual influent quality.

The concentration factor of the R of the monthly average maximum and the average value can reflect the distribution of fluctuation range which the high level value of the influent pollutant concentration moves around the mean value. This shows that the monthly average maximum value exceeds more than 20% of the average value in several situations, 50% of that in 1/4 of situations, but not more than 1-fold of the average value in most situations.

The concentration factor of the R of the average monthly maximum value and the average monthly minimum value reflects that the gap of the monthly average concentration is more than 3 times among over 80% of the water plants, and the running load of the water plants is greatly.

4 PROBLEMS AND COUNTERMEASURES

4.1 The concept of design influent quality

Statistical analysis shows that, a certain discrepancy exists between the actual influent quality and the design influent quality. Substantially, the design water quality value range is close to the average water quality, which does not reflect the actual adverse working conditions. In the design and calculations of the structures and facilities, the design quality always appeared as the same image whatever the working conditions described by the flow, and it seemingly has the implication meaning of the average value. However, in accordance with the general principle, the adverse working conditions need to be designed or checked.

The design water quality is the average value or not, as well as the basis of using the average value is abundant or not, the fundamental reason of which is the ambiguity of the concept of "design influent quality".

Be different from the detailed provisions and discussions of the influent "design flow" of the water plant, there not many related design specifications, manuals and the theoretical books involving the provisions or detailed discussions on "design influent quality" of the wastewater treatment plants. The "design flow" in the drainage system is defined as each equipment and structure assuring the passable flow during the service period, which embodies the principle of "design as unfavorable conditions". Also the Wet-weather flow, the maximum daily maximum flow, maximum daily average flow, and daily average flow, etc. are also defined in the design of water treatment structures, which characterizes different working conditions, and the "design flow" with theoretical basis and computational methods is applied in a variety of computing and checking. For the same water quality with the volatility (i.e. the design water quality), there are no terminology and definitions, no definition on whether it is the mean or maximum value, and no consideration to the relationship between it and the design flow, while linking the water quality together with working conditions of specific and practical characteristics. The ambiguity of the concept will inevitably lead to a lack of positioning basis on argumentation and calculation of the design water quality.

4.2   *Methods for determining influent quality*

The methods for determining influent quality given in the design specifications are that the actual investigation, the adjacent or similar water quality for reference and calculation according to a certain standard (GB 50014-2006). However, it is not stipulated clearly how to deal with and analyze the investigation methods and the acquired water quality data by actual measurement. Take SS design reference value (mg·L^{-1}) for example, "the water quality indicator of typical sanitary wastes" (Chui, etc. 2004; Gao, 2003) listed three levels as high, medium and low, respectively 350, 220, 100, "the water quality parameters variations range of the general municipal sewage" (Chui, etc. 2004) listed as the range of 50~330, "the sewage water quality of different drainage systems in China's southern cities" (Chui, etc. 2004; Gao, 2003) listed both triage system and combined system as 150~250 and 70~150; "the water quality indicator and significance of common sewage" (Gao, 2003) that gives an average concentration of 200 can be calculated by 40~65 g per person per day (GB 50014-2006). Due to the fuzzy concept, unknown method, broad reference value, the special demonstration on the selection and determination of the design influent quality can be found infrequently in the design file, while it is usually mentioned but vaguely or specified directly "in line with relevant data". Whatever working condition was applied in the design flow in the design, there is no change on the design water quality, leading to the similar design water quality from different designs, the consistent design water quality with the same design and different working conditions. Eventually, the design quality becomes the data that meets only the need of the design and computation, missing the functions of forecasting, planning and guiding practice.

4.3   *Countermeasures*

4.3.1   *Explicit connotation of the design influent quality*
The explicit connotation of the design influent quality, the application of the average water quality or some certain adverse water quality, or corresponding guaranteed rate, which make the "design water quality" be the exact directivity to avoid missing calculation basis and fuzzy method for determining.

4.3.2   *Emphasizing on data collection, statistics and analysis*
The analysis of the actual data reflects the defects of the traditional experience estimation method and simple statistical method. Based on the statistical distribution of the data, the method that describes the features of wastewater by the water quality data whose probability is equal to or less than 10%, 50% and 90% has been used to determine the industrial waste-water influent quality.

With the widespread establishment and operation of China's urban wastewater treatment plants and environmental monitoring stations, the actual measured data that the probability analysis requires will be much richer. Taking advantage of the actual measured data in this region or similar regions to analyze and determine the design influent quality that can meet various typical guaranteed rates (such as 50%, 80% or 85%) will enable the design value method to be more scientific and the basis more adequately.

4.3.3   *Drawing on the theories and methods of scenario analysis*
Based on putting forward various key assumptions on various important elements that affect the system, the scenario analysis method conceives every possible case in the future by the

detailed and rigorous reasoning and description. The Conceive on every scenario in the future can deepen our understanding to the regularity and predictable things that affect the system, and the uncertain things, therefore to avoid the two most common mistakes: overestimation or underestimation of the future changes and impact.

The present various design flow for the design and checking of the same structures, and its essence is to consider the scenario analysis on uncertainty of water quantity. Considering that the influent quality is an important uncertainty that can affect the water treatment effects and equipment efficiency, the rationality and credibility of the design will be greatly enhanced with the overall consideration of water quality and quantity from different levels. The targeted structures design and equipment configuration will benefit the actual operation management of water plants so that improve operational efficiency.

5 CONCLUSION

It is difficult to avoid the adverse effect that the objective water quality fluctuations on the water treatment system, however, the fuzzy concept and subjective determination method of the design influent quality further increase the differences between the operation and design. Although the water treatment systems with a certain resistance to impact loads can response to the differences between the actual water quality and design water quality by adjusting the working conditions in the operation. Actually, the differences are usually too large and go beyond the system regulation ability; therefore the design will be out of action in formulating the efficient operation program for the system in advance, which goes against the operation and management of the wastewater treatment plants. The water quality characteristics distribution values identified by statistical analysis of the actual water quality data, combined with the water quantity fluctuation characteristic values, and using scenario analysis to determine the influent quality and quantity load of the water plants are the fundamental works on improving the credibility, practical and rationality of the design. In additional, how to select the characteristics value scientifically and how to combine different water quality and quantity load scenarios with the process designs need the engineering staff to make a further exploration and practice.

REFERENCES

Chui, Yuchuan. Liu, Zhenjiang. & Zhang Shaoyi. 2004. Calculation of Design of Urban Sewage Treatment Facilities. Beijing: Chemical industry press.

Gao, Junfa. 2003. Handbook of Process Design for Wastewater Treatment Plant. Beijing: Chemical industry press.

Ju, Xinghua. Wang, Sheping. & Peng, Dangcong. 2007. Determination Methodology for Design Influent Quality of Municipal Wastewater Treatment Plant. China Water & Wastewater. 23(14):48–51.

Shanghai Municipal Engineering Design Institute. 2004. Water Supply and Drainage Design Manual—Urban Drainage. Beijing: China architecture and building press.

Zhou, Kezhao. & Zhou, Mi. 2006. Influent Quality Estimation and Effluent Quality Evaluation of Municipal Wastewater Treatment Plants. Water & Wastewater Engineering. 32(9):26–30.

Hydraulic Engineering – Xie (Ed.)
© 2013 Taylor & Francis Group, London, ISBN 978-1-138-00043-8

Application of progressive optimality algorithm to cascade reservoir optimal operation in flood control

Jinglin Qian & Jun Hou
Zhejiang Institute of Hydraulics & Estuary Hangzhou, China

ABSTRACT: According to the optimal operating problem of the river basin flood control, and to analysis of the inner relation between the inflow, discharged ability of the cascade reservoir and the flood peak discharge of the flood control point in the river, this paper puts forward the solving method combined the flood operating model and progressive optimality algorithm. The in-stance results show that this method successfully provides a set of simple, efficient and actual model and solution for real-time flood dispatching and flood control system planning.

The main task of the reservoir flood control operation is to ensure that the safety of engineering, reduce the flood peak flow, reduce the flood disaster and so on, give full play to the comprehensive benefit of the reservoir effective using of flood control storage capacity. The optimal operation of the reservoir flood control has many characteristics such as multiple constraint, high dimension and nonlinear. The main purpose of the operation that use the effective optimization algorithm to study flood regulation scheme and provide the management of the reservoir flood detention with scientific basis for decision-making when the downstream has flood control task.

At present, the conventional operating method for operating is still used of most of the reservoir flood operation in our country. Although it intuitive convenient and have certain reliability, but as a result of operating drawing has some empirical, also have the insufficiencies: (1) it is hard to solve the problem of the whole basin flood control operation. When the river basin has many reservoirs and complex hydraulic connection, it is particularly outstanding. The operating results only for local optimal solution instead of the global optimal solution; (2) it usually can only get feasible solution rather than the optimal solution. With the development of technology of optimization, people try to research the optimal operation of flood control from the whole basin and made progress. In the solving process, people often think of reservoir discharge as decision variable, and regard the reservoir water level (or pondage) as the state variables for optimal operation. But reservoir system is a dynamic nonequilibrium system. This system internal relationship is more complex. It makes the reservoir operating problem is general nonlinear constrained optimization more complicated and difficult to solve. In view of this, this paper decomposed the complicated multi-objective optimization problem into two stages to solve by constructing two stage multidimensional and gradually optimized model from the perspective of basin flood control, and ensure the feasibility and rationality of the operating decision-making.

1 THE MATHEMATICAL MODEL OF THE CASCADE RESERVOIR OPTIONAL OPERATION IN FLOOD CONTROL

1.1 *Objective function*

The optimality criterion of the reservoir flood control operation usually has three kinds of forms. To the maximum peak clipping criteria as an example, this paper reduce the flood peak flow as far as possible in the meet the flood control safety conditions of dam (or reservoir), and try to meet the requirements of downstream flood control. There is a cascade reservoir system consist of n reservoirs and n flood control points (as shown in Fig. 1), the objective function (Zhong 1995, Ye 2001) is

Not to consider the interval compensation:

$$\min \int_{t_0}^{t_d} \left[q_1^2(t) + q_2^2(t) + \cdots + q_{n-1}^2(t) + q_n^2(t) \right] dt \tag{1}$$

Consider the interval compensation:

$$\min \int_{t_0}^{t_d} \left[\left(q_1(t) + Q_{I,1}(t) \right)^2 + \left(q_2(t) + Q_{I,2}(t) \right)^2 + \cdots + \left(q_{n-1}(t) + Q_{I,n-1}(t) \right)^2 + \left(q_n(t) + Q_{I,n}(t) \right)^2 \right] dt \tag{2}$$

where $q_n(t)$ is the discharge volume of the reservoir n in the period t; $Q_{I,n}(t)$ is the internal flow from the reservoir n to the downstream in the period t; t_0 is the initial operating time; t_d is the final operating time.

1.2 *Constraint conditions*

1. Water balance constraints

$$V(t) = V(t-1) + \left[\frac{Q(t) + Q(t-1)}{2} + \frac{q(t) + q(t-1)}{2} \right] \cdot \Delta t \tag{3}$$

where Q(t) is the inflow of the reservoir in the period t, while q(t) denotes the abandoned water flow of the same reservoir in the same period. Q(t – 1) is the inflow of the reservoir in the period t – 1, while q(t – 1) denotes the abandoned water flow of the same reservoir in the same period. V(t) and V(t – 1) respectively denote the water storage of reservoir in the period t and t – 1.

2. The highest water level constraints

$$Z(t) \le Z_m(t) \tag{4}$$

where Z(t) is the water level of the reservoir at the beginning of the period t, while $Z_m(t)$ denotes the maximum limited water level of the reservoir at the same period.

3. The final level constraints at the end of the operating period

$$Z_{end} = Z_e \tag{5}$$

where Z_{end} is the water level of the reservoir at the end of the operating period, while Z_e is the control water level of the reservoir at the same period.

4. The discharge capacity of reservoir constraint

$$q(t) \le q(Z(t)) \tag{6}$$

where q(Z(t)) denotes the abandoned water flow corresponding to the water level Z(t).

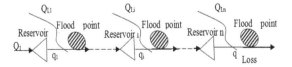

Figure 1. Cascade reservoir.

5. Hydraulic contact between reservoirs

$$Q_2(t) = q_1(t) + Q_{I,1}(t) \tag{7}$$

Where $Q_2(t)$ is the inflow of the lower reservoir in the period t, while $q_1(t)$ denotes the abandoned water flow of the upper reservoir at the same period and $Q_{I,1}(t)$ denotes the internal flow between the first reservoir and the second reservoir at the same period.

6. The nonnegative constraint of the reservoir water level

$$Z(t) \geq 0 \tag{8}$$

The above constraints reflect the relevant requirements of reservoir flood dispatching (Zhong et al. 2003, Qin et al. 2008).

2   PROGRESSIVE OPTIMALITY ALGORITHM

POA is the method of the reservoir optimal operation, provides a shortcut to solve the "curse of dimensionality" obstacles of multi-dimensional and multi-objective dynamic programming problem. POA is mainly used to solve the multi-stage dynamic decision problems. POA converges to the global optimal solution, and its solution is only, and makes computer memory small and reduces the dimension disaster of multi stage dynamic optimization (DONG 1989). This paper applied POA for reservoir flood operating, to the water level of the reservoir as decision variables. In this paper, a single reservoir as an example to describe the algorithm steps for the convenience of description. The objective function:

$$\min F = \sum_{t=1}^{T} \{q^2(t)\} \Delta t \tag{9}$$

where t = 1,2, ..., n, the initial water level Z_0 and terminate water level Z_n is known, solving steps are as follows:

1. Drawing up the initial operating line Z_1, Z_2, Z_3, ..., Z_n (as shown in Fig. 2(a)) within the allowable variation range of the water level of the reservoir.
2. Taking two periods of Δt_n and Δt_{n-1}, while fixing water level Z_n and water level Z_{n-2} and the water level before it and seeking water level Z'_{n-1}, and making the objective function of the two periods of Δt_n and Δt_{n-1} to achieve optimal, namely

$$F = \min(q^2(t-1) + q^2(t)) \Delta t \tag{10}$$

Then, the water level Z_{n-1} changed into the water level Z'_{n-1} and the new track Z_1, Z_2, ..., Z_{n-2}, Z'_{n-1}, Z_n (as shown in Figure 2(b)) is generated.

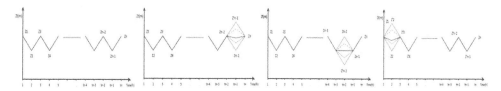

Figure 2.   The method of the POA optimization. (a) selecting the initial line (b) adjusting the Z_{n-1} (c) adjusting the Z_{n-2} (d) adjusting the water level Z_1.

3. Sliding a period of time to the left, fixing water level Z_n and water level Z_{n-1} and water level Z_{n-3} and the water level before it and seeking water level Z'_{n-2}, and making the objective function of the two periods of Δt_{n-1} and Δt_{n-2} to achieve minimum. We can obtain the optimal operating line $Z_1, Z_2, ..., Z_{n-3}, Z'_{n-2}, Z_{n-1}, Z_n$ (as shown in Figure 2(c)).
4. Similarly, in turn, sliding to the left, until obtaining the optimal operating line $Z_1, Z'_2, ..., Z'_{n-2}, Z'_{n-1}, Z_n$ (as shown in Figure 2(d)).
5. The above iterations are continued, if obtained optimal operating line with the initial operating line does not meet the accuracy requirements, then make the initial operating line for the line $Z_1, Z'_2, ..., Z'_{n-2}, Z'_{n-1}, Z_n$ and go back to step (2) to continue iteration, until a satisfactory convergence is attained.

3 APPLICATION EXAMPLE

This paper applied the theory, model and **POA** of the reservoir optimal operation, combining with the characteristics of a cascade reservoir, and calculated instance and made a detailed analysis in order to verify the correctness and validity of **POA** theories and methods, and explore for the most reasonable operating mode of the cascade reservoir optimal operating and provide a theoretical basis and the technical support for the actual production scheduling. This cascade reservoir is combine reservoir 1 with reservoir 2, and the reservoir 1 scheduling play an important role in the flood control to the downstream of a reservoir. The position of the cascade reservoir and flood control point as shown in Figure 4 shows.

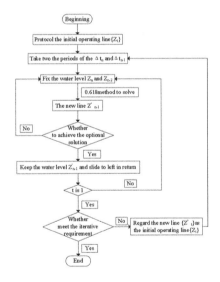

Figure 3.   The flow chart.

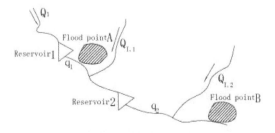

Figure 4.   Cascade reservoir and flood control points.

The reservoir 1 drainage area of 132 km², is located in a river basin. Comprehensive utilization of hydraulic engineering of a large (b) reservoir is a reservoir with the water supply, flood control combined with power generation. Flood control capacity of reservoir 1 is 22.9 million m³. It protects downstream safety by joint operating of reservoir 2 and in combination with other flood control project in downstream a river. Reservoir 2 is a large (b) reservoir of comprehensive utilization of hydraulic engineering within large flood control and irrigation and its flood control capacity of 45.83 million m³. This paper seek a relatively perfect operating rules through the POA algorithm to make the cascade reservoir to maximize develop the flood control benefit in the actual operating. The safety discharge volume of reservoir is 500 m³/s in the downstream river, and the design standard of flood control to reach a return period of twenty years (P = 5%).

1. The objective function of the cascade reservoir is

$$\min \sum_{t=1}^{T}\left[q_1^2(t) + \left(q_2(t) + q_{1,2}(t) \right)^2 \right] \Delta t \tag{11}$$

The constraint conditions of the cascade reservoir as show in the Table 1.
2. The POA calculated results as shown in Tables 2–4, Figures 5 and 7.

To select the result of the operating of reservoir 1 as inspection, this paper will compare the result calculated using the POA operating with the result calculated using the conventional

Table 1. The constraint condition of the two reservoirs.

Reservoir	The regulating water level (m)	The maximum discharge volume (m³/s)	The highest regulation level (m)	Control water level at the operating period (m)	Convergence precision (m)
Reservoir 1	227.13	280	237.89	227.13	0.001
Reservoir 2	60.18	500	77.22	60.18	0.001

Table 2. The flood regulating calculation result of the two reservoirs.

Reservoir	Reservoir 1	Reservoir 2
Flood peak flow (m³/s)	1475.56	1520.38
The hightest water level (m)	237.89	68.03
The maximum discharge volume (m³/s)	217.68	476.02

Table 3. This result of reservoir 1 is calculated using the POA and conventional algorithm. (unit: m³/s).

Time	139	140	141	142	143	144	145
Inflow	674.16	1272.04	1475.56	1296.67	1049.17	810.16	607.22
Discharge volume of the POA	214.52	214.57	214.5	214.61	214.54	214.52	214.56
The conventional algorithm	280	280	280	280	280	280	280

Time	146	147	148	149	150	151	152
Inflow	450.11	333.2	247.74	185.94	141.29	109.09	85.85
Discharge volume of the POA	214.57	214.55	214.59	178.49	148.77	101.64	101.65
The conventional algorithm	280	280	280	280	280	280	280

Time	153	154	155	156	157	158	159
Inflow	69.09	57.06	48.5	42.47	38.35	35.82	34.53
Discharge volume of the POA	101.68	101.72	101.66	101.7	101.63	101.63	101.61
The conventional algorithm	280	280	280	280	280	280	280

Time	160	161	162	163	164	165	166
Inflow	33.85	34.94	37.74	43.46	86	118.04	124.01
Discharge volume of the POA	101.55	101.53	101.49	101.46	101.46	101.41	101.37
The conventional algorithm	280	280	280	280	280	280	280

Table 4. This result of reservoir 2 is calculated using the POA (unit: m³/s).

Time	139	140	141	142	143	144	145
Inflow	824.31	1334.85	1520.38	1386.91	1190.37	995.67	829.97
Discharge volume of the POA	466.95	467.02	441.77	444.54	443.88	444.12	443.97
Time	146	147	148	149	150	151	152
Inflow	697.27	596.67	521.83	432.32	320.87	255.22	233.13
Discharge volume of the POA	442.95	441.75	440.02	394.18	323.22	70.7	269.75
Time	153	154	155	156	157	158	159
Inflow	216.77	204.68	195.86	189.51	184.95	182.01	180.29
Discharge volume of the POA	269.37	268.89	268.77	268.52	268.61	268.77	268.86
Time	160	161	162	163	164	165	166
Inflow	180.98	182.57	182.93	182.38	214.28	237.18	237.23
Discharge volume of the POA	269.03	269.26	269.25	269.5	269.66	269.8	269.92

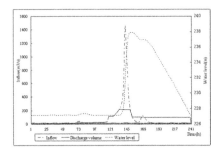

Figure 5. The flood regulating calculation chart of the reservoir 1 is calculated using the POA.

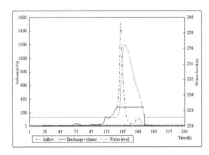

Figure 6. The flood regulating calculation chart of the reservoir 1 is calculated using the conventional algorithm.

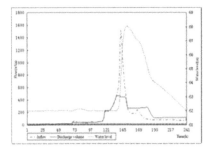

Figure 7. The flood regulating calculation chart of the reservoir 2 is calculated using the POA.

operating (Table 3, Figs. 5 and 6)under the same initial conditions to verify the efficiency of the algorithm. From Table 3, Figures 5 and 6, the values of the maximum discharge volume calculated using the POA is smaller than using the conventional method, and it appears ahead of the result of conventional operation, peak clipping effect is quite obvious, the process of discharge volume more homogeneous, flow range is small. The POA algorithm makes the reservoir discharged volume as uniform as possible to reach the largest possible clipping purpose under the highest water level to meet the requirement precondition, and makes the reservoir as large as possible discharge the flow while the flood control in satisfying the downstream watershed at the same time, which to a certain extent, lower the flood risk of the downstream flood control point and improve the flood control benefit of the reservoir. Example results show that, POA calculation speed is quicker and the result is superior compared with conventional operating method, and provides an effective way for the basin optimal operation of flood control.

4 CONCLUSION

This paper established the optimal operating model enable the reservoir maximum reduce the downstream flood and the downstream pressure of flood control in the face of the design criterion of the flood, so as to improve flood control benefit of the reservoir. By comparison with the results of the conventional flood control operation, POA is used to solve the model to obtain the optimal discharge process more uniform, so as to avoid the downstream flood loss caused by central flood discharge of reservoir. The result of the example shows that POA can get the optimal operating scheme in solving the complex reservoir operating problem such as the "curse of dimensionality" caused by many variables and constraints, nonlinear, or variable dimension increased, and provides a new way for flood control operation of cascade reservoir.

ACKNOWLEDGEMENTS

This project is supported by the Key Program of Science and Technology Department of Zhejiang Province (2009C13011).

REFERENCES

Dong Zi-ao. 1989. The Theory and Application of the Multi Reservoir Optimal Dispatchand Planning. Ji Nan: Shandong Science and Technology Press.
Qin Xu-bao, Dong Zeng-chuan, Fei Ru-jun, Liu Rui, Li Qiao, Lei Yang. 2008. Research for Optimal Flood Dispatch Model for Reservoir Based on POA. *Water Resources and Power* 26(4):60–62.
Ye Bing-ru. 2001. Planning and Dispatching of the Water Resources System. China Water Power Press.
Zhong Ping-an. 1995. Analysis of the Objective Function in the Optimal Operation of Flood Control. *Water Economy* 13(1):38–44.
Zhong Ping-an, Zou Chang-guo, Li Wei, Zhang Chu-wang. 2003. Stage trial-and-error Method and its Application to Reservoir Operation for Flood Control. *Advances in Science and Technology of Water Resources* 23(6):21–23,56.

Hydraulic Engineering – Xie (Ed.)
© 2013 Taylor & Francis Group, London, ISBN 978-1-138-00043-8

The effect of train load and water coupling to track crack growth

Xu Guihong & Wang Ji

MOE Key Laboratory of High-speed Railway Engineering, Southwest Jiaotong University, Chengdu, Sichuan, China

ABSTRACT: Numerical simulation (The FEM software ANSYS) is applied in this paper, unified governing equation and mathematical model was established in solid and fluid domain, based on the two-way coupled field computation, at the high frequency train loads action, the crack of containing water growth had been simulated in slab track. Analysis showed that: crack height and length are important factor in the crack growth. When crack length is 0.6 m and the height is 3 mm were the limit state of crack growth, when crack length L > 0.6 m, the crack will be growth; when crack height H > 3 mm, the crack will be growth; crack tip stress-intensity factor value is linear increased with the force on the crack surface, not relate to the crack height and length.

CRTS II type slab track on subgrade is composed of steel rail, elastic fastening, precast track plate, mortar adjustment layers, concrete bearing layer and so on, and it's widely used for its high stability, high smoothness and low maintenance on high speed railway. By the approaches of on-site survey, it's found that different degrees of flaw had appeared on CRTS II type slab track, and the cracking of the slab or the floating slab track and the flaw of the mortar adjustment layers are the most common. Also areas with heavy rainfall or impeded drainage, the damage rate is faster than dry areas. Water plays a vital role to the development of the flaw of slab track, so it possesses important theory significance for the improvement of the design theory of slab track and the establishment of reasonable maintenance methods and procedures by means of carrying out the research of the crack propagation of slab track made by water[1–2].

Numerical simulation (The FEM software ANSYS) is applied in this paper, according to the water crack in the bottom of mortar adjustment layers, unified governing equation and mathematical model was established in solid and fluid domain, based on the two-way coupled field computation, at the high frequency train loads action, the pressure of the crack surface produced by water had been simulated, using the calculation results, the intensity factor of crack tip are calculated, and the problems of crack propagation are analyzed.

1   MODELING AND SOLVING METHOD OF THE CRACK TIP OF THE TRACK STRUCTURE

1.1   *Computation model*

The analysis of the impact of high-speed train load and water coupling on the propagation of interlaminar crack can be modeled as under high-frequency train loads, for the fast loading speed, water in an instant too late to discharge, thus in the airtight crack internal, water produce great pressure to the surface of the track plate, that urging the crack of the crack tip. Calculation model can be created as shown in Figure 1.

1.2   *Calculated parameters and the solving method*

Plane model of II type slab tracks on linear subgrade is used in the calculation. Track plate is precasted by the ordinary reinforced concrete precast, the size is 200 × 2550 mm, mortar

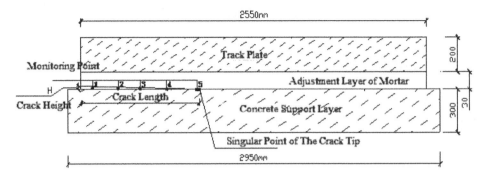

Figure 1.   Fluid-structure interaction computational model with bidirectional transient.

Table 1.   Calculated parameters of the fluid-structure interaction computational model with bidirectional transient.

Components	Items	Detailed information
Solid domain	Loading frequency	0.04 s
	Load values	10×10^5 pa
	Elasticity modulus (track plate)	36000 Mpa
	Poisson ratio (track plate)	0.2
	Elasticity modulus (base plate)	32500
	Poisson ratio (base plate)	0.2
Fluid domain	Analysis type	Transient ANSYS multi-field
	Fluid type	Water at 25°C
	Domain type	Single domain
	Time steps	0.01 S (transient)
	Reference pressure	1 [atm]
	Coupling time	1 s
	Output control	Monitor points
	Boundary conditions	Wall: mesh motion Wall: no slip wall Interface: no slip wall

adjustment layers are made in-situ casting, the size is 30 × 2550 mm, the size of the support layer is 300 × 2950 mm.

The calculation assumes the water in the crack internal is 25°C cold water, and it's better to fill the entire crack. Changes of the water pressure in the crack are monitored by the output of control points (monitoring point is arranged as shown in Fig. 1).

Based on the calculation model above, using the finite element method, solve the force state of crack water on numerical simulation under the high-frequency train loads with the commercial software ANSYS WORKBENCH13.

2   RESULTS AND ANALYSIS

2.1   Pressure of the crack surface

Obtained by simulation, when the train load is loaded by 0.04 s, crack length of 1000 mm, 800 mm, 600 mm, crack height of 2 mm, 3 mm, 4 mm, 5 mm, pressure values of the crack surface are listed as follows:

It's shown that, when the train load is constant, crack height is an important factor to effect the force of the crack surface. When the crack length is large (L = 1 m), with decreasing

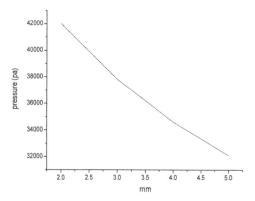

Figure 2. Relationship between the force of the 1 m long crack surface and the crack height.

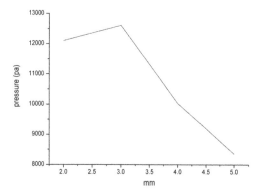

Figure 3. Relationship between the force of the 0.8 m long crack surface and the crack height.

Table 2. Intensity factor of the crack tip.

Crack length L (m)	Crack height (mm)	Maximum force acting on the crack surface (kN)	Crack tip intensity factor K_{IC} (MPa $\sqrt{m}$)	Cracking the case (K_{IC})
1	5	32.10	10.94	Cracking
	4	34.60	11.82	Cracking
	3	37.80	12.93	Cracking
	2	42.02	14.41	Cracking
0.8	5	8.35	2.00	Cracking
	4	10.03	2.44	Cracking
	3	12.60	3.11	Cracking
	2	12.10	3.15	Cracking
0.6	5	3.80	0.69	Cracking
	4	3.53	0.63	Cracking
	3	2.10	0.381	Limit state
	2	0.41	0.064	Not crack

in the height of cracks, pressure of the crack surface and the crack height rise substantially linearly. When the crack length is short (L = 0.6 m), with the increasing of the crack height, the pressure of the crack increases. For the 0.6 m long crack, while the height is 5 mm, the maximum pressure will be 3.8 KPa, while the height is 2 mm, its pressure will be 1.42 KPa.

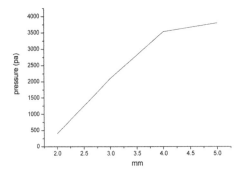

Figure 4.   Relationship between the force of the 0.6 m long crack surface and the crack height.

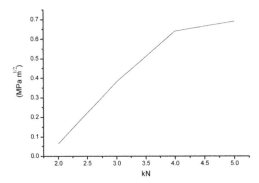

Figure 5.   Relationship between the force of the 0.8 m long crack surface and the crack tip stress-intensity factor (L = 0.6).

The length of the crack is an important factor to affect the force of the crack surface. Cracks with the same height, along with the increasing of L, the surface pressure will decrease. For the 5 mm height crack, the surface pressure is 32.1 KPa while L = 1 m, the surface pressure is 3.8 KPa while L = 0.6. For the 2 mm height crack, the surface pressure is 42.02 KPa while L = 1 m, the surface pressure is 0.408 KPa while L = 0.6.

2.2   Intensity factor of the crack tip

Obtained by calculating, under different crack lengths L = 1 m, L = 0.8 m, L = 0.6 m with different crack heights respectively as 5 mm, 4 mm, 3 mm, 2 mm, the crack tip intensity factors corresponding to crack surface pressure are shown in Table 2.

It can be inferred from the Table 2 that, the size of crack tip intensity factor has nothing to do with the crack height and the crack length. The crack tip stress-intensity factor value is linear increased with the force on the crack surface, as is shown in Figure 5.

Based on the strength toughness of the mortar adjustment layer $K_{IC} = 0.628$, it can be inferred that when crack length is 0.6 m and the height is 3 mm were the limit state of crack growth, when crack length L > 0.6 m, the crack will be growth; when crack height H > 3 mm, the crack will be growth.

3   CONCLUSIONS

1. The height of the crack is a important factor to the pressure of the crack surface which resulting the growth of the crack. When the length of the crack is big (L ≥ 1 m), the

pressure of the crack decreases with the increasing of the crack height. Namely that: cracks with larger height is hard to cracking, and the cracks with smaller height is easy to cracking. The crack with the length of 1 m, when the height is 5 mm, the pressure of the crack surface is 32100 pa; when the height is 2 mm, the pressure of the crack surface is 42019 pa. When the length of the crack is small (L ≤ 0.8 m), the pressure of the crack increases with the increasing of the crack height. The crack with the length of 0.6 m, when the height is 5 mm, the pressure of the crack surface is 3800 pa; when the height is 2 mm, the pressure of the crack surface is 408 pa.

2. The length of the crack is a important factor to the pressure of the crack surface which resulting the growth of the crack. Cracks with the same height, the pressure of the crack surface increases with the increasing of the crack length. For the 1 m length crack, when the height is 5 mm, the maximum pressure will be 32100 pa; for the 0.6 m length crack, when the height is 5 mm, the maximum pressure will be 3800 pa.

3. Crack tip stress-intensity factor value is linear increased with the force on the crack surface, not relate to the crack height and length.

ACKNOWLEDGMENTS

This is a fund project supported by the National Natural Science Foundation of China (51278431). Financial support of this work provided by the National Natural Science Foundation of China and MOE Key Laboratory of High-speed Railway Engineering are gratefully acknowledged.

REFERENCES

Antonia, R.A. Conditional Sampling in Turbulence Measurement. Ann. Rev. Fluid Mech., 13(1981), 131–156.

Bear J. Hydraulics of Groundwater [M]. London: McGraw-Hill, 1979.

Jiao Zongxia, & Hua Qing & Yu Kai. Modal analysis of the fluid-solid coupling for vibration of transmission pipeline. *Chinese journal of aeronautics* [J], 1999,20(4):316–320.

Lou Tao. Numerical simulation based on the problem of ANSYS fluid-structure coupling [D], dissertation of Lanzhou university master degree, 2008,5:7–20.

Lu Chihua & He Yousheng. Fluid-solid coupling effect of the two-dimensional elastic structure into the water in the process of impact. *Acta mechanica sinica* [J], 2000,32(2):129–140.

Management center of the ministry of Railways engineering, Beijing-Tianjin Intercity Rail Transit Engineering. Technical summary report of the slab track CRTS II.[M], 2008.8.

Pan Jiazheng. The application of fracture mechanics in hydraulic design [J], Journal of Hydraulic Engineering, 1980,(1):45–49.

Quality acceptance supplement standard of the Pack construction of passenger dedicated railway line. (Railway construction [2009] NO. 90).

Wu Zhimin & Zhao Guofan & Huang Chengkui. Fracture toughness and fracture energy of concretes with different strength levels [J], *Academic journal of dalian university of technology*, 1993.9.33(1):73–77.

Wang Guangjun & Xiong feng. Finite element study of the dynamic stress intensity factor in the three-dimensional crack propagation [J], *Applied research of the ship electric technology*, Vol. 30 No. 6 2010:55–58.

Xu Shilang. The Calculation Approaches of Double-K-Fracture Parameters of Concrete and a Possible Coding Standard Test Method for Determining Them, *Journal of three gorges university (natural science edition)*, 2002(1):1–8.

Xu Shilang & Zhou Hougui & Gao Hongbo etc. Experimental Research of Double-K-Fracture Parameters of Concrete of All Kinds of Graded Dam. *China civil engineering journal*, 2006(11):50–62.

Xu Shilang & Yu Changxiong & Li Qinghua. Determination of the Shear Fracture Process of Concrete Dam Joint Grouting and its Fracture Toughness [J], *Science of water energy*, 2007,3,38(3):300–306.

Yang Jiangang. Fluid-structure coupling dynamic research of vibration cylindrical in annular clearance. *Journal of Southeast University* [J], 2005,35(1):7–10.

Yue Baozeng & Liu Yanzhu & Wang Zhaolin. ALE fractional step finite element method for fluid-struture nolinear coupling problems. *Science of water energy* [J], 2001,19(4):43–47.

Zheng Shaohe & Yao Hailin & Ge Xiurun. Coupling analysis of the fractured rock mass seepage field and damage field [J], *Chinese Journal of Rock Mechanics and Engineering*, 2004,5,23(9):1413–1418.

Zhang Yanfeng. Research of the fracture parameters of the floating slab track[D], dissertation of Central south university master degree, 2011,5:p45–50.

Zhang Yingjun, the numerical research of the crack tip stress-intensity factor of the 3D T crack [D], 2010,4,6:6–26.

Hydraulic Engineering – Xie (Ed.)

A spatial decision support system for water resource management of Yellow River Basin in China

Yan Li
School of Geographic and Oceanographic Sciences, Nanjing University, Nanjing, China
Information Center of Yellow River Conservancy Committee of Ministry of Water Resources, Zhengzhou, China

Xiang-Lin Fang & Shi-Xing Jiao
Department of Resources and Environment and Tourism, Anyang Normal University, Anyang, China

ABSTRACT: The Yellow River is the second longest in China, the environmental changes occurring in the Yellow River Basin during the last two centuries have faced the dry-up of the Yellow River, which has become more frequent and severe since the 1990s. A Spatial Decision Support System (SDSS) for water resource management, based on the Digital Yellow River Project, was developed for the Yellow River Basin in order to gain a better understanding of the water resources management process in the basin, and to optimize planning processes and to achieve a higher effectiveness of decision-making while solving a semi-structured spatial data. The SDSS includes information subsystem that support information exchange and knowledge and model sharing from different organizations on the web, a real-time meteorological and hydrological data monitoring subsystem, a model-base subsystem for system simulation and optimization, and a graphical dialog interface allowing effective use by system operators, etc. With this SDSS, some difficult issues concerning water resource management in Yellow River Basin are explored. The SDSS is a software platform with diversity and expandability. It contains intelligent and visualization functions. The system has been put into application successfully, and satisfied effects are obtained, therefore it is worthy to be improved and popularized and has the great directive significance and applied value in optimum operation regulation and uniform management of water resources.

1 INTRODUCTION

The Yellow River is the second longest river in China which originates from Yueguzonglie basin located in Tibet plateau northern of Bayan Har Mountain. On its east is Bohai and west stretches in the inland. The mainstream is about 5464 km, the total drainage area is 794712 square kilometers. The average natural runoff for many years is 580 billion cubic meters, accounts for about 2 percent of the total river's runoff in our country. The yellow river is a very important source of water supply in northwest and northern of China, responsible for the water supply of 15 percent of arable land, 12 percent of population and more than 50 big or middle cities on its shoulders.

With the development of industrial and agricultural economic in the Yellow River Basin, some problems come into being (Wang, W.K, Liu, H. & Liu, A.R. 2002, Li, G.Y. 2002), such as the shortage of water resources and ecological environment deterioration become worse which damaging the health and life of Yellow River, affecting the harmonious development of Yellow River Basin seriously. Concentrated reflect on Firstly, the sharp contradiction between supply and demand of water resources leads to frequent cutoff in Lower Yellow River, the threaten of cutoff is faced in Yellow River Middle Reaches, cutoff situation becomes more serious in the mainstream and major tributaries of the Yellow River. Secondly, the demand of water resources is continuously increasing and the Water requirement and water consumption are close to the natural river cutoff. Besides, continued growth in water demand results in intensifying the deteriorated tendency of water quality. What's more, the intermittent cutoff and the uncoordinated between the water and sediment made the shape of river bed become worse. Thus it can be seen that the scheduling of water resources has highlighted its significance and become a key problem to water management in the Yellow River Basin. Therefore, it is a key measure of solving the great problem to strength the united management and scheduling water resources and the development and research of decision support the system.

2 OVERALL FRAMEWORK

2.1 *The goal of system construction*

A decision support system for the management of the water resources system of Yellow River is developed and constructed, based on "Digital Yellow River" project and scientific monitoring information of need water in the main water consumption areas, using contemporary theory of water resources allocation model and mature application development technology, aiming at the goal of improving the implementation of effective means and methods of the Yellow River water regulation, taking the key issues in water allocation and real-time programming schedule as the core, taking system information of water regulation as the support, taking relevant criteria of water allocation as the guide. The system can provide a complete decision support information and comprehensively management and information services of water regulation measures, improve data analysis and programming tools, provide reliable technical support of water management for sustainable social and economic development of the Yellow River Basin. We can achieve information services network of water dispatching, and monitoring visualization of operating and running status, computerized business operation of water allocation and water management, scientific water allocation scheduling and electronic water rights trading market.

2.2 *System structure*

The system is used hierarchical structure and is divided into three levels: business layer, application service platform and data layer. The business layer directly faces to end-users and provide specific business functions, the application service platform needs to extract the public infrastructure services and to provide a unified common infrastructure services for the business layer, thus, the speed and quality of system construction are greatly improved, the data layer provides information services for reading and writing from data center and sub-centers. The system structure is shown in Figure 1.

Scheduling scheme is the main contents of water regulation and management decision-making, water resources forecasting and simulation of water allocation subsystems are respectively provided support for inflow forecasting and program evaluation of water resources programming schedule, information services and comprehensive running situations of water dispatching subsystems are respectively provided basic information and feedback of program implementation of water resources programming schedule. Business process subsystem is provided business process and means of transmission of daily dispatch and upload and download of schedule programming. Diversion and return water remote monitoring subsystem is not only provided real-time information of programming schedule and evaluation, but also provided effective protection for implementation of the program scheduling. Operation and maintenance management subsystem is provided support for normal operation and fault handling of the system.

The application service platform separates the system logic business and the specific business, makes modification of the user interface, logic business and data structure independent each other, development and extension of system becomes easy. At the same time, we abstract the logic business, and publish them in the form of services in order to make easily

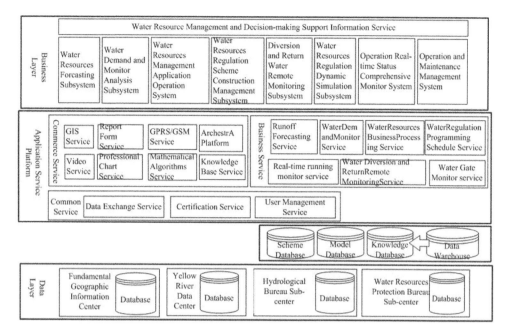

Figure 1. The logical system structure of spatial decision support system.

interact between the logic business and system. The application service platform consists mainly of commerce services, business services and common services. Commerce services includes GIS services, report forms services etc, such services need to purchase professional commercial software, and are developed based on business needs, business services are developed by the logic business services, including runoff forecasting service, simulation services, water evolution services, reservoir scheduling services; common service is to provide basic running environment for system running, including data exchange services, authentication services and users management services. From the realization of form, the application service platform is used two forms of services and components, which are used the form of services for supporting cross-boundary calls and are used the form of components for local calls.

Data layer stores all kinds of data types and is foundation of application systems and provides all kinds of data services based on standard services and distributed disposition. On the one hand, data warehouse is constructed according to different subject; knowledge base is renewed through data mining. On the other hand, data logical views are packaged into different standard services, which disposed to application service platform, to provide application support for application system, and data layer provides management services for model base.

2.3 The content and aim of system decision-making

Decision-making process is taken as the main body, method of decision-making is mainly through the consultation meeting. Decision-making process is divided into determining the target, programming schedule, choosing schedule, and implementing schedule and checking correction four stages. The mode of dynamic decision-making process is the basis of analyzing of the environment to determine what are problems and then to pick up decision-making objectives, based on the decision-making objectives, feasible schedules are searched, programmed and created, then the most satisfactory schedule is found from all kinds of feasible schedules, the effectiveness of the most satisfactory schedule is monitored in the progress of implementing, at the same time, the most satisfactory schedule will be checked and corrected when it is found faults. The content and aim of system decision-making are shown in Figure 2.

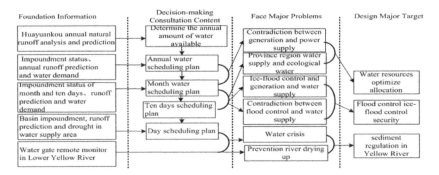

Figure 2. Decisive contents and decisive objectives of spatial decision support system.

2.4 System functions

2.4.1 Water resources prediction subsystem
Supported by the integrated database of hydrology, water and rainfall information, and In the B/S framework, the subsystem explores modules of information pick-up and feedback, hydrological analysis and simulation, result display and output, and database management etc.

2.4.2 Water demand monitor and analysis subsystem
Based on monitoring and interpreting and analyzing soil moisture contents of farm in water receiving areas and forecasting domestic water demand, the subsystem explores modules to provide spatial-temporal distribution information of water demand information of water receiving areas for water resources dispatch. So, we can realize to optimize water resources dispatch schemes based on allocation indexes, avoid no effectively dispatching water, save water resources and optimize water resources dispatch. In the end, sustainable development of water resources is realized.

2.4.3 Water regulation program compilation system
Based on ecology and benefit and water right, supported on market adjustment means water tariff leverage and water right transaction, the subsystem explores modules of water regulation careful regulation program compilation and simulation and evaluation of Yellow River Basin. At present, there are three most commonly used methods, including the currently used method, the automatically adapted method, and the moderately optimized method (Chen, L.C., Cai, Z.G. & Wang, X.L. 2003). The core of the subsystem is water resources regulation model, which is modified and optimized based on WRMM Model in water demand optimization and water quality.

2.4.4 Water diversion and return remote monitoring subsystem
The mode, which subsystem is adopted, is combined general center disposition with control in-situ, based on uniformly open platform, object-oriented and distributed control and multi-tier application technology autoimmunization and information. The subsystem explores modules of picking up water diversion information of sluice stations of all levels management departments and remotely controlling real-time running status of sluice stations and important sluice gate and pumping stations and pumps to open or close, and remotely monitoring water diversion and real-time running status of sluice stations, and remotely monitoring the process of return flow and the status of hydraulic projects.

2.4.5 Water resources dispatch simulation subsystem
Supported on strong function system software and water resources dispatch mathematic models and based on multi-scale and multi-type data integration, it is constructed the integrated platform of water resources uniform management and three-dimensional simulation subsystem for virtual environment. The subsystem give support to water dispatch visual management platform and provide real-time and exact water dispatch information.

2.4.6 Real-time running monitor subsystem

Supported by large color monitor, and based on GIS platform, the subsystem is constructed in the Yellow River water general regulation center. It is the center of the Yellow River water integrated regulation and management, scheduling in the process of programming, policy-making meeting, information (including: diversion information, water, rainfall, drought, weather information), monitor the sluice work are all performed here. The foundation of the Yellow River water general regulation center provide good working and system operating environment for water regulation, reflect the function of the regulation management system comprehensively.

3 KEY TECHNOLOGY

3.1 Model base construction

Based on software reuse and oriented-object technology and Composite design mode (Erich, G., Richard H. & Ralph, J. 2000, Yuan, F. & Li, Z.Z. 2004), the model base is standard, flexible, and expandable and lays a foundation for integration and disposition of water resources regulation management and decision-making system.

The models are programmed and used during system construction, such as runoff prediction model, runoff routing prediction model, water dispatch model, water quality prewarming and forecasting model, water quality analysis and evaluation model, and so on. In order to program and reuse, the models are abstracted, based on model base technology and expressed by standard mode.

The model base includes model base management system, model base, model dictionary and model drive system. The models are constructed, displayed, modified, deleted, quarried and printed through model base management system. Through the model base management system interface, the user can create, display, modify, delete, query or print model, combine the existing models into new, compile newly-built or modified models, implant the compiled new model or modified model to the model base and add the model attribute information to the model dictionary. The model base logical structure is shown in Figure 3.

All the various models in model base have data interaction and indirect data transmission with program database between them. The models obtain the input data from the scheme database, and then place the computing results into the program database after operation. The functions of the model management subsystem is developed to automatically achieve models running and connected between models according to the schedule types.

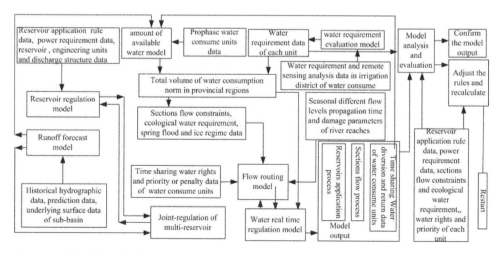

Figure 3. Model base logic structure.

Model base is a large set of business processes which directly service for all the business process, which is a collection of units programmed in accordance with sharing agreement and relevant standards. The basic principle of Model Base Construction is universal, open and practical; model base can easily connect with the database system and the business application system. Also it must have its own drive system and operational mechanism to ensure the application can be built and run through the basic models and combination models in model base.

3.2 Application service platform construction

Application service platform is the basic software supporting platform, which located in the middle layer of the system structure, based on application server, middleware and infrastructure software technology as the core technology. It offers a variety of application services for upper system and information exchange with lower data storage management system and information collection system. It is the core to achieve the information exchange, transmission, sharing among application systems and between application system and other platform.

Application service platform are adopted the unified technical work mode of centralized management and loose convenience, based on Service-oriented Architecture and Enterprise Service Bus. There are one main node and three sub-nodes in data exchange and sharing service framework structure, they communicate by communication bus and every node has a correspondingly self-governed ESB environment (Michelson, B.M. 2006, David, C. 2004, Martin, K. 2006), and they can apportion running services load and provide data exchange and sharing services and application integration among nodes. The main node can centralize to control and monitor and manage sub-node, at the same time, every sub-node keeps relative independence of running. At the same time, the system is a multi-node user service system, and that there are numerous and frequency information exchange among multiple nodes users.

Application service platform integrates heterogeneous geospatial information and natural resources database located in various regions based on Web Services technology builds an application oriented drive network, intelligent management and sharing service system (He, J.G., Huo, H. & Fang, T. 2006). It combines business processes, application software,

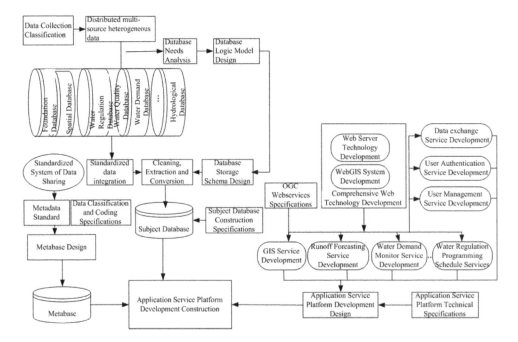

Figure 4.   The technical process of application service development and constructing.

information and a variety of standards to enable them to conduct business and share information as a whole. The application service platform follows certain rules to extract and re-organize the information from the data layer and pack in Web Services. By UUDI service in registry center, users can publish their own services and also look for the services for their own usage, and call the service to build their own applications. The realization of the technology process is shown in Figure 4. Firstly, taking to integrate and transform geospatial and natural resources data as the main line, we carry out classification, cleaning and conversion of basic data, spatial data, water dispatching and other department professional data and establish subject databases and information resources sharing directory after standardization and integration. Secondly, taking to develop and construct basin data sharing network as the core, aiming the goal of information services and following the OGC Web services and metadata standards, application service platform is built.

4 CONCLUSIONS

We keep the Yellow River from flow cutoff and try the best to give play to water resources utilization benefit, and acquire a significant ecological, social and economic benefits, and make outstanding contribution to the basin and related regions economic and social sustainable development, and raise the overall level of the Yellow River water resources management and dispatching standard, and fundamentally reflect to keep the healthy life of the Yellow River and keep healthy live of the Yellow River through the decision support system for the management of the water resource system of Yellow River development and construction. Also, it provided a full range of modern and efficient means of intelligent decision support for the basin water resources management and dispatching. Through the comprehensively using of modern information technology, automatic information collection and modernization of the monitoring management, it gradually became a digital system which adapts to the sustainable development and can support comprehensive water resources management of the Yellow River Basin, it also achieves the scientific management of Yellow River water resources, optimal dispatching and effective use of decision support system.

REFERENCES

Chen, L.C., Cai, Z.G. & Wang, X.L. 2003. Research on Yellow River Water Regulation Programming Method [J]. *Journal of Basic Science and Engineering*, vol. 11 (2), 2003, pp. 208–215.
David, C. 2004. Enterprise service bus [M]. O'Reilly Publishing, 2004.
Erich, G., Richard H. & Ralph, J. 2000. Vissides. Design Patterns: Elements of Reusable Object-Oriented software [M]. *Pearson: Addison Wesley*, 2000, pp. 107–115.
He, J.G., Huo, H. & Fang, T. 2006. Spatial Metadata Directory Service Implementation Based on MVC Model [J]. *Computer Engineering and Applications*, vol. 42 (13), 2006, pp. 165–167.
Li, G.Y. 2002. Major Problems and Countermeasures of Yellow River [J]. *Water Resources and Hydropower Engineering*, vol. 33 (1), 2002, pp. 12–14.
Martin, K. 2006. Patterns: implementing an SOA using an enterprise service bus [EB/OL]. http://redbooks.ibm.com/redbooks/pdfs/sg246346.pdf, 2006.
Michelson, B.M. 2006. Enterprise service bus Q & A [EB/OL]. http://www.ebizq.net/hot-topics/esb/features/6117.html. 2006.
Wang, W.K, Liu, H. & Liu, A.R. 2002. Countermeasure on Sustainable Utilization of Water Resources in Yellow River in 21 century [J]. *Regional Research and Development*, vol. 20 (2), 2002, pp. 62–64.
Yuan, F. & Li, Z.Z. 2004. A New Approach for Developing Model Base in DSS [J]. *Microelectronics and Computer*, vol. 21 (5), 2004, pp. 111–113.

Hydraulic Engineering – Xie (Ed.)
© 2013 Taylor & Francis Group, London, ISBN 978-1-138-00043-8

On energy dissipation of hydraulic jumps at low Froude numbers considering bubbles formation and transportation process

Shuai Chen & Junxing Wang
State Key Laboratory of Water Resource & Hydropower Engineering Science, Wuhan, Hubei, China

ABSTRACT: A hydraulic jump is the rapid transition from supercritical to subcritical flow which is associated with energy dissipation due to strong turbulence and air bubble entrainment. In the present study, some physical experiments were conducted at relatively low inflow Froude numbers ($2.01 < Fr_1 < 4.85$). The results demonstrate that the energy dissipation of a hydraulic jump significantly depends on the energy consumption during the formation and transportation process of air bubbles. The air-water flow properties exhibited a peak of energy loss in the shear layer beneath the reverse-flow region with the peak value of bubble size. Furthermore, dimensionless distributions of the diameter of bubbles were highlighted by using a micro acoustic Doppler velocimeter (microADV). An empirical formula predicting the energy consumption by bubbles formation and transportation process was developed and a good agreement was obtained.

1 INTRODUCTION

Hydraulic jumps have long been the efficient energy dissipaters which commonly encountered in hydraulic structures, wastewater treatment plants and chemical processing plants. The main characteristic of a hydraulic jump is the upper reverse-flow region which causes strong interaction between air and water, and simultaneously generates disturbances of the air-water interface leading to air entrapment (Chanson 2009). Generally, the process of air bubble formation could be reckoned as the result of turbulent stresses overcoming both surface tension and viscous forces (Ervine & Falvey 1987), while transportation process as the result of turbulence advection and diffusion. During these two processes, a certain amount of flow energy might be dissipated so as to split air mass group into bubbles and carry them downstream. From this point of view, a positive correlation between energy dissipation and bubbles formation and transportation process could be found.

Rajaratnam (1962) made the first attempt to measure void fraction in two-phase flow turbulence structure. Resch & Leutheusser (1972) highlighted some effects of the inflow conditions by conducting some hot-film probe measurements in the bubbly flow region. In the recent years, new measurement methods, such as acoustic Doppler velocimeters, have been applied to show the turbulence structure of hydraulic jumps at various Froude numbers in many researches, including Liu et al. (2004), Misra et al. (2008) and Mignot & Cienfuegos (2010). Besides, Chanson (2011) investigated the bubbly two-phase flow properties for a wide range of Froude numbers and distributions of void fraction and bubble count rate were exhibited. Moreover, the relationship between energy concept and aeration performance of hydraulic jumps got more attentions by Kucukali & Cokgor (2009).

Despite all these achievements, little is known about the energy consumption caused by the bubbles formation and transportation. Hence, in this study, laboratory investigations were conducted on hydraulic jumps at different inflow Froude numbers and Reynolds numbers. Due to the limitation of the micro acoustic Doppler velocimeter (MicroADV), the experiments had to be confined to relatively low Froude numbers. The experimental results provide detailed information on the energy consumption of bubbles and approximate distributions of energy dissipation rate are presented.

2 EXPERIMENTAL SETUP AND INSTRUMENTATION

The experiments were conducted in a rectangular glass flume, 0.5 m wide, 0.45 m high and 5.0 m long, at the State Key Laboratory of Wuhan University. A supercritical flow is enforced by a vertical sluice gate at the upstream of the flume to generate hydraulic jumps, while the location of the jumps were controlled by a tail gate located at the downstream end of the flume (Fig. 1). The water discharge was measured by an Electromagnetic Flowmeter (EMF) and the accuracy could be limited within ±1%. The flow depths were measured by pointer gauges with ±0.2 mm accuracy.

The two-phase flow properties were measured with a SonTek 16 MHz type MicroADV at 50 Hz sampling frequency for 5 minutes sampling time. It can measure flow velocities in range of 1 mm/s to 2.5 mm/s with an accuracy of ±1%. The longitudinal, vertical and lateral turbulence intensities, $\sqrt{u'^2}$, $\sqrt{v'^2}$ and $\sqrt{w'^2}$, could be determined by WinADV32 software respectively. However, the ADV will make unrealistic measurements when being used in a highly air-entrained flow (Robinson et al. 2000). Therefore, the Froude numbers of the following experiments were limited within 2.01 to 4.85 (Table 1) and the Phase-Space Thresholding Method (PSTM) was used to process the MicroADV data to get rid of the spikes taking place when air bubbles pass through the sampling volume (Goring & Nikora 2002).

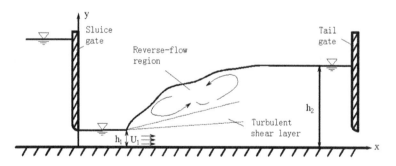

Figure 1.   Definition sketch of a hydraulic jump.

Table 1.   Hydraulic jump properties.*

ID	h_1 (m)	U_1 (m/s)	Fr_1	h_2 (m)	Q (m³/s)	Re
J1	0.05	1.41	2.01	0.14	0.0352	7.06E+4
J2	0.05	1.79	2.56	0.17	0.0448	8.99E+4
J3	0.04	2.44	3.90	0.22	0.0488	9.79E+4
J4	0.04	3.04	4.85	0.27	0.0608	1.22E+5

*h_1 = supercritical water width; U_1 = inflow velocity; Fr_1 = inflow Froude number; h_2 = subcritical water width; Q = discharge; Re = Reynolds number.

3 RESULTS AND DISCUSSION

3.1 *Energy dissipation rate*

The turbulence structure of a hydraulic jump will become complicated and the energy dissipation will change accordingly after air bubbles being entrained into the water mass. In order to demonstrate the effect of interaction between bubbles and water mass, the distribution of energy dissipation rate at certain sections were put forward (Fig. 2). According to some former researches, the dissipation rate can be estimated by a two-step approximation (Liu et al. 2004). Firstly, the Kolmogorov theory of local isotropic turbulence can be used to obtain the spectrum, $G_i(k)$, when the sampling frequency of MicroADV is greater than 21. Therefore, $G_i(k)$ is described as

$$G_i(k) = \alpha_i \varepsilon_i^{2/3} k_i^{-5/3} \tag{1}$$

where $G_i(k)$ = spectrum of the ith velocity component (i = 1, 2, 3); α_i = Kolmogorov constant with values of 0.53, 0.71 and 0.71 for longitudinal, vertical and lateral turbulence, respectively (Sreenivasan 1993); ε_i = the ith energy dissipation rate in three directions; and k_i = the ith wave number in three directions. Secondly, Taylor's frozen turbulence hypothesis can be used to approximately transfer the spectrum from the frequency domain to the wave number domain by

$$k_i = 2\pi f_i / u_i \tag{2}$$

$$G_i(k) = \frac{u_i}{2\pi} G_i(f) \tag{3}$$

on condition that the velocity fluctuations are much smaller than the streamwise mean velocity, where u_i = the ith velocity component. Substituting Equations (2) and (3) into Equation (1), we note that

$$\varepsilon_i = 2\pi \alpha_i^{-3/2} u_i^{-1} f_i^{5/2} G_i^{3/2}(f) \tag{4}$$

Then the energy dissipation rate of certain sections within hydraulic jumps could be calculated from the sampling data of MicroADV using Equations (4). Figure 2 shows the distribution of energy dissipation rate along the jumps in the dimensionless form of $\varepsilon h_1 / U_1^3$, where ε is the Root Mean Square (RMS) of ε_i in three directions, while x and y being defined

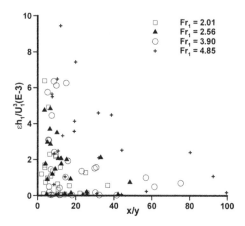

Figure 2.   Variation of energy dissipation rate in central plane with x/y.

as the longitudinal distance from the jump toe and the vertical distance from the flume bottom, respectively. It was found in Figure 2 that most of the dimensionless values appeared to cluster near the corner regardless of Froude numbers. It should be noted that the vertical distance of the sampling position from the flume bottom would increase as x/y decreased. Therefore, the energy dissipation rate took its highest value as $x/y < 20$, which means the energy dissipation took place mostly in the shear layer beneath the reverse-flow region. When compared with earlier experimental data by Chanson (2011), a good agreement was observed between the distribution of energy dissipation rate and bubble count rate. Moreover, the dissipation rate gradually increased with the inflow Froude numbers, mainly owing to more bubbles entrained into water mass at larger inflow Froude numbers.

3.2 Energy consumption of bubbles

After a cloud of air being entrapped into hydraulic jumps, the air mass will encounter a group of forces, including water pressure, viscous force and surface tension. Therefore, within the jump length, air bubbles will experience two kinds of process, formation and transportation, simultaneously. On one hand, air bubbles will be split from air mass into smaller ones and the average bubble size will vary obviously with the physical property of the flow. This process herein can be named as air bubbles formation process. During the process, a part of energy of the flow was extracted to form the boundary of bubbles, which can be regarded as a certain part of energy dissipation of hydraulic jumps. To identify the quantities of such part of energy consumption, the bubble size should be determined first as (Hinze 1955)

$$d_{95} = 0.725\left[(\sigma/\rho)^3 \varepsilon^2\right]^{1/5}$$

(5)

in which, d_{95} = diameter of bubbles accounting for more than 95% in the sampling area; σ = coefficient of surface tension with a value of 0.0728 N/m; ρ = density of water; ε = energy dissipation rate. Hinze (1955) pointed out that almost 95% of total bubbles would finally become stable in a certain size after dispersion processes. Substituting the results of energy dissipation rate in Equation (5) leads to the distribution of diameters of bubbles, shown in Figure 3. Apparently, the diameter closely corresponded with the energy dissipation rate so that the peak value appeared at the same location in the shear layer, where $1 < y/h_1 < 2$. Although the results showed some increase in the bubbles size near the flow bottom, some rapid decrease took place in the reverse flow region of the jumps at all different inflow Froude numbers.

On the other hand, air bubbles will be carried downstream along with the main flow during the formation process. When bubbles are transported to a certain position within the jump, the inner pressure of bubbles will rise to the corresponding level to balance the pressure from outside. During this transportation procedure, another portion of energy will be consumed on compressing bubbles. Besides, there should be a minimal quantity of energy loss resulting from bubbles overcoming viscous force of water. Since there had been evidence that the proportion of energy consumption on overcoming viscous force to that on compressing bubbles ranges from 0.01% to 1% (Guo 2002), the former one could be ignored in predicting total energy loss.

Assuming that the bubbles size would keep unchanged after the formation process, the energy consumption from both formation and transportation process of bubbles in a certain position could be estimated as (Ni 2000):

$$W = \frac{4}{3}\pi R^3 \left(p_w + \frac{2\sigma}{R}\right) \ln\left(\frac{p_w + \frac{2\sigma}{R}}{p_0}\right) + 4\pi\sigma R^2$$

(6)

where p_0 = absolute atmospheric pressure (ATA); p_w = water pressure on the bubble surface; R = radius of bubbles. On the right side of Equation (6), the first part represents the energy loss caused by bubbles compression during transportation process while the second part demonstrate the energy dissipation produced by bubbles formation. Figure 4 presents the

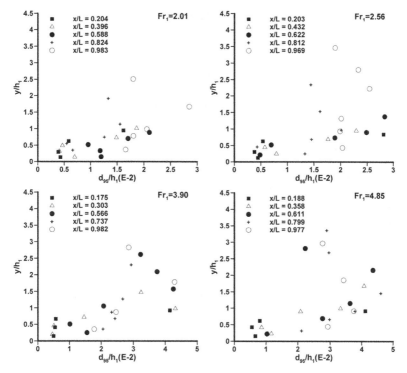

Figure 3. Dimensionless distribution of the diameter of bubbles accounting for more than 95% in various sections (L = length of hydraulic jumps).

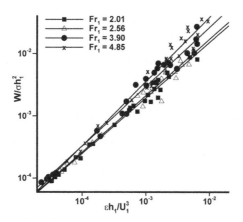

Figure 4. Bubbles energy consumption W as a function of energy dissipation rate ε.

evolution of dimensionless bubbles energy consumption $W/\sigma h_1^2$ with the energy dissipation rate. The data showed some effect of the inflow Froude number since they were scattered around four different lines, which could be described by the equation

$$
\begin{aligned}
W/\sigma h_1^2 &= A(\varepsilon h_1/U_1^3)^B \\
A &= 0.5741 Fr_1^2 - 2.7426 Fr_1 + 4.2111 \\
B &= 0.0159 Fr_1^2 - 0.0614 Fr_1 + 0.9554
\end{aligned}
\tag{7}
$$

with the normalized coefficient of correlations ranging from 0.972 to 0.994. The results highlighted some monotonic increase in energy consumption by bubbles with increasing energy dissipation rate for a given inflow Froude number. The rates of increase were about the same for both inflow Froude numbers $Fr_1 = 2.01$ and 2.56. For a larger inflow Froude numbers, however, the growth of dimensionless energy consumption $W/\sigma h_1^2$ was relatively slow as the dissipation rates increasing. Therefore, the energy consumption of bubbles is more significant to energy dissipation of jumps at small Froude numbers than at larger ones.

4 CONCLUSIONS

The research herein was to study the energy dissipation of hydraulic jumps with low Froude numbers of 2.01, 2.56, 3.90 and 4.85. A MicroADV probe was used to obtain the distributions of energy dissipation rate in different section within the jumps. It was found that the dissipation rate reached its peak value in the turbulence shear layer beneath the reverse-flow region regardless of the inflow Froude numbers and gradually decreased along the direction to the surface or the flume bottom, mainly due to a maximum bubble count rate showed in Chanson's study. Moreover, the energy consumption by bubbles formation and transportation process, predicted by estimation of the diameter of bubbles within jumps, has been revealed and it is implied that energy consumption by bubbles represents a certain degree of the turbulence energy dissipation rate. Finally, an empirical formula of estimating energy consumption by bubbles has been developed and verified.

REFERENCES

Chanson, H. & Murzyn, F. 2008. Froude similitude and scale effects affecting air entrainment in hydraulic jumps. *World Environment and Water Resources Congress*. Ahupua'a.
Chanson, H. 2011. Bubbly two-phase flow in hydraulic jumps at large Froude numbers. *Journal of Hydraulic Engineering*. 137(4): 451–460.
Chachereau, Y. & Chanson, H. 2011. Free-surface fluctuations and turbulence in hydraulic jumps. *Experimental Thermal and Fluid Science*. 35(6): 896–909.
Chachereau, Y. & Chanson, H. 2011. Bubbly flow measurements in hydraulic jumps with small inflow Froude numbers. *International Journal of Multiphase Flow*. 37(6): 555–564.
Goring, D.G. & Nikora, V.I. 2002. Despiking acoustic Doppler velocimeter data. *J. Hydraul. Eng.* 128(1): 117–126.
Guo, L.J. 2002. *Two-phase and Multi-phase flow dynamics*. Xian: Jiaotong University Press.
Hinze, J.O. 1955. Fundamentals of the hydrodynamic mechanism of splitting in dispersion processes. *AM. Inst. Chem.* 1(3): 289–295.
Kucukali, S. & Cokgor, S. 2006. Aeration performance of a hydraulic jump. *World Environment and Water Resources Congress*. Nebraska.
Kucukali, S. & Cokgor, S. 2009. Energy concept for predicting hydraulic jump aeration efficiency. *Journal of Environmental Engineering*. 135(2): 105–107.
Kumar, B. & Rao, A.R. 2009. Oxygen transfer and energy dissipation rate in surface aerator. *Bioresource Technology*. 100(11): 2886–2888.
Liu, M.N., Rajaratnam, N. & Zhu, D.Z. 2004. Turbulence structure of hydraulic jumps of low Froude numbers. *Journal of Hydraulic Engineering*. 130(6): 511–520.
Mignot, E. & Cienfuegos, R. 2010. Energy dissipation and turbulent production in weak hydraulic jumps. *Journal of Hydraulic Engineering*. 136(2): 116–121.
Ni, H.G. 2000. *Effective energy dissipators*. Dalian: University of Technology Press.
Rajaratnam, N. 1962. An experimental study of air entrainment characteristics of the hydraulic jump. *Journal of Instn. Eng. India*. 42(7): 247–273.
Resch, F.J. & Leutheusser, H.J. 1972. Le ressaut hydraulique: mesure de turbulence dans la region diphasique. *La Houille Blanche*. (4): 279–293.
Sreenivasan, K.R. 1993. On the universality of the Kolmogorov constant. *Phys Fluids*. 7(11): 2778–2784.

Hydraulic Engineering – Xie (Ed.)
© 2013 Taylor & Francis Group, London, ISBN 978-1-138-00043-8

Research on time-varying resistance attenuation law of hydraulic steel gate members

Shuhe Wei, Yushan Ren, Zhigang Yin & Yi Wang
Changchun Institute of Technology, Jilin Province, China
Jilin Water Project Safety and Disaster Prevention Engineering Center, Jilin Province, China

ABSTRACT: Corrosion on hydraulic steel gate is a natural process, however, the assumption of constant corrosion rate in previous researches lead to linear change of member thickness. Based on non-linear description of corrosion process, non-linear resistance attenuation model of hydraulic steel gate members is developed in this paper. Attenuation process of resistance and modulus of steel gate structural members are calculated and studied according to real corrosion data.

1   INTRODUCTION

Resistance attenuation of hydraulic steel gate structures and members has notable influence on structural reliability. Resistance of hydraulic steel gate members and whole structure attenuates with time. Time variation of structural resistance had notable influence on structural reliability index, however, Unified standard for reliability design of building structures (GB5068-2001) and various code of design of building structures do not explicitly take the time variation of structural resistance R into consider which should be improved (Li 2001 & Ren 2005).

In order to predict the service life of existing hydraulic steel gate structures, resistance attenuation law must be mastered. Generally, the influencing factors of resistance attenuation can be summarized as load effect and material effect and environment effect (Li 2001 & Zhou 2003). Environment effect is the most important factors for hydraulic steel gate and manifested as steel gate corrosion. So, resistance attenuation law of hydraulic steel gate caused by corrosion is mainly analyzed in this paper.

2   CORROSION LAW OF HYDRAULIC STEEL GATE

Corrosion of steel gate belongs to electrochemical corrosion. Due to impurities in the steel, impurity is cathode for its higher electric potential and steel is anode for its lower electric potential, thus lots and lots of corrosion micro-cells are formed. Corrosion process is formed with the dissolution of iron ion in anode region.

Corrosion rate of steel is controlled by the electrode process of corrosion cells, that is, anodic process and cathodic process. The electrode process is directly or indirectly influenced by various internal and external factors. Internal factors include surface condition of steel structure and internal force and deformation and so on. External factors include water chemical compositions, various ion concentration and distribution, water temperature distribution, flow velocity, stray current, aquatic animals, and so on.

Based on the available information and data about steel corrosion, the main factors influencing the decaying resistance of steel gate structural members are analyzed and a law of corrosion is derived. A model of linear decaying resistance of gate structural members is proposed by Zhou (2003).

In fact, corrosion of steel structure in natural environment is nonlinear process (Guedes Soares 1999 & Xia 1989 & Zheng 1997). The time-dependent model of corrosion degradation could be separated into three phases. In the first phase there is in fact no corrosion because the protection of the metal surface works properly. The second phase is initiated when the corrosion protection can not work properly and corresponds really to the existence of corrosion, ($t \in$ [O, B] in Fig. 1). The process last a period about 4–5 years in typical ship plating by Maximadj AI (1982). The third one corresponds to a stop in the corrosion process and the corrosion rate is close to zero ($t >$ B in Fig. 1). Corroded material stays on the plate surface, protecting it from the contact with the corrosive environment and the corrosion process stops.

The model proposed by C. Guedes Soares (1999) can be described by the solution of a differential equation of the corrosion wastage.

$$d_\infty \dot{d}_L(t) + d_L(t) = d_\infty \tag{1}$$

where d_∞ is the long-term thickness of the corrosion wastage, $d_L(t)$ is the thickness of the corrosion wastage at time t, and $\dot{d}_L(t)$ is the corrosion rate.

The solution of Equation (1) can have the general form

$$d_L(t) = d_\infty \left(1 - e^{-t/\tau_t}\right) \tag{2}$$

and the particular solution leads to

$$d_L(t) = \begin{cases} 0 & t \le \tau_c \\ d_\infty \left(1 - e^{-(t-\tau_c)/\tau_t}\right) & t > \tau_c \end{cases} \tag{3}$$

where τ_c is equal to the time interval between the painting of the surface and the time when its effectiveness is lost, which could be called the coating life, and τ_t is the transition time.

Based on the corrosion data, research on nonlinear corrosion description proposed by C. Guedes Soares (1999) and linear model (Zhou 2003) is illustrated in Figure 2. Corrosion wastage $d_L(t)$ increased to 2.55 mm after 22 years for linear model, $d_L(t)$ stabilized at about 1.60 mm in the 22nd year for the nonlinear model. In correspondence to the models, parameter values are $\tau_c = 5$ a, $d_\infty = 1.6$ mm, $\tau_t = 4.5$ a, $v = 0.15$ mm/a. Nonlinear model better fitted the corrosion data obviously and reflected the nonlinear characteristics of $d_L(t)$ variation with time.

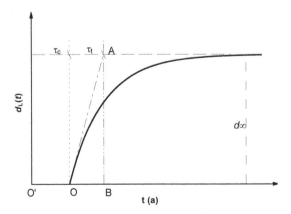

Figure 1. Thickness of corrosion wastage as a function of time.

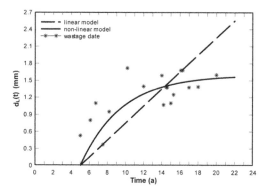

Figure 2. Time-dependent corrosion wastage for linear and nonlinear model.

Furthermore, taking into consideration periodic preventive maintenance in gate operation process and practical corrosion process, nonlinear description of corrosion law is adopt and resistance attenuation nonlinear model of hydraulic steel gate members is proposed in this paper.

3 TIME-VARIANT RESISTANCE MODEL OF HYDRAULIC STEEL GATE MEMBER

Generally speaking, structure resistance is one-dimensional or multidimensional non-stationary random process varying with time. To introduce a practical approach for structure reliability, the non-stationary random process is stabilized and relatively simple stochastic process model is adopted (Zhao, 2000).

$$R(t) = \phi(t) R_0 \tag{4}$$

where $R(t)$ is structure resistance at time t and R_0 is the structure resistance at time $t = 0$, $\phi(t)$ is attenuation function. The determination of $R(t)$ comes down to the determination of $\phi(t)$ for the reason that R_0 is known.

Based on working environment and corrosion of hydraulic steel gate, resistance attenuation caused by corrosion has been mainly studied in this paper, furthermore the question comes down to the determination of $\phi(t)$ according to the law of corrosion.

In fact, $\phi(t)$ is a random variable and a random process, more strictly speaking. In order to reduce the complexity of the problem and introduce a practical approach for structure reliability, only deterministic function $\phi(t)$ is discussed in the paper.

Taking hot rolled H-beam section of gate structure for example, the determination of intensity attenuation function $\phi(t)$ has been illustrated in tension-compression state and flexural state. It is assumed that structure resistance use the representation of internal forces and the deformation of the section is according with the plane hypothesis after corrosion, and member corrosion in width direction is ignored.

3.1 Axial tension-compression members

Resistance $R(t)$ under tension-compression condition is given as

$$\begin{aligned} R(t) = A(t) f_y &\approx \left[2bd_2(t) + hd_1(t) \right] f_y \\ &= \left\{ 2bd_{20} \left[1 - \frac{d_L(t)}{d_{20}} \right] + hd_{10} \left[1 - \frac{d_L(t)}{d_{10}} \right] \right\} f_y \end{aligned} \tag{5}$$

where $A(t)$ is H-beam section area at time t and f_y denotes design value of steel strength, b and h is respectively breadth and depth of steel H-beam section, $d_1(t)$ and $d_2(t)$ is respectively

99

the web and flange thickness at time t, d_{10} and d_{20} are corresponding thickness values at $t=0$, which are called initial thickness, $d_1(t)$ is corrosion wastage thickness at time t.

It is assumed that the initial thickness is equal, that is, $d_{10} = d_{20}$, so

$$R(t) = (2bd_0 + hd_0)\left[1 - \frac{d_L(t)}{d_0}\right]f_y$$

$$= A_0 f_y\left[1 - \frac{d_L(t)}{d_0}\right] = R_0\left[1 - \frac{d_L(t)}{d_0}\right] \tag{6}$$

Where A_0 is member initial area, and $R_0 = A_0 f_y$ is member resistance at $t=0$.

Equation (3) is substituted into Equation (6), and the expression of attenuation function is introduced in comparison with Equation (4)

$$\phi(t) = 1 - \frac{d_L(t)}{d_0} = \begin{cases} 1 & 0 \le t \le \tau_t \\ 1 - \dfrac{d_\infty\left(1 - e^{-(t-\tau_c)/\tau_t}\right)}{d_0} & t > \tau_t \end{cases} \tag{7}$$

3.2 Flexural members

Modulus of bending section illustrated in Figure 3 can be approximated to be

$$W(t) = \frac{h^2 d_1(t)}{6} + \frac{bd_2^3(t)}{3h} + bhd_2(t)$$

$$\approx \frac{h^2}{6}d_{10}\left[1 - \frac{d_L(t)}{d_{10}}\right] + bhd_{20}\left[1 - \frac{d_L(t)}{d_{20}}\right] \tag{8}$$

when $d_{10} = d_{20} = d_0$

$$W(t) = \left(\frac{h^2}{6}d_0 + bhd_0\right)\left[1 - \frac{d_L(t)}{d_0}\right] = W_0\left[1 - \frac{d_L(t)}{d_0}\right] \tag{9}$$

where W_0 is initial section modulus in bending, and the resistance is given by

$$R(t) = W(t)f_y = W_0\left[1 - \frac{d_L(t)}{d_0}\right]f_y = R_0\left[1 - \frac{d_L(t)}{d_0}\right] \tag{10}$$

and expression of $\phi(t)$ developed from Equation (11) is the same as that given by Equation (7).

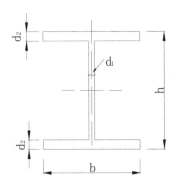

Figure 3.   Hot rolled H-beam section size.

4 ANALYSIS OF MODELS

According to $R_0 = A_0 f_y$ and deterministic function, initial resistance was discussed. Taking hot rolled H-beam 200×200 mm² for example, it is assumed that design value of Q235 steel strength is $f_y = 210$ N/mm², section size $h = 200$ mm and $b = 204$ mm and $d_1 = d_2 = 12$ mm and $A_0 = 7296$ mm², so initial resistance $R_0 = A_0 f_y = 1.53$ MPa, initial section modulus in bending $W_0 = 569,600$ mm³ are received.

Based on the linear resistance model given by Zhou (2003) and non-linear resistance model given in this paper, resistance attenuation curves of H-beam member were calculated and illustrated in Figure 4. For linear model, resistance was equal to 1.53 MPa and remained unchanged in the first 5 years, and linear attenuation was observed after 5a and the resistance value was reduced to 1.20 MPa at the 22nd year. For non-linear model, resistance was equal to 1.53 MPa and remained unchanged in the first 5 years, rapid attenuation with decreasing rate was observed after the first 5 years, the resistance curve became flat and was eventually kept stable at 1.33 MPa.

Based on the linear model given by Zhou (2003) and non-linear model given in this paper, section modulus $W(t)$ curves of H-beam member were calculated and illustrated in Figure 5. For linear model, section modulus was equal to 569,600 mm³ and remained unchanged in the first 5 years, and linear attenuation was observed after 5 years and the section modulus value was reduced to 448,560 mm³ by 21.25% at the 22nd year.

For non-linear model, rapid attenuation with decreasing rate was observed after the first 5 years, and the non-linear curve intersected the linear curve at 14th year, resistance curve became flat and section modulus was reduced by 13.00% and eventually kept stable at 49,5300 mm³.

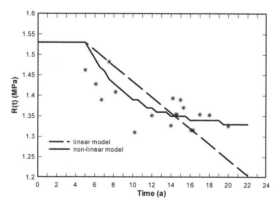

Figure 4.   Resistance attenuation curves of H-beam member.

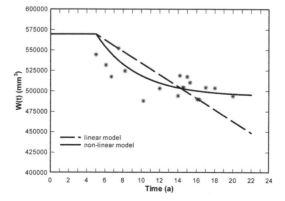

Figure 5.   Section modulus $W(t)$ curves of H-beam member.

For actual process of corrosion, corroded material stays on the plate surface, protecting it from the contact with the corrosive environment and the corrosion rate will slow down gradually. Therefore this article hold that resistance attenuation law of hydraulic steel gate caused by corrosion can be more realistically decrypted with non-linear model.

5 CONCLUSIONS

Based on non-linear description of corrosion process, resistance attenuation function of hydraulic steel gate H-beam members is derived and resistance attenuation non-linear model is further developed. Attenuation law of resistance and modulus of steel gate structural members is calculated and tested through real corrosion data. The result comparison between the models and real corrosion data shows that better fitness of non-linear attenuation model of steel gate members to data is observed in comparison to the linear model, steel structure resistance attenuation law expressed by non-linear description is more coincident with practical corrosion process.

Taking into consideration gate periodic maintenance and anti-corrosion treatment, non-linear resistance attenuation model proposed in this paper could fit the real corrosion process more flexibly. Non-linear resistance attenuation model of hydraulic steel gate members is a profitable attempt at the resistance attenuation law of hydraulic steel gate structure and deepens the research to time-varying reliability practical calculation method.

ACKNOWLEDGEMENTS

The model and formulation presented here has been developed in the project "Research on degradation mechanism and durability evaluation theory for existing hydraulic steel gates", which was financed by the National Natural Science Foundation of China through the contract No. 51179011 and Jilin Province Science and Technology Development Planning Project (No. 20100707).

REFERENCES

Guedes Soares, C. & Garbatov, Y. 1999. Reliability of maintained, corrosion protected plates subjected to non-linear corrosion and compressive loads. *Marine Structures* 12(6): 425–445.

Li, G.Q. & Li, Q.S. 2001. Time-varying reliability theory and application to engineering structures. Beijing: Science Press.

Maximadj, A.I. & Belenkij, L.M. & Briker A.S. & Neugodov, A.U. 1982. *Technical Assessment of Ship Hull Girder*. Petersburg: Sudostroenie.

Ren, Y.S. 2005. Study on estimation of remaining life of members of existing hydraulic steel gates. *Engineering Journal of Wuhan University* 38(6):62–65.

Xia, N.L. 1989. The service life and protection design of steel gates. *Metal Structure* (3):10–18.

Zhao, G.F. & Jin, W.L. & Gong, J.X. 2000. *Reliability theory of structures*. Beijing: Chinese Building Industry Press.

Zheng, S.Y. & Yuan, Y.Q. 1997. Safety examination and assessment of arc-type head of water outlet in Yuecheng reservoir. *Metal Structure* (4):33–35.

Zhou, J.F. & Li, D.Q. 2003. Time-variant resistance model and reliability analysis of steel gate structures. *Engineering Mechanics* 20(4):104–109.

Hydraulic Engineering – Xie (Ed.)
© *2013 Taylor & Francis Group, London, ISBN 978-1-138-00043-8*

Near-dam reach propagation characteristics of daily regulated unsteady flow released from Xiangjiaba Hydropower Station

Ai-Xing Ma
Nanjing Hydraulic Research Institute, State's Key Laboratory of Hydrology Water Resources and Hydraulic Engineering Science, Nanjing, China

Si-En Liu
Delft University of Technology, Delft, The Netherlands

Xiu-Hong Wang & Min-Xiong Cao
Nanjing Hydraulic Research Institute, Nanjing, China

ABSTRACT: During the operation of Xiangjiaba Hydropower Station, the unsteady flow released from the Station's daily regulation will change the channel flow condition of near-dam reach (Shuifu-Yibin river stretch). A two-dimensional flow mathematical model concerning 33 km downstream reach is applied to calculate and analyze propagation characteristics of daily regulated unsteady flow released from Xiangjiaba Hydropower Station. The result shows that during the operation of the Station, daily regulation release wave gradually flattens as it propagates downstream, the varying amplitude of discharge decreases, and the amplitude of water level shows a generally falling trend, but rises when flow section shrinks. Stage-discharge (*Z-Q*) relationship shows anticlockwise looped curve. As it propagates, the discharge wave will cause additional water surface slope, which is positive in rising periods and negative in falling periods. The unsteadiness of unsteady flow in narrow-deep reach is stronger than that in wide-shallow reach.

1 INTRODUCTION

Along with the development of hydroelectric station cascade, currently the majority of rivers are regulated by hydraulic junctions. During the medium and low water periods, flow is influenced by power generation and the originally steady flow shows apparent ups and downs (daily regulated unsteady flow). The unsteadiness is very strong and exerts influence on downstream channel flow condition, sediment movement and evolution of shoals and channels (De Sutter 2001, Kuhnle 1992, Ma 2012). Huang (2004) calculated the influence of daily regulated wave from Three Gorges Reservoir on downstream water surface slope and projected that when the station increased or decreased its loads, the downstream water surface slope is steeper or milder than that in steady condition. Cao (2011) studied the unsteady water and sediment characteristics based on the physical model of Heshangyan rapids below Xiangjiaba Station and showed maximum water surface slope steepens with the increase of discharge gradient before the arrival of peak wave and is irrelevant to the amplitude of discharge. Ma (2012) experimented on unsteady flow flume and found that in the propagation of unsteady flow, water surface slope J, shear velocity U^*, depth averaged velocity U, discharge Q and water depth h occur asynchronously. The order of hydraulic elements reaching the peak is $J_{max} \rightarrow U^*_{max}$, $U_{max} \rightarrow Q_{max} \rightarrow h_{max}$. The asynchronous variations of hydraulic elements result in the looped curve relationship. Xiangjiaba Hydropower Station, impounding and operating in October, 2012, is the last cascade in the downstream cascade development of Jinshajiang River. Its installation capacity is smaller than those of Three Gorges

Station and Xiluodu Station, ranking the third in China. The operation of the Station will certainly changes the downstream natural flow condition below the dam. This paper uses two-dimensional flow mathematical model concerning 33 km downstream reach to calculate and analyze near-dam reach (Shuifu-Yibin river stretch) propagation characteristics of daily regulated unsteady flow released from Xiangjiaba Hydropower Station.

2 DESCRIPTION OF XIANGJIABA HYDROPOWER STATION AND THE NEAR-DAM REACH

2.1 Description of Xiangjiaba Hydropower Station

Xiangjiaba Hydropower Station is the last cascade in the downstream cascade development of Jinshajiang River. The dame site lies at 3 km upstream Shuifu Harbor which is at the boundary of Sichuan and Yunnan Province. It is 157 km to Xiluodu Hydropower Station in the upstream, and 1.5 km to the Shuifu County, 33 km to Yibin City in the downstream, the geographic location as shown in the Figure 1. The normal reservoir level of the station is 380 m, dead level 370 m, reservoir area 95.6 km^2, total reservoir capacity 5.163 billion m^3, backwater length 156.6 km, controlled drainage area 458,800 km^2. It is considered as ravine type reservoir and occupies 97% of the total basin area of Jinshajiang River. There are two types of typical daily regulation operating mode in low flow stages (Liu 2012) (Fig. 2), operating mode I is the most disadvantageous guaranteed output in the design low flow years under the condition that hour amplitude of water level below the dam is less than 1.0 m/h, and operating mode II is the mean annual output of each month in low flow stages. Thus, daily regulation discharge range of Xiangjiaba Hydropower Station changes greatly (1200~5589 m^3/s), and the maximum amplitude is 4.66, there is one peak flow (18:00~19:00) in 24 hours, the load of the station decreases from 24:00 to 6:00 am in the following day, correspondingly the release discharge is at the bottom during the whole day.

2.2 Overview of Shuifu-Yibin river stretch

The river reach from Shuifu to Yibin covers 30 km with an average surface slope of 0.27%. The stream bed is mainly composed of bedrock or sand-pebble, and can fall into the category the raw river without any systematic regulation in the history. The branch of Hengjiang River converges in the upstream and Minjiang River injects in the downstream. The Xiangjiaba Hydropower Station, 3 km upstream from Shuifu Harbor, initiated to generate power with its first hydropower unit in October, 2012. There are 13 rapids and shoals, including Niupi Rapids (NPR), Hengjiangkou Rapids (HJKR), Mapibao Rapids (MPBR), Heshangyan Rapids (HSYR), Erlang Rapids (ELR), Zhanqiao Rapids (ZQR), Shuimi Rapids (SMR), Jichibang

Figure 1. Geographic location of Xiangjiaba Hydropower Station.

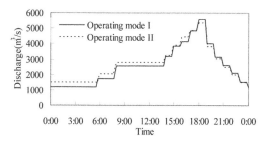

Figure 2. Daily regulation hydrographs of Xiangjiaba Hydropower Station.

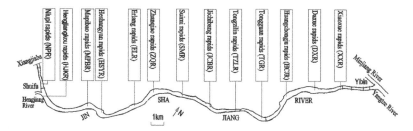

Figure 3. Rapids and shoals in Shuifu-Yibin river reach.

Rapids (JCBR), Tongzilin Rapids (TZLR), Tongguan Rapids (HGR), Huangchongju Rapids (HCJR), Daxue Rapids (DXR) and Xiaoxue Rapids (XXR), Figure 3, with an average distance of 2.3 km in the river reach, characterized by rapidity, shoal and hazard.

3 NEAR-DAM REACH PROPAGATION CHARACTERISTICS OF DAILY REGULATED UNSTEADLY FLOW FROM THE STATION

Two-dimensional flow mathematical model is applied to calculate near-dam reach propagation process of daily regulated unsteady flow released from Xiangjiaba Hydropower Station. The model adopts the 1:3000 underwater topography data (from Xiangjiaba to Hejiangmen in Yibin) measured in March, 2008, with the total length 33 km. The computational grid is made up of structureless triangular grid unit. The size of the grid unit along the flow direction is generally 30 m, perpendicular to the flow direction 28 m. The total amount of computational grid nodes is about 44,890, the total units about 86,625. The model uses the finite volume method for solving. The roughness of the model's main pool is 0.025~0.029, and the roughness of shoal or point bar side is 0.032~0.05. The model is verified with field data during low, flood and average flow periods of steady and unsteady flow (Ma 2010), and the result shows that the verification of water-level, velocity correspond to the filed data, and the model can simulate the flow motion law in the downstream of Xiangjiaba Station.

From the perspective unfavorable to channel flow condition, the operating mode I, whose minimum discharge is at the lowest and maximum at the highest, is chosen to simulate its process of daily regulation discharge flow. Meanwhile, the discharge from branch of Hengjiang River is considered as 50 m³/s, and the lower boundary of the model is provided by the one-dimensional mathematical model from Xiangjiaba to Luzhou reach (Ma 2009).

3.1 Water level propagation characteristics and variation of water surface slope

After the operation of Xiangjiaba Station, the water level from Shuifu to Yibin becomes flatting (Fig. 4), the water level variation in the dam is correspond to the hydrograph of the release wave, showing the shape of steps. The step-like changes in river reach gradually disappear and become flatting and the amplitude of water level decreases along with the unsteady flow moving to the downstream. Through the analysis of daily amplitude of water level along the way (Fig. 5), it can be seen that as the unsteady flow propagates downstream, the overall trend of water level daily amplitude gradually reduces, but water level daily amplitude from Mapibao rapids to Suimi rapids remain stable during the fluctuations; judging from water level daily amplitude of regional rapids, it increases first and then decreases. The analysis shows this phenomenon is related to the changes of landscape along the way. Sixty cross sections are selected to calculate the minimum flow section (A-min) during the daily regulation and also plotted in Figure 5, which indicates the flow section from Mapibao rapids to Suimi rapids is relatively narrow compared with other rapids. When the unsteady flow propagates into this reach, it is restricted by the sidewalls and the flow energy concentrates, leading to increases of amplitude. Flow section downstream Suimi rapids increases, the consumption of flow energy also increases due to the bed friction, thus amplitude falls. Some regional rapids, such as Mapibao rapids, Zhanqiao rapids, Suimi rapids, Daxue rapids, experience similar changes. According to statistics, daily water level amplitude is between 3.28 m and 5.16 m, and the maximum daily level amplitude of variation is 5.16 m in the dam section.

Figure 6 reflects the variation of water level hour amplitude in the part of rapids. It can be seen that the water level hour amplitude less than 1 m, the maximum value occurs in 19:00~24:00, the amplitude is large near the Station.

As daily regulation unsteady flow moves downstream, the water surface slope is not only related to the characteristics of rapids and shoals (the water surface slope of low water rapids is large in small discharge condition, while that of average and flood rapids is steep in large discharge condition), but also relevant to the Station's discharge hydrograph. For example, Niupi rapids, 1.5 km away from the Station, is an average and flood rapids, mean water surface slope increases with rapids reach discharge and additional water surface slope will occur when influenced by unsteady discharge flow (Fig. 7). To be more specific, when the discharge increases and the release wave move forward, water surface slope soars with flow increase rapidly, and then restores to the original discharge-slope relation; when the discharge

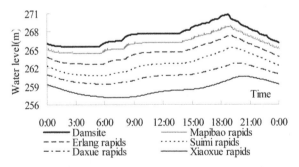

Figure 4.   Water level variation with the release wave.

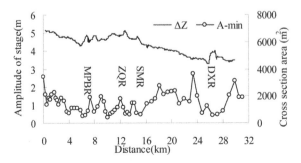

Figure 5.   Variation of water level daily amplitude.

106

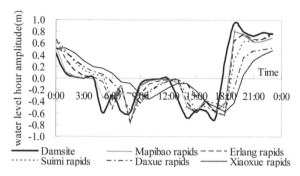

Figure 6.   Variation of water level hour amplitude.

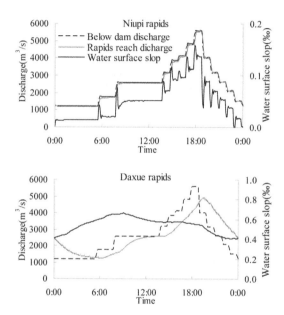

Figure 7.   Variation of average water surface slope in partial rapids reaches.

decreases markedly and recession wave move forward, water surface slope plunges, and then restores to the original discharge-slope relation again. The Variation of water surface slope in rapids reach corresponds to the changes of discharge. According to the statistics, the water surface slope in Niupi rapids reaches −0.098‰~0.056‰. When unsteady flow propagates to Daxue rapids, 29 km away from the Station, discharge variation flattens, additional water surface slope weakens obviously, and water surface slope is much steeper in small discharge because this rapids is low water rapids.

3.2 Discharge propagation characteristics and stage-discharge relation

Since daily regulation release wave is affected streamwise by friction and channel storage, the step-like changes of discharge in river reach gradually disappear and become flatting (Fig. 8), the maximum discharge below the dam decreases while the minimum discharge increases, so the flow amplitude decreases (Fig. 9). Unlike the amplitude of water level, discharge amplitude is irrelevant to flow section. Under the operating mode I, the maximum discharge is 5589 m³/s, and the minimum 1200 m³/s. When flow reaches Yibin, 30 km away from the dam in the downstream, the maximum flow declines to 4734.6 m³/s, whereas the minimum flow increases to 1347.6 m³/s. Discharge rate of variation falls to 3387 m³/s in Yibin from 4389 m³/s, and the average discharge amplitude decreases at the rate of 33.4 m³/s/km.

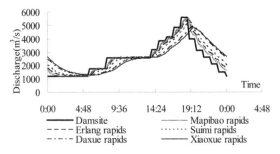

Figure 8.   Discharge variation with the release wave in partial rapids reaches.

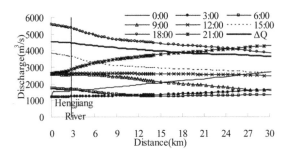

Figure 9.   Downstream variation of discharge.

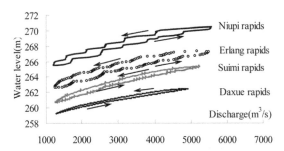

Figure 10.   Variation of stage-discharge relation.

When release wave propagates downstream, variations of water level and discharge are not synchronous, which can be obtained from the analysis of the stage-discharge relationship. Normally, for the river reach where variations of flow rate are gentle, its rating curve displays single-valued cure. When influenced by the strong unsteadiness of daily regulation wave, the rating curve shows anticlockwise looped curve (Fig. 10), that is, discharge attains its maximum value before the water level peak appears, which is non-synchronous and reflected in the following: at the same discharge, water level in rising limb is lower than that in falling limb, and discharge capacity in rising limb is greater than that in falling limb at the same water level. Water level and discharge hydrograph of Niupi rapids is directly influenced by release wave, and its rating curve is step-like looped rating. The looped rating curve is relatively wide due to strong unsteadiness. As the unsteady discharge wave propagates downstream, the wave loses its power, and the steps vanish. Simultaneously, looped rating curve becomes narrow.

3.3   *Variation of unsteadiness of the flow along the way*

As can be seen from the above, during the propagation of Station's daily regulated unsteady flow, the hydraulic elements (such as water level and discharge) change constantly, the change rates of water level and discharge directly reflect the unsteadiness of flow and its variation along the way. Dimensionless unsteadiness parameter P is proposed on the basis of change rates of water level and discharge:

$$P = \begin{cases} \dfrac{1000\,BT_r}{Q_p - Q_b}\left(\dfrac{Z_p - Z_b}{T_r}\right)^2 & Q_p \neq Q_b \\ 0 & Q_p = Q_b \end{cases} \tag{1}$$

where B is the river width (m), which can be obtained from the average river width corresponding to the highest and the lowest water level, T_r is the lasting time (s) of the unsteady flow in rising period, Q_p and Q_b represents the peak flow (maximum discharge, m³/s) and basic flow (minimum, m³/s) of unsteady flow respectively, and Z_p and Z_b indicate the highest, lowest water level (m) of unsteady flow; as for steady flow ($Q_p = Q_b$), the unsteadiness is zero, then $P = 0$. As can be seen from the formula (1), the influence of the rate of change of the water level on the parameter P is sensitive to the change rate of the unit width discharge. Change rate of water level has a more sensitive influence on the unsteadiness parameter P than change rate of discharge does.

Figure 11 reflects the variation of unsteadiness for near-dam reach unsteady flow when Xiangjiaba Hydropower Station runs daily regulation module I. Overall, the unsteadiness does not show obvious decline along the way. Although the amplitude of water level in dam's cross section is maximum in near dam reach, but the discharge amplitude is also large, which makes the unsteadiness here is not the strongest in near dam reach. Stronger unsteadiness appears in the reach from Hengjiangkou rapids to Suimi rapids. Statistics indicates unsteadiness parameter P fluctuates between 0.01 and 0.05 in near dam reach.

As can be seen from Equation (1), unsteadiness is related to change rate of water level, unit width discharge, and variations of these parameters are linked with the section configuration along the way under the premise that the discharge wave remains unchanged. In order to analyze the relationship between unsteadiness and section configuration, morphological relationship ξ can be applied to represent the variation of section configuration corresponding to the minimum discharge:

$$\xi = \frac{\sqrt{B_{min}}}{h_{min}} \tag{2}$$

where B_{min} corresponds to the river width in the lowest water level, and h_{min} is the average water depth of the corresponding section in the lowest water level.

Figure 12 shows the relationship between unsteadiness parameter P and morphological relationship ξ. It can be seen that P becomes stronger as ξ increases, that is to say, unsteadiness caused by discharge wave propagating in narrow deep reach is stronger than that in wide shallow reach.

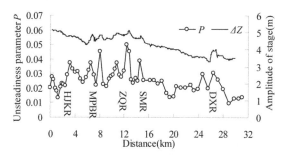

Figure 11. Downstream variation of unsteadiness parameter P.

109

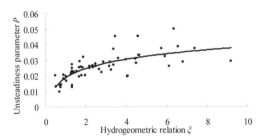

Figure 12. Relationship between unsteadiness parameter P and morphological relationship ξ.

4 CONCLUSION

The operation of Xiangjiaba Hydropower Station will exert influence on the channel flow condition below the dam from Shuifu to Yibin. A two-dimensional flow mathematical model concerning 33 km downstream reach is applied to calculate and analyze near-dam reach (Shuifu-Yibin river stretch) propagation characteristics of daily regulated unsteady flow released from Xiangjiaba Hydropower Station. The main conclusions are listed as following:

1. During the operation of Xiangjiaba Hydropower Station, daily regulation release wave gradually flattens as it propagates downstream, the amplitude of water level shows a general falling trend. Its magnitude is not only related to the discharge hydrograph of the Station, but also to the flow section of the cross section along the reach: the decrease of flow section will result in the increase of water level amplitude. During its propagation, discharge wave will lead to additional water surface slope, positive in rising periods and negative in falling periods.
2. Release wave is affected by factors like friction and channel storage, the maximum discharge below the dam decreases while the minimum discharge increases, so the discharge amplitude decreases. Discharge amplitude is irrelevant to flow section.
3. When release wave from Xiangjiaba Hydropower Station spreads downward, water level and discharge in Shuifu-Yibin river reach do not synchronize. Stage-discharge (Z-Q) relationship show anticlockwise looped curve. The unsteadiness of Niupi Rapids (NPR) is much larger, so the loop is wider, and the downstream loop becomes narrower.
4. The unsteadiness of flow is relatively strong from Hengjiangkou rapids to Suimi rapids, unsteadiness parameter P increases with morphological relationship ξ, i.e., unsteadiness of release wave propagating in narrow deep reach is stronger than that in wide shallow reach.

REFERENCES

Cao M.X. & Pang X.S. & Wang X.H. 2011. Experimental study on unsteady flow and sediment characteristic downstream of Xiangjiaba power station. *Hydro-Sciencce and Engineering* (1):28–34.
De Sutter R. & Verhoeven R. & Krein A. 2001. Simulation of sediment transport during flood events: laboratory work and field experiments. *Hydrological Sciences Journal* 46(4):599–610.
Huang Y. & Li Y.T. & Han F. 2004. Influences of the Three Gorges Hydropower Station's daily regulation on water surface slope of downstream reach. *Hydro-Sciencce and Engineering* (3):62–66.
Kuhnle R.A. 1992. Bed load transport during rising and falling stages on two small streams. *Earth Surface Processes and Landforms* 17(2):191–197.
Liu Z.H. & Ma A.X. & Cao M.X. 2012. Shuifu-Yibin channel regulation affected by unsteady flow released from Xiangjiaba Hydropower Station. *Procedia Engineering* 28:18–26.
Ma A.X. & Lu Y. & Lu Y.Y. 2012. Advances in velocity distribution and bed-load transport in unsteady open-channel flow. *Advances in Water Science* 23(1):134–14.
Ma A.X. 2012. Mechanism of unsteady water-sediment process from reservoirs on sand-gravel movement. Nanjing: Nanjing Hydraulic Research Institute.
Ma A.X. & Cao M.X. 2010. Study 2-D flow mathematical model on Shuifu-Yibin waterway regulation in Yangtze mainstream. Nanjing: Nanjing Hydraulic Research Institute.
Ma A.X. & Cao M.X. 2009. Study 1-D mathematical model on Shuifu-Yibin waterway regulation in Yangtze mainstream. Nanjing: Nanjing Hydraulic Research Institute.

Hydraulic Engineering – Xie (Ed.)
© *2013 Taylor & Francis Group, London, ISBN 978-1-138-00043-8*

Study on flood storage and water logging control in Huainan mining subsidence area

Jinming Li, Zuhao Zhou, Ziqi Yan, Huaxiang He & Qingyan Sun
China Institute of Water Resources and Hydropower Research, State Key Laboratory of Simulation and Regulation of Water Cycle in River Basin, Beijing, China

ABSTRACT: Huainan mine located in Huai River basin, coal resources are abundant, coal mining brings the surface subsidence problem. At present China for the subsidence area management are limited. And the use of subsidence area storage area and exert the flood control, water logging prevention case are rarely to happen in China. In this paper the 3 main subsidence area in Huainan analyzed preliminarily. The subsidence area of the dynamic change law, and at the same time, predicts the future subsidence area storage capacity. Greatly alleviate the local and even the Huai River basin detention water logging prevention of pressure. It will make full use of rainfall flood resources to improve reliability of water supply. It's great significance to the region's flood control and water logging elimination work.

Huaibei plain is one of rich coal and flood disaster area. Chinese scholars in subsidence area management put forward a variety of measures and ideas. As for subsidence area with buildings, the water body of subsidence area, a linear building subsidence area and so on the different kind of subsidence area put forward different management methods. Some innovation methods emergence like Huainan "lake model", "Xin Sen logistics park model", help the mining area residents live and work in peace and contentment (Xu et al., 2012). At the same time in subsidence area management technology is also launched a wide range of discussion (Jiang et al., 2010; Li et al., 2011; Zong et al., 2010; Xu et al., 2010). In recent years, subsidence area impoundment and comprehensive utilization of water resources research have rise. Chinese scholars analyzed subsidence area impoundment feasibility and diversion way. The obtain of subsidence area and storage capacity index have been confirmed by these scholars (Zhang et al., 2010; Tan et al., 2009). Taking Huainan mining area as an example, it have the dynamic change characteristics of quantity and tend to be integrated into a big subsidence area. Subsidence area to implement storage scheme for feasibility are analyses. Estimated future subsidence area water storage capacity, this paper discusses the region detention water logging prevention and the positive role of water resources allocation.

1   GENERAL SITUATION OF THE STUDY AREA

The study area is located in the middle reaches of Huai River northern plains, Figure 1.

Huaibei plain in different four seasons, sufficient sunlight, rainfall is appropriate, hot summer and rainy, drought in autumn. Years of average rainfall is 886.5 mm, from June to September, the flood season precipitation throughout the year accounted for 55% of the precipitation. Average annual temperature is between 14.3 °C to 16.4 °C, 80% of the year concentrated in 14.8 °C to 15.8 °C. The study area has no large water conservancy engineering, wet period flood no way out in the plain, lead the area flood disaster seriously and water resources development and utilization efficiency is not high. Because of the enormous amount of coal mining there are three subsidence area located in these 3 rivers. At present, this subsidence area is gradually merged into one big subsidence area. It can storage rainfall

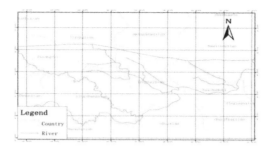

Figure 1.   Research scope.

Table 1.   2009 subsidence area and related rivers.

Number	Mine name	Rivers	Subsidence area name	2009 subsidence area (km²)
1	Guqiao	XiFei Yong Xing	Yong Xing	8.5
2	Gubei	XiFei Yong Xing	Yong Xing	2.7
3	Dingji	Yong Xing	Yong Xing	9.9
4	Zhangji	XiFei and Ji	XiFei	20.5
5	Xieqiao	XiFei and Ji	XiFei	14.6
6	Panyi	Ni	XiFei	19.2
7	Paner	Ni	Ni	9.4
8	Pansan	Ni	Ni	21.4
9	Pansi	Ni	Ni	3.01

Figure 2.   2010 subsidence area.

flood resources, river water and the surrounding drainage. The 3 main subsidence area are, Yong Xing subsidence area, Ni River subsidence area, XiFei River subsidence area. Related mine and the rivers and 2009 subsidence area see Table 1.

By 2010, the area of subsidence area of 123.38 km², subsidence area distribution see Figure 2.

2   COAL MINING SUBSIDENCE AREA AQUIFEROUS FEASIBILITY ANALYSIS

2.1   *Subsidence area impoundment water analysis*

The quantity of water source in subsidence area are huge, drainage are well developed, XiFei River, Ji River, Ni River. Also surrounding groundwater recharge, their water production, atmospheric precipitation, etc. Area rainfall is rich, the study area 1956~2000 annual average precipitation 874.47 mm, flood season average precipitation 516.18 mm. Precipitation station 1956~2000 annual maximum, minimum, and average precipitation in the area see Table 2.

Table 2. Part of the study area precipitation station 1956~2000 material.

Station name	City	Maximum annual precipitation (mm)	Year	Minimum annual precipitation (mm)	Year	Average annual precipitation 1956~2000 year (mm)
Yue Zhangji	Huai Nan	1502.0	1991	478.7	1978	947.8
Feng Tai	Huai Nan	1573.1	1991	446.9	1966	860.3
Huai Nan	Huai Nan	1522.6	1956	450.3	1978	921.9
Gu Dian	Huai Nan	1392.9	1991	443.3	1986	842.7
Ding Ji	Huai Nan	1418.2	1991	360.8	1978	857.2
Pan Ji	Huai Nan	1426.6	1991	328.1	1968	816.9

In 2007, the whole Huai River basin occurred mass of great waters. Anhui province take a heavy toll again, there are 4 obvious rainfall process. Yue Zhangji station annual precipitation reached 1242.9 mm, a day of maximum rainfall occurs in July 8, rainfall 220.2 mm, 3 days maximum rainfall reached 454.2 mm. And for 1956~2000 years of Huainan average precipitation 886.5 mm, flood season 6~9 month precipitation throughout the year accounted for 50%. It is notable that because of Huai bei plain low-lying, Huai River upstream water level head big, middle and lower reaches head is small, "close the door flooded" phenomenon seriously, is easy to appear the flood control pressure. Huai River main stream flood will also become one of the important sources of subsidence area, to alleviate the shortage of the Huai River middle reaches water storage space, slow down in the middle and lower reaches of the Huai River flood control situation grim. So in wet years, subsidence area water more kinds of water, the water is rich, make subsidence area have higher storage guarantee.

2.2 Subsidence area impoundment impact on the local environmental analysis

The situation of present study area river water quality: XiFei zha at the annual I~II class water 41.67%, IV~V class water 58.33%, the rest of the river system on the whole in III~V class between standard. GaoTang lake water quality for the whole year, III~IV class water is given priority to, III class water 16.67%, IV class water 75.00%, V class water accounted for 8.33%. According to the surface water environment quality standard (GB3838-2002), every river has been polluted to varying degrees, the main pollution index is fecal coli form fall, BOD and COD. Subsidence area water environment pollution sources are mainly from the surrounding agricultural non-point, industrial point source influence. With the increasing of water area and depth, subsidence area near the original farmland will gradually succession for waters, form terrestrial ecosystem and aquatic ecosystem coexist pattern.

2.3 Subsidence area water leakage loss analysis

Subsidence area belong to northern Anhui province formation area in Huai River stratigraphic division, from top to bottom is the distribution of formation: The fourth system (Q), Tertiary (N + E), Trias (T), Permian (P), Carboniferous (C), Ordovician (O), Cambrian system ($\in$), Snian system (Z). The Permian and carboniferous area for coal seam, ordovician below is the base of coal seam. So subsidence area need seepage loss minimum of water-resisting layer. The fourth system mainly consists of silty clay, heavy silty loam, sandy loam, in fines and layer of composition, thickness of about 100~400 m. Tertiary constitute of clay rock, powder sandstone and argillaceous sandstone, thick about 50~200 m; Triassic system in the research area appears very little. The fourth is thicker, tertiary mainly by the rock composition, leakage loss is very small, for regional water storage is extremely advantageous.

3 COAL MINING SUBSIDENCE AREA WATER STORAGE CAPACITY ANALYSIS

3.1 *Coal mining subsidence area and river hydraulic contact*

The study area of coal seam group mining, underground coal resources of the large area mining lead to ground subsidence area, with the increase of mining depth, the ground subsidence depth and scope of influence further increase. The research of coal mining subsidence area can through the XiFei River, and Yong Xing River, Ni River lower segment and Huai River main stream connected. These rivers have sluice control with the connection of Huai River main stream. By 2030 or so, due to coal mining will affect the river formed large area subsidence area, and the main drainage connected, could develop into a large storage area group.

3.2 *Water storage volume estimate*

The size of the water resources was decided by subsidence area water storage capacity, therefore calculated in the subsidence area water storage capacity is to determine subsidence area can be poundage primary task.

3.2.1 *Calculation method*

To study area in 2009 for 30 meters spatial resolution of Digital Elevation Model (DEM) as the basic material, Use of China university of mining subsidence by probability integral method is expected to generate mining subsidence value and forecast data. Through the arcgis software platform, the use of Spatial Analyst Tools etc. The tool can be calculated in 2010, 2030, and 2030 of the Discrete Element Method (DEM) numerical, see Figure 3. Calculate subsidence area, the application of GIS Calculate Geometry, and other functions, calculation subsidence isoline each value of the area, and with contour multiplication, can wait until water storage capacity.

3.2.2 *Coal mining subsidence area subsidence area is expected*

Set the subsidence zone height in 22 m the following areas as a storage area. By 2020, the regional subsidence area of 194.25 km², water area of 157.16 km²; By 2030, the regional subsidence area of 276.14 km², water area of 205.45 km². In 2020, in 2030 subsidence range as shown in Figure 4.

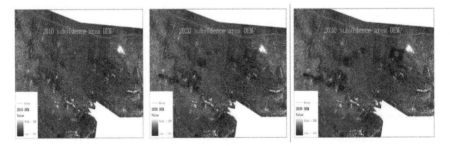

Figure 3.   Subsidence area DEM value.

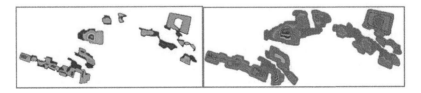

Figure 4.   2020, 2030 subsidence range.

Table 3. Subsidence area water storage capacity.

Subsidence area name	2010 volume (m³)	2020 volume (m³)	2030 volume (m³)	Location
XiFei River	0.71	2.5	3.4	XiFei River
Yong Xing River	0.36	1	2.4	Yong Xing River
Ni River	0.74	2.4	3.5	Ni River
Total	1.81	5.9	9.3	

3.2.3 *Mining subsidence area water storage capacity is expected*

The next 20 years of mining subsidence area water storage capacity in Table 3.

Table 3 shows, in 2010 is expected to the city of subsidence area water storage capacity for 181 million m³. Along with the subsidence area subsidence depth deepening, subsidence area further expanded, the subsidence area water storage capacity will be increased. Huainan in 1956~2000, the average total water resources for 824 million m³, the average surface water resources amount for 576.3 million m³, surface water rate for 513.3 million m³. If the subsidence area all used to hysteresis storage local water logging water, can greatly improve the except water logging drainage ability. And hysteresis storage of rainfall flood resources can also be used to participate in the local water resources configuration, so as to improve the reliability of water supply. For 2030 years above three subsidence seeper area will gradually become a whole, subsidence area average concave in surface 10 m, subsidence area of 276.14 km², and at the same time, integrated peripherals of drainage depression, will form the more large areas of water area, subsidence area water volume is expected to 930 million m³.

4 CONCLUSION

In this paper, the Huaibei plain mining subsidence area aquiferous feasibility and importance are discussed, the subsidence area in Huaibei plain and Huai River basin can play detention water logging prevention role of a preliminary analysis. After analysis think: give full play to the subsidence area detention effect, can effectively alleviate the local and Huai River main stream flood control and waterlogged elimination pressure; Make full use of subsidence area water within the scope of the rainfall flood resources, can effectively improve the surrounding natural ecological environment, and gradually improve people, water, and the contradiction between, the realization of social economy and the ecological environment of common coordinated development.

REFERENCES

Jiang, F.H. 2010. Discusses on carrying out comprehensive governance of the Two Huais' mining coal depressed area combining harnessing the Huai River. *China Water Resources* 22:61–63.

Li, F.M. 2011. Current Status and Development Tendency of Coal Mining Subsidence Area Treatment Technology in China. *Coal Mining Technology* 16(3):8–10.

Su, J.M. 2008. Coal Mine Depresses Area Preventing and Controlling and Control Technology Discussion. *Coal Technology* 27(7):154–155.

Tan, B.B. & Shen S.L. & Wu Z.H. 2009. Application of Neural Networks in Project Optimization of Water System Management in Mining Subsidence. *Journal of Anhui University of Science and Technology* 29(2):13–16.

Xu, L. 2012. Huainan Coal Mine on mining subsidence control new ideas. *Energy Environment Protection* 26(1):42–44.

Xu, S.J. & Liu, J. & Zhang, S.J. 2010. Study of comprehensive development and utilization of water resources in coal mining subsidence area. *Water Resources and Hydropower of Northeast China* 8:29–31.

Zhang, S.J. & Xu, S.G. & Gao, Y. 2010. Study on Unconventional Water Resources Utilization of Coal Mining Subsidence Areas in Huaibei City. *Water Resources and Power* 28(7):27–30.

Zong, Y.F. 2010. Huainan Coal Mine Subsidence Area Construction (Structures) Treatment Project of Building Materials to Protect. *Energy Environment Protection* 24(1):36–3.

Hydraulic Engineering – Xie (Ed.)
© 2013 Taylor & Francis Group, London, ISBN 978-1-138-00043-8

Research of CCS Hydropower Station Bieri sand-flushing facility test

Chen Junjie, Ren Yanfen, Li Yuanfa & Wu Guoying
Yellow River Institute of Hydraulic Research, Zhengzhou, China

ABSTRACT: This paper has made experimental study on Bieri sand-flushing facility of CCS hydropower station and got the following conclusion that original scheme has better desilting effect but sedimentation rate of coarse-size sediment in silting tank cannot meet design requirements. After 5 scheme modification of silting tank, we bring forward recommended scheme that simplifying inlet channel shape, lengthening the length of silting tank by 30 m, and adding 3 honeycombs so as to make sedimentation rate of coarse-size sediment basically meet design requirements, and suggest that desilting work shall be started when front sediment deposition is high to 1,270 m in silting tank.

1 INTRODUCTION

CCS hydropower station is located in the intersection of Napo Province and Sucumbios Province of northern Ecuador in South America. Water collection downstream of first hub adopts Bieri sand-flushing facility—continuous flushing silting tank with 222 m³/s designed diversion flow and 2 kg/m³ silt concentration. There are 6 silting tanks with gentle slope at the bottom, 13 m tank room in width, 6 m working depth and 120 m effective working length. 4 sets of 30 m desilting sections in length are set at the bottom of silting tank with 58 $0.19 \text{ m} \times 0.2 \text{ m}$ holes in each set. There is one desilting corridor in the lower part of each silting tank, and two desilting corridors converge into three larger desilting corridors in the upper reaches of silting tank to the lower reaches of the hub. Static tank is connected with the rear part of silting tank, and water can enter into water conveyance tunnel through energy consumption in static tank. Sedimentation rate of d $\geqq$ 0.25 mm sediment is designed to be 100%.

The test is mainly to survey silt concentration, size distribution of inlet and outlet of silting tank, search silting effects of silting tank, deposition shape of sediments in the tank and desilting effects of desilting corridor at the bottom of silting tank, analyze rationality of continuous water-supply while sand-flushing, and optimize design of silting tank.

2 ORIGINAL DESIGNED SILTING AND SAND-FLUSHING TEST

The test result shows that front deposition is high to 1,268.80 m with 3.4 m deposition thickness in the silting tank, 2.1 in the middle part and 1.5 m in the rear part when diversion flow is 222 m³/s and silt concentration is 2 kg/m³ for 10 h. Sand sedimentation rate is 32%–38% primarily in silting tank and that of d > 0.25 mm sediment is 80%–91%. After 10 h, deposition becomes thicker and sand sedimentation rate is reduced to 32%–38% gradually and that of d $\geqq$ 0.25 mm sediment is 80%–91%.

When reservoir level is 1,275.5 m and diversion flow is 222 m³/s, two sets of desilting holes are opened so as to take desilting test. Silt concentration is up to 90 kg/m³ primarily while desilting and desilting stops and there is no sediment deposited in the corridor when deposited sediment in desilting section is discharged in silting tank after about 0.25 h. Silt Concentration Change of Desilting Corridor Outlet While Desilting refers to Figure 2.

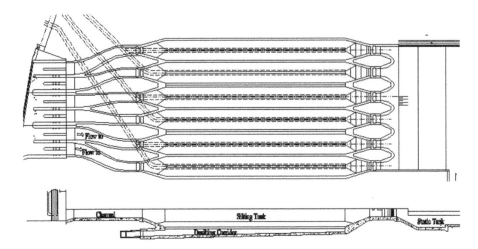

Figure 1.   Layout diagram and section plan of silting tank.

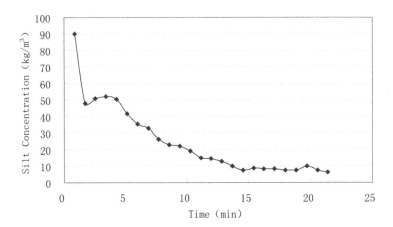

Figure 2.   Silt concentration change of desilting corridor outlet while desilting.

Additionally, water level is slightly reduced in silting tank while desilting with little change of flow pattern, which does not influence normal operation of silting tank.

Desilting method is good enough in original design to meet design requirements. However, sedimentation rate of d ≥ 0.25 mm sediment is 80%–91%, which does not meet design requirements. It is analyzed that the reason for unsatisfied sedimentation effect is that: after water enters into silting tank, flow rate is uneven with fierce mixture and the maximum flow rate is up to 0.76 m/s–2.2 m/s in the front part of silting tank, which is bad for silting and shall be modified for shape of silting tank.

3   MODIFIED TEST OF SILTING TANK

5 scheme modification tests have been taken for silting tank.

1. Add 1 honeycomb in the front part of silting tank of original design scheme: water intake flow is reduced by about 6% after adding honeycomb, but vertical flow rate in cross-section is still uneven;
2. Add 4 honeycombs in the front transition of silting tank of original design scheme: cross-section flow rate of silting tank is still uneven and slightly larger on the surface and bottom;

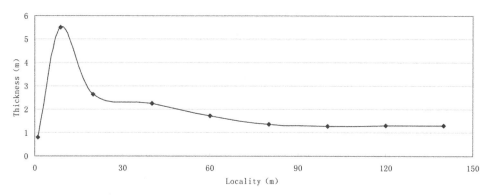

Figure 3.   Longitudinal deposition thickness distribution in the center of silting tank.

3. Adjust structure and layout of 4 honeycombs: flow rate becomes stable in silting tank, cross-section flow rate becomes even and water intake flow is reduced by about 4.8%. Sedimentation rate of d ≧ 0.25 mm sediment is 94%–97%;
4. Lengthen the length of silting tank by 30 m on the base of (3): Sedimentation rate of d ≧ 0.25 mm sediment is up to 99%–100%, water intake flow is reduced by about 4.8%, and sand sedimentation rate is averagely 46%;
5. Based on lengthening the length of silting tank by 30 m, reduce the bottom height of intake channel of silting tank to 1,270, change the curved side-wall into straight side-wall on both sides of the channel with the same width as silting tank, and set 2 honeycombs in connection section: primarily, intake flow rate is 1.3% larger than design value and sedimentation rate of d ≧ 0.25 mm sediment is up to 95.1%–97.4%. However with longer operation time, there are more and more sediments deposited in intake channel with maximum 2.1 m deposition. Intake flow is reduced 4.3% than design value after deposition in channel, which cannot meet intake requirement.

4   OPTIMIZED SCHEME TEST OF SILTING TANK

According to 5 modification test results of silting tank, make optimized design for upper channel shape of silting tank, reduce the width of upper channel to 7.5 m and set 3 honeycombs in connection section.

Test result shows that there are little deposited sediments in the channel in the primary stage of water intake with 0.5 m maximum deposition thickness and unobvious influence of channel deposition on flow capacity, intake flow is 1.2% larger than design value and sedimentation rate of d ≧ 0.25 mm sediment is up to 97%–99.7%. With longer operation time, a bunker occurs in 8.8 m to front part of silting tank. The bunker exceeds 1,270 m than channel bottom height, takes certain water-prevention effects on water flow and makes flow capacity reduced. Longitudinal Deposition Thickness Distribution in the Center of Silting Tank refers to Figure 3. Intake flow is near to design value after cleaning the bunker, so it is suggested that desilting work shall be started when the bunker is high to 1,270 m.

5   CONCLUSION

1. Sedimentation rate of d ≧ 0.25 mm sediment is only 80%–91% in original scheme and sedimentation cannot meet design requirements. Sediments in silting tank can be discharged after opening desilting hole, and there is no sediments deposited in the corridor after desilting, so desilting corridor meets design requirements.

2. Seen from results of 5 modification schemes, adding honeycombs in upper transition, lengthening the length of silting tank and simplifying inlet channel shape can increase sedimentation rate of d ≧ 0.25 mm sediment on the condition that intake flow shall not be reduced after shape simplifying.
3. Sedimentation rate of d ≧ 0.25 mm sediment can be up to 97%–99.7% in final optimized scheme. It is suggested that desilting work shall be started when front sediment deposition is high to 1,270 m in silting tank.

REFERENCES

Dou Guoren. 1978. Changjiang River Scientific Research Institute, Wuhan University of Hydraulic and Hydropower. Similarity law of total sediment transport modeling and design example collection of sediment model reports 1.

Qu Menghao et al. 1980. Yellow River Institute of Hydraulic Research. Test Report on Sediment Prevention Model in Hydropower Station of Xiaolangdi Reservoir of Yellow River.

Wu Caiping et al. 2005. Yellow River Institute of Hydraulic Research. Test Report on Warping Model for Lianbotan of Lower North Main-Stem of Yellow River.

Zhang Junhua et al. 2002. Yellow River Institute of Hydraulic Research. The Report on Model Verification Test and Evaluation of Xiao Langdi Reservoir.

Zhang Junhua et al. 1997. Yellow River Institute of Hydraulic Research. The Report on Model Validation Test of San Menxia Reservoir Area.

Hydraulic Engineering – Xie (Ed.)
© 2013 Taylor & Francis Group, London, ISBN 978-1-138-00043-8

Model design comparative experiment of reservoir sediment transport piping suction head

Wu Guoying, Ren Yanfen, Chen Junjie & Song Lixuan
Yellow River Institute of Hydraulic Research, Zhengzhou, China

ABSTRACT: Suction concentration control device is designed on the self-suction sediment transport piping suction head of sludge cleaning piping system to solve the question of worsening reservoir sedimentation. Yellow River Institute of Hydraulic Research has experimented sludge-cleaning sediment transport piping system in reservoir area and compared four types of suction heads in details by experiments, and the experimental result shows that vertical horn mouth suction head is suitable for unconsolidated sediment treatment in reservoir sediment transport piping system.

1 INTRODUCTION

Four different types of suction heads are designed totally and their advantages and disadvantages have been compared via experiments.

1.1 *Position control type suction head*

Sketch map of position control type suction head refers to Figure 1. Suction head moves horizontally on the sediment surface with pulling rope and takes suction operation in the different positions.

1.2 *Vertical expanding horn mouth suction head*

Vertical expanding horn mouth suction head, which is designed to reduce suction head hoist influence on sediment concentration sensitiveness, expands 6cm vertically in the outlet and connects with pipe at the tail gradually.

Vertical expanding horn mouth suction head has enlarged effective vertical suction scope and reduced suction head sensitiveness of vertical change, as well as adjusted suction head

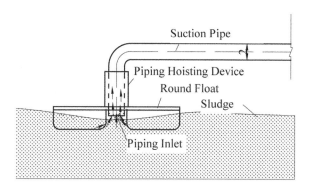

Figure 1.   Import clear water and muddy water control system.

depth immerged in deposited sediment only with pulling rope and then accordingly controlled sediment concentration stably. Sketch map of vertical expanding horn mouth suction head refer to Figure 2.

1.3 *Protective cover type suction head and counter weight type suction head*

The other two suction heads are protective cover type suction head and counter weight type suction head. Sketch map of protective cover type suction head refers to Figure 3, while sketch map of counter weight type suction head refers to Figure 4.

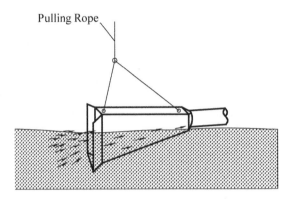

Figure 2.   Sketch map of vertical expanding horn mouth suction head.

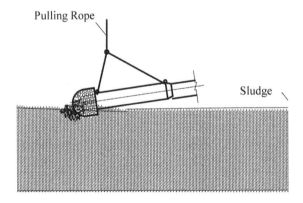

Figure 3.   Sketch map of protective cover type suction head.

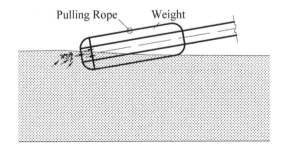

Figure 4.   Sketch map of counter weight type suction head.

2 SUCTION HEAD MODEL EXPERIMENTAL RESULT

2.1 *Model experimental result of position control type suction head*

Sucked sediment concentration control model experiment of position control type suction head has taken piping concentration observation continuously (piping inlet adopts manual control in model experiment). When experimental predicted control target concentration is 350 kg/m^3, sucked sediment concentration change of the suction head refers to Figure 5.

Sediment designed concentration is 350 kg/m^3 in design scheme, according to designed concentration is 70% maximum controlled concentration while minimum controlled concentration is 50% maximum concentration, experimental controlled concentration shall be 250~500 kg/m^3.

As to this suction head, average measured concentration is 345 kg/m^3, maximum concentration is 530 kg/m^3, and minimum concentration is 254 kg/m^3 in experimental period. Seen from sediment sucking up experiment with suction head, average measured sediment concentration is close to predicted value, while controlled concentration scope is a little larger than predicted concentration scope.

According to observation of model experimental operation, it is analyzed that piping inlet is downwards sludge and sediment concentration in the pipe is very sensitive to the distance from pipe to sludge, therefore, sediment concentration in the pipe shall be controlled in a smaller scope, and pipe hoisting operation shall be more precise and mechanical facility, automatic or semi-automatic hoisting device shall be used to improve piping hoisting control and reduce concentration controlled scope.

Float to control position of this suction head shall be suspended on the surface of sludge, so it is suitable in application of high-concentration sludge and sediment.

2.2 *Model experimental result of vertical expanding horn mouth suction head*

Experimental target concentration of horn mouth suction head model is controlled at 350 kg/m^3 and the controlled concentration shall be 250~500 kg/m^3. Average concentration is 344 kg/m^3 in the pipe and the changeable concentration scope is 255~387 kg/m^3.

Seen from experimental result, only 2% deviation between average concentration and controlled target concentration meets design requirements. Measured average concentration is 89% maximum controlled concentration which is 19% higher than design concentration. Sucked sediment concentration change of the suction head refers to Figure 6.

In the process of sucking operation of suction head, sucked sediment concentration change scope, which is only 40%~54% required value, is quietly different from design requirement. Average sediment concentration is 86%~89% while sucking operation of suction head, which is 16%~19% higher than adopted concentration of sediment transport volume in design scheme.

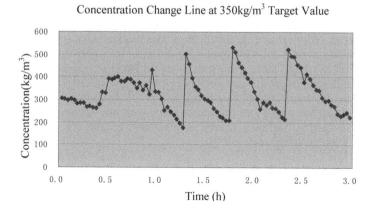

Figure 5. Sucked sediment concentration change of position control type suction head.

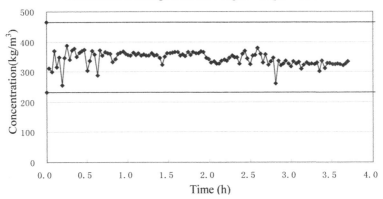

Figure 6. Sucked sediment concentration change of vertical expanding horn mouth suction head (target concentration is 350 kg/m³).

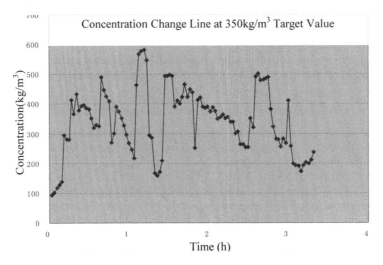

Figure 7. Sucked sediment concentration change of counter weight type suction head.

Seen from experimental data, sucked sediment concentration change of this suction head can be controlled within designed sucked sediment concentration scope. It can control the suction head to suck up the sediment, meet piping system demands for sucked sediment concentration of the suction head, effectively ensure the effects of cleaning sludge of sediment transport piping system as design, even is good for optimizing sediment transporting concentration design parameters, as well as realizes more efficient sediment transporting operation of piping system.

Vertical expanding horn mouth suction head possesses simple structure and much more stable piping inlet, therefore, except that it is good for stably controlling sediment sucking, the suction head can be hoisted and transported only with one suspending pulling rope, for the suction head sucking operation is very simple and is suitable for self-suction sediment transport piping system in the reservoir.

2.3 Model experimental result of counter weight type and protective cover type suction head

Experimental sediment concentration line of counter weight type and protective cover type suction head refers to Figures 7 and 8 when target value is 350 kg/m³.

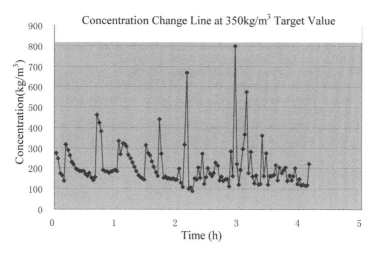

Figure 8. Sucked sediment concentration change of protective cover type suction head.

The above two types of suction heads possess simple structure and simple sediment sucking operation, but it is found in the experiments that the above two suction heads get very lower concentration controlled precision. It is analyzed that inlet of the two suction heads is not expanding type, so elevation change can cause great change of pipe concentration for sucked sediment concentration is sensitive to the inlet height of suction head, while single rope gets relatively lower height precision of suction head, concentration controlled scope is hard to control. So the above two types of suction heads cannot be considered as recommended suction head of piping system.

3 SUCTION HEAD LECTOTYPE CONCLUSION

The following can be seen from the sucked sediment concentration control experiment of suction head:

1. Position control type suction head must be installed with piping hoisting machinery on the float, such as piping automatic hoisting device, to control precisely elevation of the suction head and accordingly to meet demands of sediment transport piping system for suction head concentration controlled scope. But the suction head structure is very complex, piping inlet elevation shall be operated by underwater piping automatic hoisting device on the floating box of suction head, sucking operation precision shall be higher. Furthermore, position control type suction head shall rely on floating box to suspend on the surface of underwater deposit (instead of relying on rope to adjust elevation, only change sucking location with rope), so it is hard to design stabilizing suspension.
2. Vertical horn mouth suction head is much simpler compared to position control suction head in terms of structure. Its sucking operation can work with rope. Target concentration can be controlled precisely and sucking sediment concentration scope is lower than design requirements, so it can meet the demands for suction head of different concentration operation in piping system.
3. Seen from sucked sediment concentration change line, counter weight type and protective cover type suction head gets sucked concentration that is hard to control.

To sum up, vertical horn mouth suction head is advantageous and quite suitable in application of sediment treatment and sediment transport piping system of the reservoir area.

REFERENCES

Ganshu Research Institute for Water Conservancy. 1987. Experimental Research Results Collection of Sediment Discharge and Desilting in Gansu Province Reservoir.
Han Qiwei. 2003. Science Press. Reservoir sedimentation. Beijing.
Jiao Enze. 2004. Yellow River Conservancy Press. Reservoir sediment of the Yellow River.
Qu Menghao. 2005. Yellow River Conservancy Press. Model test theory and method on movable bed model test in the Yellow River.
Wang Chenjiang, Wang Yanchun, Bai Tieliang. 2005. Heilongjiang Science and Technology of Water Conservancy, *Discussion on the Sedimentation of North-Diversion Project Reservoir* 33(2):87–89.

Hydraulic Engineering – Xie (Ed.)
© 2013 Taylor & Francis Group, London, ISBN 978-1-138-00043-8

Study on the key technique of the control system adjustment for the Yellow River estuary model

Zhu Chao, Zhao Lianjun & Wang Jiayi
Yellow River Institute of Hydraulic Research, Zhengzhou, China

Qin Xin
North China University of Water Resources and Electric Power, Zhengzhou, China

ABSTRACT: Tide model control system is an important part of the construction of the yellow river estuary model, the way of tidal generating between the yellow river estuary and other large rivers estuary obviously different, the tidal generating equipment and its control technology intended to be used should be subject to a small-scale generalized model test for further verification of the entire system stability condition, in order to adapt to the unique requirements of the yellow river estuary large model to tidal range, tidal shaped and coastal streams.

1 INTRODUCTION

Tide model control system is an important part of the construction of the yellow river estuary model, which is key to whether tail water level and the corresponding amount of flooding and ebbing tide's flow can be controled by the tide level change process. The yellow river estuary tidal type is complex, and most of the sea area belongs to irregular semidiurnal tidal. The tide range is less than 2 m, therefore it is categories as weak tide. The ocean dynamics is relatively weak, the runoff tidal strength changes frequently, and the tide reciprocating flow feature is obvious. Therefore, the yellow river estuary model test equipment requirement is high. According to these characteristics, how to improve the measurement and control precision of the equipment arises as a problem which needs urgent solution.

At present the Yellow River Institute of Hydraulic Research, YRCC (YRIHR) is carrying out the generalization yellow river estuary physical model construction, focusing on tidal current control and hoping to reproduce the flow field of the yellow river estuary and sand bar. Considering the facts that the trend of the yellow river coastal current is strong, the tidal range is small, and the tide generating mode differs from the other rivers, the present tidal generating equipment and its control technique should be reformed and upgraded, in order to adapt to the unique requirements of the tidal range, tidal shape and coastal current for the yellow river estuary model. The tidal generating equipment and its control technology intended to be used should be subject to a small-scale generalized model test for further verification of the entire system stability condition.

2 GENERALIZED PHYSICAL MODEL OF TIDAL GENERATING CONTROL SYSTEM

2.1 *The design of the model*

According to the needs of the tide simulation model, the building experience of the current domestic and international river movable bed model and the estuarine tidal model, and considering the function of the yellow river estuary model should include the evolution of the channel part and flood control, the Yellow River Estuary generalized physical model adopts horizontal scales $\lambda_L = 600$ and vertical scale $\lambda_H = 60$.

In the river area, the inverter one-way control flow is used. In the sea area, the inverter is used to control pumps directly and adjust the speed of submersible pumps so as to control the flow into the model sea to complete the water pump running status in different flow control, the pump output attenuation rate, flood tide and ebb tide pump discharge flow, in order to control the north and south flow rate and velocity, simulate the reciprocating movement of the tide and achieve automatic similar simulation in tidal shape, tidal level and flow velocity.

The main task of the model is to achieve the water level automatically follow given water level process with the mehod of open loop control or closed loop control. The data exchange between the inverter, water level meter, current meter measurement and control equipment and the central control computer is achieved by RS-485 communication interface implementation, the control devices associate through the levels of the independent controller and the method of communication, after calculating by the measured digital PID control algorithm, the error digital PID control algorithm between the water level given values and the water level measured values, the output controlling value controls the inverter frequency, thereby controlling the pump motor speed, real-time adjustment of the controlled variable so that the measured value of the water level is in conformity with the water level given value, thus completing the process of measurement, calculation, comparison and implementation.

The test is based on the existing water level data, automatically tracking water level with the automatic control system, simulating the tide generating and ebb process, and generating the flow velocity in the corresponding time. The upstream of the model is located in the tide river reach. A clean water pump and mud pump is arranged in the upstream of the model for the simulation of the Yellow River into the sea. The downstream of the model is sea area with strong two-way flow, therefore, a steady flow tank is arranged in the north and south sides of the downstream whose outflow is controlled by submersible pumps. The flow field of the Yellow River Estuary is simulated together with the upstream control system.

The Yellow River estuary entity model in the sea area set up two level control station. Input the water level hydrograph of these two stations $h_0 \sim t$ into microcomputer, compare water value h of the timing (usually 1s) acquisition control station in tide generating process and given value, get a difference $\Delta h = h_0 - h$, then calculate by the control algorithm, and the difference is sent to the site machine, obtain the control of the controllers output, by adjusting the inverter to change the pump speed to control flow into the model in order to adjust sea area water level, so that the control station water level is closer to the given water level until the difference between the control station water level and the given level of the level is zero, and thus the tidal phenomena similar to the prototype can be produced in the model.

2.2 *The control system of the model entry*

In order to control sediment concentration of the model entry, two sets of independent circulation systems of clear and muddy water are used, which respectively controls the clear and muddy water flow into model. In the muddy water stirred pool prepare muddy water of certain concentration, input flow numeral of muddy water and the corresponding clear water during various periods into industrial control computer, by controlling the inverter and electromagnetic flow meter, regulating the water pumps and muddy water pumps outflow, clear water flows into the pool before the mixture of clear and muddy water through clear water pipe, muddy water flows into the pool before the mixture of clear through muddy water pipe, clear and muddy water fully mixed evenly the pool before the mixture of clear and muddy and enters into the model river. Referring to Fig. 1.

2.3 *Model arrangement and existing problems*

The first Yellow River Estuary generalized physical model design uses an inverter, submersible pumps as tide generating equipment. The submersible pump is installed in the reservoir

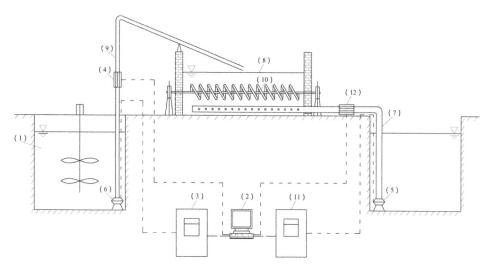

Figure 1. Entry clear water and muddy water control system. (1) muddy water stirring pool (2) industrial control computer (3) muddy water inverter (4) muddy water electromagnetic flow meter (5) clear water pump (6) muddy water pump (7) clear water pipe (8) the pool before the mixture of clear and muddy water (9) muddy water pipe (10) clear water and muddy water mixer (11) clear water inverter (12) clear water electromagnetic flow meter.

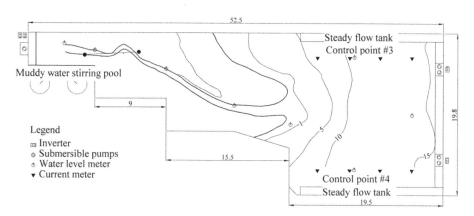

Figure 2. Plane figure of the model.

rear side, connected to the steady flow tank by the water supply pipe. In the model test, water flows into the sea from the steady flow tant through the partition wall. Figure 2 shows the plane layout of the model.

The first scheme model layout causes charged object lag seriously due to the reasons that the control of north-south tide is through a long pipe water to the north and south sides of the overflow sink, the water enters from the pool into the model by long pipeline with many bends, and the surplus flow of the water pump is insufficient. As the the north-south physical distance of the model is short, the terrain is almost no slope changes, the water level of the north-south side influence and interfere each other, the control accuracy of the model is restricted. When the two control points are doing closed loop control according to the given level process curves, one control point control output can not immediately spread to the other control points, which makes it difficult for the model to control water level process. In addition, the two control points are quite close and the terrain is flat, so effective transverse slope cannot be produced, and velocity of flow is only in one direction of flowing from the original control flow to the other control point at a fast flow rate, which obviously cannot meet the need of simulating tide for the the model test.

2.4 The improved scheme layout of the model

To solve the serious lag of the first scheme, scheme two focuses on improvement from two aspects. On the one hand, replace the long and winding water pump by the pump being connected to the pipe, then gradually change the pipe into rectangular outlet, where the waterflow is divided into four outlet, so that the effluent can be even; on the other hand, increase the number of the water pumps to solve the problem of water shortage. The plane layout of the model is shown in Figure 3.

2.5 Test results of the improved model scheme

Generalized model revised scheme closed-loop control has completed water level tracking, but the flow velocity was not well controlled. Therefore, the control is changed to mixed control, namely the open loop control on one side and the closed loop control on the other. Final water level tracking curve is shown in Figure 4. In the process of flood tide and ebb tide, because the frequency of the inverter has been increasing or decreasing, so the water level tracking is quite good. But in the conversion period form rising tide to ebb tide and ebb tide to rising tide, and the inverter frequency changes from increasement to decreasement, the water level in a given curve has a slight deviation. Figure 5 is the flow velocity trace graph. Test curve can basically meet the given curve velocity value.

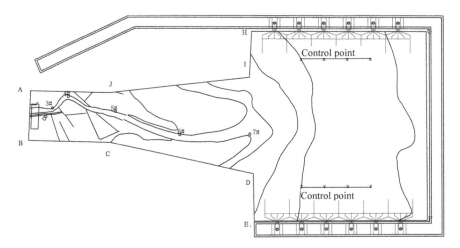

Figure 3.　The plane layout figure of the second scheme model.

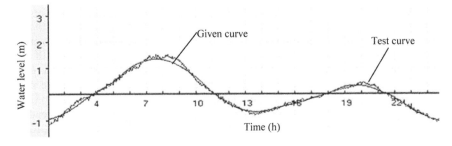

Figure 4.　Hybrid control water level trace graph.

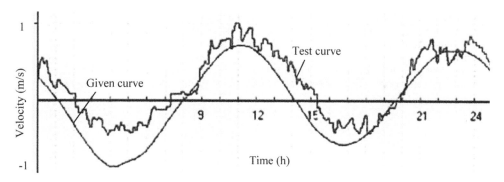

Figure 5. The flow velocity trace graph.

After the adjustment of the revised scheme, the model adopts hybrid control effect which can basically meet the needs of the model test, but there still exists some shortage for future reference for entity model control system: Due to the water pump is one-way flow, when the water level height needs to reduce frequency, water can only be discharged through the pump outlet and drainage holes without restraint; when the water level is low, the inverter frequency increase need a time, during when the pump outlet and drainage holes are still sluicing, making the control accuracy difficult to guarantee. There is a waterhead gap between generalized model of sea area and reservoir water level, making water level tracking lags behind, which fails to track a given curve immediately.

3 CONCLUSION

Because of the particularity of the Yellow River estuary type of tide, and large scale, large variable ratio of the Yellow River estuary physical model, the tidal generating equipment and its control technology intended to be used should be subject to a small-scale generalized model test for further verification of the entire system stability condition.

Model entry control system structure is clear and concise with automatic control, and it has great practical value. Muddy water electromagnetic flow meter is installed in the vertical segment of the muddy water pipes, the muddy water pipes slope to the pool begore the mixure of muddy and clear water from the highest point, solving the muddy water piping precipitation and plugging problems. System automation level is relatively good, the accuracy is high, the proper amount of clear water and muddy water flow required by the model entry flow process is guaranteed and the control is in accordance with the water and sediment required, which can completely meet the needs of the river model test entry control.

The pump chosen to be used in the present test is reasonable, and the variable speed pump to be used for examination and the working condition of the layout in different flow amount can basically meet the test needs. Waste pipe is replaced by waste hole and the number of pumps is increased so that the model can achieve similar and automatic simulation of the tide shape and tidal level. If the bidirectional pump is used to replace the present one-way submersible pumps to make the tidal control more flexible, the water level tracking lag problem can also be solved.

It is proposed that HMMC2000 water conservancy physical model of computer control system (including software) which is intended to be used can satisfy the Yellow River estuary physical model test requirements.

REFERENCES

China Water Power Press. 2006. The regularity study on the interaction between the evolution of the Yellow River downstream channel and the estuarine.

Jiang Niachang. 1998. China Building Industry Press, Pumps and pump.

Li Zhixi, Chen Zhijuan. 2003. The estuary model construction and estuarine research.

Water Transport Engineering. 1999. The tidal model design and validation of Changjiang river estuary.

Yang Jin. 2008. Huadian Technology, Digital PID control of the integral saturation problem.

Yellow River Conservancy Press. 2003. The Yellow R iver Estuary problems and Countermeasures Seminar Expert Forum.

Hydraulic Engineering – Xie (Ed.)
© 2013 Taylor & Francis Group, London, ISBN 978-1-138-00043-8

Long-term water temperature simulation for Miyun Reservoir

Zhongshun Li, Dejun Zhu, Yongcan Chen & Zhaowei Liu
State Key Laboratory of Hydro Science and Engineering, Tsinghua University, Beijing, China

Xing Fang
Department of Civil Engineering, Auburn University, Alabama, USA

ABSTRACT: The thermal module of the one-dimensional (vertical) water quality model CE-QUAL-R1 was modified to include subroutines simulating ice and snow covers and heat exchange with bottom sediments that make it possible to simulate more reasonable water temperature profiles during the ice cover period. The modified model was calibrated and validated using data in Thrush Lake, MN, USA and then applied to long-term water temperature simulation for Miyun Reservoir. The results were analyzed and compared with those obtained with the original CE-QUAL-R1. The results show that the water level decline in Miyun Reservoir starting from 1997 has different impacts on the temperature distribution in the epilimnion and hypolimnion.

1 INTRODUCTION

Miyun Reservoir, located in the northeast of Beijing, is the largest artificial lake in Asia. As the biggest water source of Beijing, Miyun Reservoir can have a great impact on the health for millions of people. The water quality is of therefore great importance. As an important parameter of water quality, temperature has a significant impact on its ecosystem (Chen et al. 1998).

The water temperature profile of a lake or reservoir is determined by the heat source and transfer. Heat source includes the solar radiation, heat exchange through the water surface (long wave radiation, latent heat and sensible heat), heat exchange with the bottom sediments, and heat exchange through inflow and outflow streams. Heat transfer is controlled by the wind-induced mixing, advection, and turbulent diffusion. The heat source and transfer in the ice cover period is quite different from those in the open water season. In the ice cover period, heat exchange with the atmosphere is insulated by the ice covers. Solar radiation reaching the ice surface is reflected first and then penetrates through the snow and ice cover layers. The amount of solar radiation absorbed by the water body after reflectivity and attenuation is therefore much less than those in the open water season. Heat exchange between water and bottom sediments may not be important in the open water season, especially in deep lakes, while it becomes an important source of heat during the ice cover period. Wind-induced mixing in the surface layers becomes zero due to the isolation of ice. The vertical diffusion coefficient in an ice-covered lake should be quantified using a different formula in the absence of wind (Fang et al. 1996).

The original thermal module of CE-QUAL-R1 can be applied to thermal simulation during the open water seasons, but it doesn't contain the heat process about ice cover and bottom sediments described above (Li et al. 2012; EL 1995). It just simply sets the surface water temperature equal to zero when the surface water temperature predicted from the heat balance equation is less than zero. In this paper, modifications were made to the thermal module of CE-QUAL-R1 to develop a year-round vertical thermal model, and then its applications to Miyun Reservoir are presented and discussed.

2 MODEL

2.1 Basic equation

The numerical simulation model for water temperature solves the one-dimensional (vertical along depth) unsteady heat transport equation

$$\frac{\partial T}{\partial t} = \frac{1}{A}\frac{\partial}{\partial z}\left(K_z A \frac{\partial T}{\partial z}\right) + \frac{H_W}{\rho c_p} \tag{1}$$

where $T(z,t)$ = water temperature as function of time t and depth z, t = time, $A(z)$ = lake horizontal area as function of depth z, K_z = vertical turbulent heat diffusion coefficient, H_W = heat source term resulted from solar radiation and other factors, ρ is the density of water, and c_p is the specific heat capacity of water.

2.2 Model modification

The modification made in this study includes ice thickness prediction in the winter period, snow thickness prediction in the winter period, couple of the ice/snow subroutine to open water temperature model, and bottom sediment heat transfer prediction.

Ice grows at the ice-water interface; ice decays at the snow-ice interface, ice-water interface, and within the ice layer. In the ice subroutine, the mass balance equation of ice is solved,

$$\frac{dz_i}{dt} = \dot{z}_{ic} - \dot{z}_{is} - \dot{z}_{ir} \tag{2}$$

where z_i is the ice thickness, $\dot{z}_{ic}$ is the rate change (growth or reduction) of ice thickness due to the conduction/convection, $\dot{z}_{is}$ is the ice thickness reduction rate due to absorbed solar radiation, $\dot{z}_{ir}$ is the ice thickness reduction rate due to rainfall in early spring. The formulae used to compute each term in the equation (2) were adopted from the literature (Gu & Stefan 1990; Fang et al. 1996; Fang & Stefan 1996a).

Snow accumulates due to precipitation (winter snowfall) accompanied by a compaction process; snow melts due to solar radiation, conduction, evaporation, and rainfall (Fang & Stefan 1996b). The mass balance equation to predict snow thickness is as follows,

$$\frac{dz_{sw}}{dt} = C_{sw}P_{sw} - \dot{z}_s - \dot{z}_c - \dot{z}_e - \dot{z}_r \tag{3}$$

where z_{sw} is the snow thickness above the ice cover, P_{sw} is the snowfall from given weather data (model input), C_{sw} is the snow compaction coefficient, $\dot{z}_s$ is snow depth reduction rate of melting due to solar radiation, $\dot{z}_c$ is snow depth reduction rate of melting due to conduction heat transfer, $\dot{z}_e$ is snow depth reduction rate of melting due to evaporation, $\dot{z}_r$ is snow depth reduction rate of melting due to rainfall. The formulae to quantify all terms in the equation (3) can refer to the literature (Hondzo & Stefan 1994; Fang & Stefan 1996b; Fang & Stefan 1998).

In the sediment heat transfer subroutine, the one dimensional, unsteady heat conduction equation

$$\frac{\partial T_s}{\partial t} = K_s \frac{\partial^2 T_s}{\partial z^2} \tag{4}$$

is solved first. $K_s = k_s/\rho c_p$ is the thermal diffusivity of the sediment. The flux of heat exchange between water and bottom sediments is then calculated by integration of bottom sediments temperature profiles (Fang & Stefan 1996b). Sediment temperatures at 10 m below the littoral waters are assumed to be constant and equal to mean annual air temperature at the study site (Fang & Stefan 1998). The first-type boundary condition is applied to the water-sediment

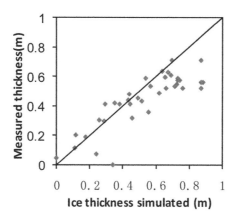

Figure 1. Simulated versus measured ice thickness in Thrush Lake. The solid line would indicate perfect agreement.

interface, i.e., equal to water temperature above. The second-type (zero gradient) boundary condition is applied to the 10 m below of the water-sediment interface.

After the three equations above are solved, the amount of heat exchange between water body and ice cover and bottom sediments can be calculated and substituted into equation (1).

2.3 *Model calibration*

The expanded model described above extends the model to year-round simulation from the open water season in summer to the ice cover period in winter. Calibration parameters for the new subroutines include snow compacting coefficient, thermal conductivity of the sediment and the material properties (albedo, absorption coefficient, extinction coefficient). Lack of the measured data of ice, snow thickness and sediment temperature profiles from Miyun Reservoir, the model was first calibrated and validated using the data of Thrush Lake (Fang & Stefan 1996a), MN, USA and then applied to Miyun Reservoir keeping the parameters unchanged. The ice thickness and water temperature in Thrush Lake were simulated well (Figure 1) using the expanded model.

3 RESULTS

Using the calibrated model, water temperature distributions in summer and winter periods in Miyun Reservoir from 1998 to 2007 were simulated.

Figure 2 shows the daily change of water level in Miyun Reservoir from 1997 to 2009. As can be seen from the graph, ever since 1997, water level of Miyun Reservoir has been decreased year by year. The lowest water level was reached in June 2004, approaching to 130 m above mean sea level. The drawdown had been 25 m, while the water depth in Miyun Reservoir was less than the half of the normal level.

Figures 3–5 show the time-series of simulated water temperatures in Miyun Reservoir at three different water depths. The results from the modified model which includes the ice/snow/sediment subroutines are more reasonable than those from the original model. In the surface layer, the difference is not significant, because the thermal balance is dominated by weather condition. In the middle layer, the original model predicts water temperatures during the ice cover periods having a large decline and being less than 4 °C in winter, but the modified model predicts a gradual decrease in water temperature in each winter period, which is more reasonable. This is because the ice cover increases its thickness when the air temperature continues to drop below 0 °C during the ice cover period, and most of heat loss due to colder air temperatures and low solar radiations does not directly cool down water below the ice-water interface. In the bottom layer, the fluctuations of water temperature are apparently reduced by sediment heat

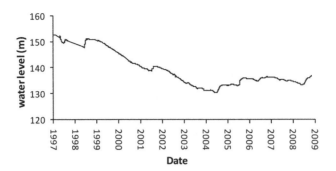

Figure 2.   Water level hydrograph of Miyun Reservoir from 1997 to 2009.

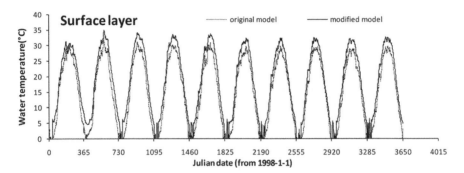

Figure 3.   Time-series plot of simulated water temperatures in Miyun Reservoir at surface layer.

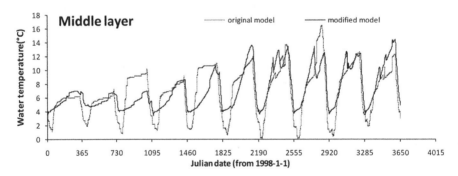

Figure 4.   Time-series plot of simulated water temperatures in Miyun Reservoir in the middle layer.

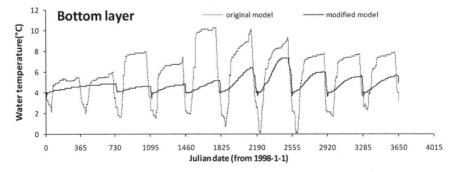

Figure 5.   Time-series plot of simulated water temperatures in Miyun Reservoir in the bottom layer.

storage when the expanded model was used. In summer, heat transfers from water into bottom sediments while in winter heat transfers from bottom sediments into water.

Both of the original model and modified model show that water level decline has different effects on water temperature of the surface layer, middle layer, and bottom layer respectively. For the surface layer (epilimnion), the decline has little influence on the temperature, because surface temperature depends on the surface heat flux and the turbulent mixing by wind, while the annual cycle of the surface temperature is still quite obvious. For the middle layer, the decline of water level leaded to substantial rises of water temperature during the summer open water seasons starting from the year 2001, but winter cooling brought water temperature down to below 4 °C no matter what the water level was in each year. Additionally, the lower water level was the more temperature in the middle layer rose since 2001. For the bottom layer (hypolimnion), water level decline resulted in a certain amount of temperature rise after 2003, but the increase was small.

4 CONCLUSIONS

In this paper we have developed a one-dimensional (vertical) year-round thermal model and applied it to long-term simulation in Miyun Reservoir. The results from the modified model which includes the ice/snow/sediment subroutines are more reasonable than those from the original model. The large temperature decline in the middle layer during the ice cover periods vanished due to the ice cover isolation and the fluctuation of temperature in the bottom layer reduced due to sediment heat storage. Water level decline in Miyun Reservoir has little influence on the temperature dynamics in the surface layer, where the temperature varies periodically with seasons. For the middle layer, the decline of water level affected water temperature significantly, additionally, the lower water level was, the more temperature in the middle layer rose since 2001. For the bottom layer, water level decline resulted in a certain amount of temperature rise also after 2003.

ACKNOWLEDGMENTS

The authors gratefully acknowledge the financial support from the National Natural Science Foundation of China (No. 51039002) and Tsinghua University Initiative Scientific Research Program (No. 20121088082).

REFERENCES

Chen, Y.C., Zhang, B.X. & Li Y.L. 1998. Study on model for vertical distribution of water. *Journal of Hydraulic Engineering* (9):15–21.

Environmental Laboratory, 1995. CE-QUAL-R1: A Numerical One-Dimensional Model of Reservoir Water Quality; Users Manual, Instruction Report E-82-1 (Revised Edition), US Army Engineer Waterways Experiment Station, Vicksburg, MS.

Fang, X., Ellis, C.R. & Stefan, H.G. 1996. Simulation and observation of ice formation (freeze-over) in a lake. *Cold Regions Science and Technology* 24(2):129–145.

Fang, X. & Stefan, H.G. 1996a. Long-term lake water temperature and ice cover simulations/ measurements. *Cold Regions Science and Technology* 24(3):289–304.

Fang, X. & Stefan, H.G. 1996b. Dynamics of heat exchange between sediment and water in a lake. *Water Resources Research* 32(6):1719–1727.

Fang, X. & Stefan, H.G. 1998. Temperature variability in lake sediments. *Water Resources Research* 34(4):717–729.

Gu, R. & Stefan, H.G. 1990. Year-Round Temperature Simulation of Cold Climate Lakes. *Cold Regions Science and Technology* 18(2):147–160.

Hondzo, M. & Stefan, H.G. 1994. Riverbed Heat Conduction Prediction. *Water Resources Research* 30(5):1503–1513.

Li, Z.S., Chen, Y.C., et al. 2012. Water temperature distribution in the Miyun Reservoir. *J Tsinghua Univ (Sci &Tech)*, 52(6):798–803.

Hydraulic Engineering – Xie (Ed.)
© 2013 Taylor & Francis Group, London, ISBN 978-1-138-00043-8

Regulated deficit irrigation effects on Tomato growth and irrigation water use efficiency

Zhang Jinxia
College of Forestry, Gansu Agricultural University, Gansu, China

Cheng Ziyong, Li Xiaoli & Zhang Rui
College of Engineering, Gansu Agricultural University, Gansu, China

Li Aizhuo
Station of Water and Soil Conservation of the 1st Division, Xinjiang Production and Construction Corps, Xinjiang, China

ABSTRACT: This study aims at Regulated Deficit Irrigation (RDI) effects on growth and Irrigation Water Use Efficiency (IWUE). Field experiment with tomato was conducted in Qinwangchuan Irrigation Area, Gansu Province, China. The results show that: irrigation of full irrigation (CK) was the largest, but its growth, yield and IWUE were not the best. Compared with CK, appropriate deficit water could shorten tomato growth period, was conducive to the growth of plants, and increased yield and IWUE. The height, yield and IWUE of the mild deficit in seeding (MH-60) were all the highest; and the mild deficit in later fruiting (GH-60) followed by. Those of severe deficit in flowering (KH-50) were the lowest, the severe deficit in fruit setting (JH-50) followed. It could clearly be seen that MH-60 would be beneficial to the growth of plants, have the effect of saving-water and improving-yield. So it is the best deficit period and degree of RDI in greenhouse tomato. Thus, our results suggest that local farmers should consider adopting mild deficit in seeding because of benefits to root growth, saving-water, yield and IWUE.

1   INTRODUCTION

In the mid-1970s, RDI concept was put forward by Center Tatura of Agricultural Research Institute of Australia Continued Irrigation (e.g. Kang & Cai 2001). Since then, many domestic and foreign scholars had carried out extensive experimental research about it (e.g. Chalmers et al. 1981, Kang et al. 1998, Zhang et al. 1999, Cai et al. 2000, Wery 2005, Intriglio & Castel 2005, Goldhamer 2006, Zhang et al. 2012). The results showed that: RDI could improve water use efficiency and quality of crop (e.g. Shi et al. 2004, Ma et al. 2006, Guo et al. 1999). It was also reported that RDI reduced fruit yield (e.g. Liu et al. 2001, Liu et al. 2005). Reducing the upper limit of soil moisture increased the yield in appropriate growth stages of tomato, especially in flowering and fruiting (e.g. Liu & Chen 2002, Li et al. 2000). The mild deficit increased crop yield, however, severe deficit reduced production in the greenhouse tomato (e.g. Guo 2007, Guo et al. 2007). It is visible that moderately reducing water consumption will increase water use efficiency.

In Qinwangchuan irrigation area, water resources is seriously shortage, soil layer is thin, leakage phenomenon is universal, and soil is serious salinization. Therefore, to improve agricultural ecological environment and increase crop yield, it is an urgent need to develop water-saving irrigation technology in this region. And experimental zone will put the vegetables as an emerging industry. On the basis of the actual situation, this project would study the effects of RDI on growth and IWUE in greenhouse tomato, and find the best deficit period and degree of RDI in greenhouse tomato. These will provide scientific basis for the better implementation of RDI in this area, and have important guiding significance to solve agriculture water-saving problem in arid and semi arid region of Northwest China.

2 MATERIALS AND METHODS

2.1 General situation of test area

The experiment was conducted at Qinwangchuan Irrigation Area, Gansu Province, China. It belongs to typical continental arid climate zone, where less precipitation and large amount of evaporation. The annual average precipitation is only 287 mm. However, the annual evaporation capacity is 1950 mm. The annual sunshine time is up to 2688.7 h. The average frost-free period is around 150 d. The average annual temperature is 6.2°C. Soil texture is in the main of powder sand clay loam and clay loam, and the thickness of soil layer is only 0.4 m in local area, which soil dry bulk density is 1.24 g/cm³. No perennial surface runoff in irrigation area, groundwater is meager and poor water quality, buried depth of 70–90 m. At present, the water resource is mainly dependent on the water content from Irrigation Engineering Datong River into Qinwangchuan, which is mainly used for agricultural irrigation and drinking water of resident in irrigation area (e.g. Zhou 2009, Zhang et al. 2012).

2.2 Experiment design

The test crop was tomato, whose breed was Appollo F1. The growth period of tomato was divided into four stages. There were severe deficit, mild deficit and full irrigation treatment respectively in each stage. Their lower limit of soil relative water content (the percentage of field capacity) was 50%, 60% and 75%, respectively. There were 9 treatments with 3 replicates, a total of 27 plots. The plot area was 5.6 m². When the soil moisture content reduced to the lower limit, irrigation would be done, and times not limited. Farming, fertilization, pest control, et al. was the same as the traditional field planting in each treatment. The experiment was a random block design, provided with a protection zone. The situation about experiment design of seed maize is shown Table 1. Irrigation water quota is 18 m³/667 m².

3 RESULTS AND ANALYSIS

3.1 Effect of RDI on growth process in the tomato greenhouse

Effect of RDI on growth process in the tomato greenhouse could be seen in Table 2. The data showed that: water deficit would shorten the growth period of the tomato, and the more serious water deficit, the longer shortened time. However, full irrigation could make vegetative growth vigorous, and relatively postpone growth process. Compared with CK,

Table 1. Experiment design of RDI in the tomato greenhouse.

Code*	Treatment	Seeding	Flowering	Fruit setting	Later fruiting
CK	Full irrigation	75	75	75	75
MH-60	Mild deficit in seeding	60	75	75	75
MH-50	Severe deficit in seeding	50	75	75	75
KH-60	Mild deficit in flowering	75	60	75	75
KH-50	Severe deficit in flowering	75	50	75	75
JH-60	Mild deficit in fruit setting	75	75	60	75
JH-50	Severe deficit in fruit setting	75	75	50	75
GH-60	Mild deficit in later fruiting	75	75	75	60
GH-50	Severe deficit in later fruiting	75	75	75	50

*Data in the table is the lower limit of soil relative water content (the percentage of field capacity).

Table 2.   Effect of RDI on growth process in the tomato greenhouse.

Treatment	Seeding		Flowering		Fruit setting		Later fruiting	
	Start m-d	Stop m-d	Start m-d	Stop m-d	Start m-d	Stop m-d	Start m-d	Stop m-d
CK	5-17	6-6	6-7	7-6	7-7	8-5	8-6	9-4
MH-60	5-17	6-4	6-5	7-2	7-3	8-1	8-2	8-30
MH-50	5-17	6-3	6-4	7-1	7-2	7-30	7-31	8-28
KH-60	5-17	6-6	6-7	7-4	7-5	8-2	8-3	8-29
KH-50	5-17	6-6	6-7	7-3	7-4	8-1	8-2	8-27
JH-60	5-17	6-6	6-7	7-6	7-7	8-3	8-4	8-31
JH-50	5-17	6-6	6-7	7-6	7-7	8-2	8-3	8-28
GH-60	5-17	6-6	6-7	7-6	7-7	8-5	8-6	9-2
GH-50	5-17	6-6	6-7	7-6	7-7	8-5	8-6	9-1

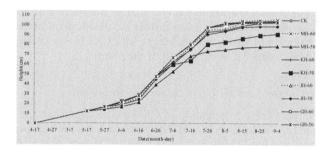

Figure 1.   Effect of RDI on plant height in the tomato greenhouse.

growth process of deficit treatment was shortened. That is respectively 5 days on MH-60, 7 days on MH-50, 6 days on KH-60, 8 days on KH-50, 4 days on JH-60, 7 days on JH-50, 2 days on GH-60 and 3 days on GH-50. Because picking tomato in the greenhouse according to the regularly scheduled batch, delaying growth period in full irrigation would lead to increase the number of no effect fruit. This caused full irrigation treatment to reduce yield and economic benefits.

3.2   Effects of RDI on height in the tomato greenhouse

Figure 1 showed that, plant height had an increasing trend in the whole growth period. In seeding stage, the growth was mainly focus on root, so slow in plant height. There was no significant difference of the plant height in seeding ($P > 0.5$), mainly because of soil initial conditions were substantially the same in the greenhouse tomato. It is flowering and fruit setting that were rapid growth periods of plant height. During these periods, the maximum of growth rate was up to 2.7 cm/d, and the minimum was 1.3 cm/d. In later fruiting, plant height was almost no longer increase, turning most of biological nutrient in tomato to fruit ripening. Plant height of the water deficit was significantly lower than CK, especially severe deficit treatment: MH-50 < KH-50 < JH-50 < GH-50 < KH-60 < GH-60 < JH-60 < CK < MH-60. There was very significantly difference of the plant height between MH-50 and CK ($P < 0.1$), it was because that MH-50 was affected by drought in seeding too long to reach the level of CK until harvest. However, MH-60 was suitable for water deficit in seeding, which made root growth more developed; Together with timely irrigation in flowering, the difference quickly was narrowed in fruit setting; And in later fruiting, its plant height had exceeded the level of CK. Since deficit, the plant height of KH-50 had been disadvantaged until to later fruiting, because of severe deficit in flowering. In later fruiting due to suitable deficit, plant height of

Table 3. Effects of RDI on yield and IWUE in the tomato greenhouse.

Treatment	Irrigation (m³/hm²)	Yield (kg/hm²)	IWUE (kg/m³)
CK	2430[aA]	177,492[aA]	73.04
MH-60	2160[abA]	222,930[aA]	103.21
MH-50	1890[abA]	157,966[abA]	83.58
KH-60	2160[abA]	158,322[abA]	73.30
KH-50	1890[bA]	107,382[bA]	56.82
JH-60	2160[abA]	160,453[abA]	74.28
JH-50	1890[bA]	137,911[abA]	72.97
GH-60	2160[abA]	189,392[abA]	87.68
GH-50	1890[bA]	154,950[abA]	81.98

KH-60 was not obvious difference from CK. Although JH-50 had longer time of deficit and lower plant height, there was no significant difference with CK ($P > 0.5$). And there is no significant difference between JH-60 and CK ($P > 0.5$), as mild deficit in fruit setting. In later fruiting, because that plant height almost no further increased, plant height of water deficit (GH-60 and GH-50) was not significant difference with CK. It was visible that appropriate water deficit would be conducive to the growth of the plant.

3.3 Effects of RDI on yield and IWUE in the tomato greenhouse

The results showed that (Table 3) irrigation of CK was the largest but the yield was not the highest, its value for 177,492 kg/hm². It is MH-60 that yield was the highest, GH-60 followed by, their value respectively for 222,930 kg/hm² and 189,392 kg/hm². It is KH-50 (severe deficit in flowering) that yield was the lowest, JH-50 (severe deficit in fruit setting) followed, their yield respectively for 107,382 kg/hm² and 137,911 kg/hm². It could be visible that MH-60 (mild deficit in seeding) and GH-60 (mild deficit in later fruiting) had a significant effect on increased yield, their increase rate of 24.5% and 6.7%, respectively. While the yield of MH-50 (severe deficit in seedling), KH-50 (severe deficit in flowering), JH-50 (severe deficit in fruit setting) and GH-50 (severe deficit in later fruiting) had different degrees of reduce, their cut rate respectively of 11%, 39.5%, 22.3% and 12.7%. Especially KH-50 (severe deficit in flowering), which cut effect was most obvious.

IWUE fully reflected water supply source of the crop—the utilization of irrigation water, which had a close relationship with the actual production. Table 3 also showed that: IWUE of CK was 73.04 kg/m³, and not the highest. It is MH-60 that IWUE was the highest, GH-60 followed by, whose value respectively for 103.21 kg/m³, 87.68 kg/m³. IWUE of KH-50 (severe deficit in flowering) was the lowest, JH-50 (severe deficit in fruit setting) followed by, whose value were respectively 56.82 kg/m³ and 72.97 kg/m³.

In summary, MH-60 was the best treatment to not only increase yield but also saving water, GH-60 secondly. Therefore, in order to achieve the purpose of water saved and yield increased, in the actual production mild deficit in seeding or later fruiting could be taken. This was mainly because that the appropriate deficit in seeding or later fruiting was conducive to the roots growth in tomato greenhouse, could promote the tomato branch, increased yield and IWUE, thereby saved the limited water resources.

4 CONCLUSION AND DISCUSSION

1. Water deficit would shorten the growth period of the tomato, and the more serious water deficit, the longer shortened time. Those could meet the requirements of picking in the regularly scheduled batch, which reduced the increase of the number of no effect fruit.

2. Plant height had an increasing trend in the whole growth period. Appropriate water deficit was conducive to the growth of the plant. It was the highest to mild deficit treatment in seeding, full irrigation flowered. Plant height of severe deficit treatment in seeding was the lowest, severe deficit treatment in flowering flowered.
3. It is full irrigation that irrigation was the largest but the yield and IWUE were not the highest. Mild deficit in seeding had the best effects on water saved and yield increased, which could increase not only IWUE but also economical benefits. Therefore, it was the best deficit period and deficit degree of RDI in greenhouse tomato.

Therefore, in order to achieve the purpose of water saved and yield increased, in the actual production mild deficit in seeding or later fruiting could be taken. This was mainly because that the appropriate deficit in seeding or later fruiting was conducive to the roots growth in tomato greenhouse, could promote the tomato branch, increased yield and IWUE, thereby saved the limited water resources. But water deficit must not be in flowering or fruit setting, because these were the reproductive growth stage and water demand peak. Water deficit in these two stages would lead to reduced yield, thereby reduced the economic benefits.

REFERENCES

Cai, H.J., Kang S.Z., Zhang, Z.H., Chai H.M., Hu, X.T., & Wang J. 2000. Proper Growth Stages and Deficit Degree of Crop Regulated Deficit Irrigation. *Transactions of the CSAE* 16(3): 24–27.
Chalmers, D.J., Mitchell, P. & Heek, L. 1981. Control of peach tree growth and productivity by regulated water supply, tree density, and summer pruning. *J. Am. Soc. Sci.* 106(3): 307–302.
Goldhamer, D.A., Viveros, M. & Salinas, M. 2006. Regulated deficit irrigation: Effects of variation in applied water and stress timing on yield and yield components. *Irrigation Science* 24(2): 101–114.
Guo, H.T. 2007. Effect of RDI on physiology characters and yield, quality of tomato. Northwest A & F University.
Guo, H.T., Zou, Z.R., Yang, X.J., & Zhou, H.F. 2007. Effects of regulated deficit irrigation (RDI) on physiological indexes, yield, quality and WUE of tomato. *Agricultural Research in the Arid Areas* 25(3): 133–137.
Guo, X.P., Liu, C.L. & Shao, X.H. 1999. The effect of regulated deficit irrigation on waterconsumption regulation and water production efficiency of maize. *Agricultural Research in the Arid Areas* 17(3): 92–95.
Intriglio, D.S. & Castel, J.R. 2005. Effects of regulated deficit irrigation on growth and yield of young Japanese plum trees. *Journal Horticultural Science and Biotechnology* 80(2): 177–182.
Kang, S.Z., Shi, W.J., Hu, X.T. & Liang, Y.L. 1998. Effects of Regulated Deficit Irrigation on Physiological Indices and Water Use Efficiency of Maize. *Transactions of the CSAE* 14(4): 82–72.
Kang, S.Z. & Cai, H.J. 2001. *The theory and practice of alternative irrigation and regulated deficit irrigation in the crop root.* Beijing: China Agriculture Press.
Li, J.M., Zhou, Z.R. & Fu, J.F. 2000. Study on Water-saving Irrigation Index for Greenhouse Tomato. *Journal of Shenyang Agricultural University* 31(1): 110–112.
Liu, M.C. & Chen, D.K. 2002. Effects of Deficit Irrigation on Yield and Quality of Cherry Tomato. *China Vegetables* (6): 4–6.
Liu, M.C., Kojima, T., Tanaka, M. & Chen, H. 2001. Effect of Soil Moisture on Plant Growth and Fruit Properties of Strawberry. *Acta Horticulturae Sinica* 28(4): 307–311.
Liu, M.C., Zhang, S.H. & Liu, X.L. 2005. Effects of different deficit irrigation period on yield and fruit quality of tomato. *Transactions of the CSAE* 21(S): 92–95.
Ma, F.S., Kang, S.Z., Wang, M.X., Pang, X.M., Wang, J.F., & Li, Z.J. 2006. Effect of regulated deficit irrigation on water use efficiency and fruit quality of pear-jujube tree in greenhouse. *Transactions of the CSAE* 22(1): 37–43.
Shi, W.J., Kang, S.Z. & Song, X.Y. 2004. Physiology of growth control of cotton under regulated deficit irrigation. *Agricultural Research in the Arid Areas* 22(3): 91–95.
Wery, J. 2005. Differential effects of soil water deficit on the basic plant functions and their significance to analyze crop responses to water deficit in indeterminate plants. *Australian Journal of Agricultural Research* 56(11): 1201–1209.

Zhang, J.X., Cheng, Z.Y. & Zhang, R. 2012. Regulated Deficit Drip Irrigation Influences on Seed Maize Growth and Yield under Film. *2012 International conference on Modern Hydraulic Engineering. Procedia Engineering* 28: 464–468.

Zhang, S.B., Li, S. & Niu, K.P. 2012. Agricultural Water-saving Approaches in Qinwangchuan Irrigation Area, Gansu, China. *Journal of Desert Research* 32(1): 270–275.

Zhang, X.Y., You, M.Z. & Wang, X.Y. 1999. Preliminary study on the regulated deficit irrigation system of winter wheat. *Ecological Agricultural Research* 6(3): 33–36.

Zhou, X.F. 2009. Study on the Scientific Water Resources Allocation Model of Large Key Water Conservancy Project. *Journal of Lanzhou University (Social Sciences)* 37(5): 101–105.

Hydraulic Engineering – Xie (Ed.)
© 2013 Taylor & Francis Group, London, ISBN 978-1-138-00043-8

Reservoir self-suction sediment transport piping practice and research progress

Li Yuanfa, Chen Junjie, Wu Caiping & Jiang Enhui
Yellow River Institute of Hydraulic Research, YRCC, Zhengzhou, China

ABSTRACT: As to the situation of reservoir sedimentation in China, this paper has summed up hazards of reservoir sedimentation and pointed out that prevention and cleaning reservoir sedimentation is one of key issues of water resources workers. This paper has analyzed several sludge cleaning measures in the reservoir, summed up application in actual engineering of reservoir self-suction sediment transport piping method and its advantages of energy saving, environmental protection, high sediment transport efficiency, low sediment transport cost and simple structure, so it is worthy of promotion mainly. The scholars have taken deep research on Xiaolangdi Reservoir self-suction sediment transport piping method in every aspect in recent years, in view of complexity and urgency of Xiaolangdi reservoir sediment treatment, research on the pilot test shall be taken early so as to promote development and perfection of such a sediment transport method effectively.

1 RESERVOIR SEDIMENTATION INFLUENCE

Hydraulic engineering was mostly built in 1950s to 1960s in China, unitl 1989, Yellow River drainage area got 601 small (I) reservoirs, with 5,221.5 billion m^3 and 1,091.0 billion m^3 taking up 21% total volume. Until 1992, upper Yangtze River got 11,931 reservoirs with 20.5 billion m^3 water reserves, and annual sedimentation is about 11.4 billion m^3, annual sedimentation rate is about 0.68%. Until the end of 2007, there are about 87,000 built-up reservoirs including 37,000 dangerous reservoirs, which took large part in accidents for sedimentation.

Main problems resulted by reservoir sedimentation: (1) volume reduced and benefit reduced; (2) upper breaches get higher water level due to backwater sedimentation, and accordingly make severe flood protection pressure and worsen ecological environment; (3) scouring and silting change in backwater area results in adverse influence on shipping; (4) Sedimentation before the dam influences safe operation of the hub and worsens erosion of unit flow parts; (5) discharge clear water of the reservoir influences lower river erosion and deformation; (6) pollutant adhering to the sediment influences water quality of the reservoir, etc.

Yunnan, Guizhou, Guangxi, Chongqing and Sichuan in the southwest China got large size sustainable drought around 2010, which influenced people's production and life. Water conservancy is the lifeblood of agriculture, while the reservoir shall be the heart of hydraulic system, and reservoir sedimentation makes the "heart" worsen. So how to prevent and clean sedimentation in the reservoir and how to maintain long term sustainable application of the reservoir shall be considered as one of key issues of water resources workers.

2 RESERVOIR SELF-SUCTION SEDIMENT TRANSPORT PIPING APPLICATION PRACTICE

Self-suction piping sediment suction device for sediment transport adopts water level difference between upper and lower reservoir as energy, and discharge the sediment through pipe. The location refers to Figure 1. Such a sediment transport system is of low

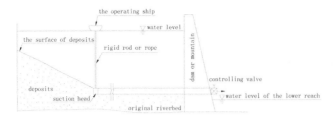

Figure 1.   Sketch map of reservoir self-suction sediment transport piping system.

energy consumption, low sediment transport cost and good future development. Assumption of reservoir self-suction sludge cleaning through pipe was brought forward by M. Zaherting from France before 100 years and applied in the reservoirs with severe sedimentation and invalid precipitation scouring in Africa of 1970s, such as Bogivan reservoir, Sidy Mohammed reservoir, Vofda reservoir in Algeria and Jose Bintafy reservoir from Morocco. Vofda reservoir dam is 90 m in height, and sedimentation thickness before the dam is up to 50 m. It uses two of four sediment transport holes as sediment transport channel. As to 400 m suction pipe diameter, 1~2 km the length and 4,000 h annual work, it can transport 3 million m^3 sediments, with transport sediment volume surpassing the annual charged sediment volume. Sediment transport pipe adopts steel pipe with one flexible rubber joint each 100 m and flexible tube joint connected with sediment transport hole.

Tianjiawan reservoir is a small reservoir with 9.42 million m^3 total volume and 29.5 m dam height. Shanxi Research Institute for Water Conservancy took self-suction sediment transport piping experiment in Tianjiawan reservoir in 1975. Sediment transport device consists of suction pipe and operational ship with "dustpan" suction head as inlet, and the sediment transport piping concentration is up to 760~800 kg/m^3. Sediment can be transported through sediment transport pipe connected with irrigation inclined pipe.

Beichaji reservoir, which is located in Jingning County of Gansu Province, was built in 1958 and got 3.95 million m^3 total volume after heightening the dam in 1981. Beichaji reservoir set self-suction sediment transport device in 1978 with main devices including suction operation ship, suction head, and D410 mm sediment transport steel pipe. It adopted blowing-suction type suction head, namely, install several water nozzles in suction head, crush reservoir bottom sludge from the plow with high pressure water, then discharge it from the pipeline, sucked sediment concentration is 484.25 kg/m^3, and sediment transport volume is 550 t/h. It only operated 392.5 h and transported 135,000 t sediments in 1984, which surpassed the annual sediment charged volume. It obtained third prize for scientific progress of Gansu Province with its sediment suction device getting national patent certificate in 1986.

Xintian reservoir, which is located in Huining County of Gansu Province, was built in 1960 with 24 m dam height, 5.56 million m^3 total volume and 169,000 m^3 average annual sediment charged volume. Xintian reservoir self-suction sediment transport piping system adopted 300 mm steel pipe as sediment transport pipe with blowing-suction type suction head, put into operation in 1981, and average sediment content in winter irrigation was up to 188.41 kg/m^3 in 1983 with 485.81 kg/m^3 the highest content. Average sediment transport rate is 176.3 t/h. Muddy water discharged is of higher fertility, which can improve soil fertility after introducing into farmland and accordingly take good economic effects.

Xiaohuashan reservoir, which is located in secondary tributary Jinduiyu River of Weihe River in Huaxian County of Shaaxi Province, was built in 1960, and got 1.768 million m^3 total volume and 1.392 million m^3 effective volume after heightening 33 m on the dam in 1977. Sedimentation became severe after application water conservancy of the reservoir. In order to keep the reservoir volume, it adopted density current sediment transport, emptying scour and manual sludge cleaning method. Effective volume became 558,000 m^3 in 1976. Self-suction sediment transport piping cleaning sludge experiment is taken to solve sedimentation of Xiaohuashan reservoir since 1976 and sediment transport piping diameter was 300 mm. Total charged sediment volume was 368,400 t, sediment discharged through sediment transport piping device was 47,830 t and discharged sediment volume was 129.88% total charged volume

146

during 1981~1986. Discharged muddy water entered into the channel and flow in the farmland to improve soil fertility, and was used for irrigation and sediment transport or store water and trap sediment, which can be made full use of water and sediment resources.

Danghe reservoir, which is located in mountain exit of Danghe River in southwest Dunhuang City, got 44.60 million m³ total volume and took irreplaceable effects in comprehensive application of irrigation, flood protection, power generation and urban water. Danghe reservoir made the first self-suction sediment transport piping device in 1993, which was the first device in medium reservoir promotion based on small reservoir experimental results of Gansu Province. Sediment transport pipe diameter was 400 mm, cleaning sludge capacity was 216 t/h and annual sludge cleaning volume was 690,000 t. Due to long term reservoir sedimentation, sediment become harden heavily, and discharged sediment even became block, so sediment content was hard to control, with the highest 1100 kg/m³ and the lowest about 220 kg/m³.

In addition, this sediment transport method were adopted by Quhe reservoir of Linfen in Shanxi Province, Hongqi reservoir of Pinglu, Hushan reservoir of Fanzhi, Xiaohe reservoir of Taigu and Youhe reservoir of Weinan in Shaanxi. But such a sediment transport method disappears and abandoned due to national change of investment on water conservancy, no further funds support of the reservoirs and no extra funds of reservoir management department to maintain normal operation. So far, these reservoirs have to adopt the measures like adding dam height or empty discharging sediment to maintain reservoir operation, and accordingly increase reservoir reinforcement fees and reduce comprehensive benefits.

3 RESEARCH OF RESERVOIR SELF-SUCTION SEDIMENT TRANSPORT PIPING METHOD

"Central No. 1 Document" pointed out in 2011 that to develop water conservancy and to relieve water disaster is cardinal issue for national security. We shall make average investment on water conservancy in the following ten years one time more than 2010 in the whole society. Furthermore, we shall pay attention to combination of water conservancy and water disaster, as well as treat both the head and the root, recover flood prevention and strengthen water resources regulation capacity. Aim to Xiaolangdi reservoir sedimentation of Yellow River, Yellow River Water Resources Committee held sediment treatment key technique and equipment meeting at Xiaolangdi reservoir in 2006. Director of Yellow River Water Resources Committee Li Guoying pointed out that Xiaolangdi reservoir sediment shall be treated relying on natural power, while sediment transport device under research shall be featured by little investment, high efficiency, few operational fee and strong endurance. The papers including "Sediment Transport Piping Technology and Equipment Research in Xiaolangdi Reservoir Area" of Gao Hang from Yellow River Institute of Hydraulic Research, "Feasibility of Sediment Transport at High Water Level in Xiaolangdi Reservoir" of Huang Ziqiang from Yellow River Committee and "Reservoir High Efficient Sediment Transport Technology Research" of Lian Jijian from Tianjin University have discussed sediment transport pipe from different angles (they can be considered as self-suction sediment transport piping method seen from sediment transport principle), and all of them think such a sediment transport method is of high sediment transport efficiency, large sediment transport scope, much economic than sludge cleaning by machine, low operational cost, simple structure as well as convenient control.

Later, Yellow River Research Institute for Water Conservancy designed self-suction sediment transport piping large and small schemes in Ministry of Water Resources Public Industrial Research Funds Project "Comparative Research of Xiaolangdi Reservoir Area Sediment Start and Transport Scheme" according to Xiaolangdi reservoir sedimentation condition and building location. The large scheme designs 4 m pipe diameter, sludge cleaned within 25 km before the dam and 0.3 billion t annual average sludge cleaning volume; while small scheme designs 2 m pipe diameter, sludge cleaned within 5 km before the dam and 0.06 billion t annual average sludge cleaning volume, comprehensive sludge cleaning cost is about RMB 1 Yuan/t. It takes detailed research of pipe dam location, construction method, pipe lectotype, pipe laying method as well as suction heat lectotype, and considers self-suction sediment transport piping method adopted is feasible in Xiaolangdi reservoir. In view of complexity of the issue and urgency of Xiaolangdi sediment treatment as well as demands for future Guxian reservoir and Yellow River water and sediment regulation, self-suction sediment transport piping method site pilot studies shall be taken earlier so as to effectively promote the technical work development and perfection. As to Dakupan reservoir sedimentation, College of Water Resources and Civil Engineering from Xinjiang Agricultural University has taken analyzed physical model experiment and certified the advantages like more sediment transportation, less water consumption and low energy consumption of self-suction sediment transport piping system.

4 CONCLUSION

1. Reservoir sedimentation has reduced reservoir volume and benefits from water conservancy, aggravated flood prevention pressure and sharpened ecological environment. How to make the reservoir that loses function for sedimentation work sustainably for a long term has become one of key issues of water resources workers.
2. There are various measures to relieve and clean reservoir sludge, and self-suction piping sediment transport suction device adopts water level difference between upper and lower reservoir as energy, and discharge the sediment through pipe. It is of low energy consumption, low sediment transport cost and good future development.
3. Self-suction piping sludge cleaning method of the reservoir was promoted and applied in some small and medium foreign reservoirs in 1970s. When the water is abundant, it can discharge the coming sediment of this year as well as the deposited sediment before. Due to neither funds support of the reservoir nor extra funds of reservoir management department to maintain normal operation, such sediment transport method disappear gradually.
4. Aim to Xiaolangdi reservoir sedimentation of Yellow River, Yellow River Research Institute of Water Resources has made detailed research of pipe dam location, construction method, pipe lectotype, pipe laying method as well as suction heat lectotype in terms of self-suction sediment transport piping method under the idea that Xiaolangdi reservoir sediment shall be treated relying on natural power. Self-suction sediment transport piping scheme designed makes only RMB 1 Yuan/t comprehensive cost for sludge cleaning. Self-suction sediment transport piping method adopted is feasible in Xiaolangdi reservoir.
5. In view of complexity of the issue and urgency of Xiaolangdi sediment treatment as well as demands for future Guxian reservoir and Yellow River water and sediment regulation, self-suction sediment transport piping method site pilot studies shall be taken earlier so as to effectively promote the technical work development and perfection.

REFERENCES

Han Qiwei, Yang Xiaoqing. 2003. Journal of China Institute of Water Resources and Hydropower Research, *A Review of the Research Work of Reservoir Sedimentation in China.* (12): 169–178.
Lanzhou, Ganshu Research Institute for Water Conservancy. 1987. Experimental Research Results Collection of Sediment Discharge and Desilting in Gansu Province Reservoir.
Liao Yiwei, Xue Songgui, Shang Hongqi. 2006. Yellow River Conservancy Press. Key Technique and Equipment Research Paper Collection of Yellow River Xiaolangdi Reservoir Sediment Treatment.
Shanxi Research Institute for Water Conservancy. 1976. Hydraulic Suction Device Primary Experimental Research, *Tianjiawan Reservoir Hydraulic Sediment Suction Device Experimental Team.*
Shanxi Research Institute for Water Conservancy, Shanxi Huaxian Water Conservancy Bureau, Huaxian Guapo Reservoir Administrative Station. 1987. Experimental Summary of Xiaohuashan Reservoir Hydraulic Suction Sediment and Sludge Cleaning.
Ye Zhiqiang. 1988. Water Conservancy & Electric Power Machinery. *A New Method of Sludge Cleaning in the Reservoir-Hydraulic Suction Device for Sludge Cleaning.* (3): 26–28.
Zhang Li, Xia Xinli, Chen Chenglin, et al. 2010. Journal of Hydroelectric Engineering, *Experiment Study on Self-Pressure Pipe Desilting in Dakupan Reservoir.* 36(4): 92–94.

Hydraulic Engineering – Xie (Ed.)
© 2013 Taylor & Francis Group, London, ISBN 978-1-138-00043-8

Application of steel fiber reinforced shotcrete in dam strengthening project

Jianhe Peng & Lei Zhu

Anhui and Huaihe River Water Resources Research Institute, Bengbu Anhui, China

ABSTRACT: According to field construction conditions, mix proportion of steel fiber reinforce shotcrete is proposed through laboratory and field experiments. The results of the study show that the steel fiber reinforced shotcrete under the optimized design can meet the requirements of compressive strength, tensile strength, flexural strength, bending temper index, temper coefficient, bond strength with surrounding old concrete and other technical indexes.

1 INTRODUCTION

Steel Fiber Reinforced Shotcrete (SFRS), composed of aggregate, cement and the radom-orientd steel fiber, is a kind of multiphase and inhomogeneous composite material. SFRS is a useful engineering material and is used extensively for the Hydraulic Engineering, since it has good physical and mechanical properties.

Properties of SFRS exert great influence on the construction quality and safety of dam strengthening protect. As water cement ratios of dry-jeting and moist-jetting are easy to subject to change, problem of SFRS like large discreteness or low impermeability would probably appear. With the development of material technology, SFRS and other additional materials are gradually applied into concrete to improve the performance of concrete. SFRS has excellent performance of tensiling, bending, shearing, crack arresting and toughening.[1–5] This paper studies the performance of SFRS under different mix proportions and different volume ratios of steel fiber. The result of the study aims to provide reference for engineering application.

2 TEST MATERIAL AND MIX PROPORTION

2.1 Materials

Cement: apply 42.5 ordinary Portland cement whose early strength is comparatively high, with qualified routine determination on fineness, volume soundness, setting time, etc. Fine and coarse aggregates: for the fine aggregates, the fine modules is 3.2, with 0.10% mud content, 1% water content and 0.4% water-soluble sulfide content; for the coarse aggregates, the maximum grain size is 15 mm. Accelerator: apply powder accelerator with qualifies initial and final setting time, which is easy to meet the requirement of wet spraying. To reduce shrinkage, improve the strength of SFRS and lower water cement ratio, low-alkali and highly-effective expansive agent and water reducing agent which are produced domestically are chosen. Steel fiber: cutting steel fiber is used, with its aspect ratio as 64.

2.2 Mix design

As the concrete mix system cannot be installed near dam strengthening work site and the mixing plant must be set at least 1 km far away, there is certain distance for the transportation

Table 1. Mix proportion of SFRS.

	Mix proportion of concrete (kg)											
Num	Water-binder ratio	Water	Cement	Cly ash	Silica fume	W	PJ	Steel fiber	Medium sand	Fine sand	Pebble	Slump (mm)
1	0.400~0.420	220~231	370	110	25	–	45	60	600	260	700	150~170
2	0.411~0.435	226~239	370	110	25	–	45	60	600	260	700	150~200
3	0.381	200	380	70	–	30	45	60	770	330	600	170~220
4	0.400	220	370	110	25	–	45	60	600	260	700	140
5	0.400	220	370	110	25	–	45	60	600	260	700	160

Table 2. Mix proportion of SFRS.

Mix proportion of concrete (kg)											
Water-binder ratio	Water	Cement	Cly ash	Silica fume	PJ	Steel fiber	Medium sand	Fine sand	Pebble	S3	Slump (mm)
0.418	230	370	110	25	45	60	600	260	700	11.1	170~210

of concrete mix. In the process of engineering, there are still some other uncertain factors that would disrupt the immediate spraying or jetting of concrete mix and time intervals are unavoidable. Therefore, tests on the slump degree of SFRS are carried out. Considering the frozen time and variety and content of cementing materials are mainly responsible for the slump of SFRS, while variety and content of steel fober exert less influence, results of tests are shown in Table 1.

Field tests show, applying SFRS mix with the following mix proportion (Table 2) can ensure effective spraying, and the performance, compressive strength, tensile strength, bending strength, bond strength as well as flexural toughness of SFRS mix can meet the design and construction requirement.

3 MECHANICAL PROPERTIES OF SFRS

3.1 Strength of SFRS

Under different mix proportion, as water cement ratio differs, the distribution of steel fiber in concrete may also be influenced, which causes further alteration in compressive strength and tensile strength. The paper just takes the concrete mix proportion suggested in filed tests as example. And the detailed test results can be seen in Table 3.

3.2 Adhesive axial tensile strength of SFRS

The adhesive axial tensile strength of old and new SFRS plays vital role in dam strengthening project. To increase adhesive axial tensile strength, the surface of old concrete should be roughened through pneumatic drill.

Apply the suggested concrete mix proportion and hold tests on adhesive performance of SFRS in site, and we can get the results shown in Table 4.

From Table 4 it can be seen that when SFRS is sprayed on the inclined arch, the adhesive axial tensile strength of the old and new concrete interface is between 0.3 Mpa and 0.9 Mpa, which indicates relatively large discreteness, and the average adhesive axial tensile strength is only 0.5 Mpa; the adhesive axial tensile strength of concrete surface alters between 1.2 Mpa and 3.9 Mpa, with large discreteness, and the average adhesive axial tensile strength is 2.4 Mpa. When SFRS is poured on side wall, the adhesive axial tensile strength of the old and new concrete interface is between 0.4 Mpa and 1.5 Mpa, with large discreteness, and the

Table 3. Strength of SFRS.

Compressive strength at 28 days (MPa)	Flexural strength at 28 days (MPa)	Axial tensile strength at 28 days (MPa)	Tensile elastic modulus at 28 days (104 MPa)	Splitting tensile strength at 28 days (MPa)		
				Core sample	Large panel	Trying die
49.0~52.2	5.92~7.28	3.41	2.73	3.56~4.01	6.22	5.68

Note: Specimen size of compressive strength and splitting tensile strength is 100 mm × 100 mm × 100 mm; specimen size of compressive strength is 100 mm × 100 mm × 400 mm; results time conversion coefficient respectively.

Table 4. Adhesive axial tensile strength of SFRS.

Position	Adhesive tensile strength (MPa)		Adhesive axial tensile strength (MPa)	
	Range	Average	Range	Average
Side wall	1.9~2.8	2.3	0.4~1.5	1.0
Inclined arch	1.2~3.9	2.4	0.3~0.9	0.5

Table 5. Flexural toughness of SFRS.

Deflection (mm)	Initial cracking toughness (kN·mm)	Bend toughness index η_{10}	Bend toughness index η_{30}	Toughness factor $R_{30\text{-}10}$
0.05	0.50~0.55	7.31~8.04	20.59~22.04	66.41~69.97

average adhesive axial tensile strength is 1.0 Mpa; the adhesive axial tensile strength of concrete surface alters between 1.9 Mpa and 2.8 Mpa, with large discreteness, and the average adhesive axial tensile strength is only 2.3 Mpa.

The destruction of combined structure of old and new concrete also occurs on the interface, which is due to the damage on the old concrete surface in engineering. This is in accordance to the filed test results. Therefore, it is suggested to strengthen the treatment on old concrete surface when toughening to increase the adhesive axial tensile strength of old and new concrete interface.

To ensure enough bond strength of old and new concrete interface, appropriate anchor bars are added in the tests, which is proved sound according to tests results.

3.3 Flexural toughness of SFRS

Apply the suggested concrete mix proportion and hold tests on flexural toughness in site, and we can get the results shown in Table 5.

Table 5 shows when the suggested mix proportion of SFRS is applied, initial cracking toughness, bend toughness index and toughness factor can meet the design requirement.

4 CONCLUSION

Through field tests and long period of observation and examination after dam strengthening, conclusions can be drawn as follows:

1. Application of suggested mix proportion of SFRS and control of the slump degree in the range of 170 mm–210 mm can ensure effective spraying and meet engineering requirement.

2. In 28 days after dam strengthening project, compressive strength, splitting tensile strength, flexural toughness and other indexes can all meet the requirement of design.
3. The axial tensile strength of SFRS and the adhesive tensile strength of surface are a bit lower than design index. The adhesive axial tensile strength of old and new concrete interface is far from design requirement, but field tests prove appropriate addition of anchor bars is available.
4. The dam has been tried for more than 6 years after strengthening project, which shows the above scheme is sound and technology measures are effective. Control measures also meet the expectation.

REFERENCES

Guoping, DU. & Xinrong, LIU. 2008. Performance of steel fiber reinforced shotcrete in tunnel and its engineering application. *Journal of Rock Mechanics and Engineering* 27(7): 1448~1454.
Jeng, F. & Lin M. 2002. Performance of toughness indices for steel fiber reinforced shotcrete. *Tunnelling and Underground Space Technolog* 17(9): 69~82.
Minggao, LIU. & Wenxue, GAO. 2006. Steel-fibre shotcrete and its application in tunnelling. *Railway Engineering* (2): 43~45.
Sing, S. & Madan S. 2006. Flexural fatigue strength of steel fiber concrete containing mixed steel fibers. *Zhejiang Univ SCIENCE A*7 (8): 1329~1335.
Xiuyun, CHEN. & Jiangang, FEI. 2009. Experimental studies on the mix proportion design for steel fiber reinforced concretes. *Concrete* (7): 111~116.

Hydraulic Engineering – Xie (Ed.)
© 2013 Taylor & Francis Group, London, ISBN 978-1-138-00043-8

An experiment about diversion recharge in Sibaishu Haize Lake of Jingbian county

Yang Tao
Department of Geology/State Key Laboratory of Continental Dynamic, Northwest University, China
Shijiazhuang University of Economics, China

Wang Jia-Ding
Department of Geology/State Key Laboratory of Continental Dynamic, Northwest University, China

ABSTRACT: On the base of field diversion recharge experiment in Sibaishu Haize Lake, the infiltration coefficient is 0.01 m/d, and the minimum diversion with water all the year round is confirmed. Variable amplitude of groundwater level in the current condition of exploitation is predicted through the establishment numerical simulation model. The results shows water quantity of diversion recharge should account to 200,000 m^3/d, and the amount of groundwater infiltration is 160,000 m^3/a in order to ensure that Sibaishu Haize has water resource all the year around, increasing area of Sibaishu Haize by 42,623 m^2, the increasing amplitude of groundwater level is more than 1 m. Obviously, it is no doubt that he diversion recharge is an efficient path of improving the whole zone's environmental and ecological effect, on the two aspects of pumping and recharge of groundwater developing benignly, fulfilling the resume of oasis ecology and making human settlement better.

1 INTRODUCTION

Diversion recharge usually needs four basic conditions. The first is a place that is useful to stream water recharge, the second is the water resource that can be recharged, the third is the underground space which can contain water, and the last is the good underground extracting condition.

On the base of the field recharge experiment of Sibaishu Haize Lake, the possibility of water environment improved through diversion recharge is studied emphatically and numerical simulated mode is used to proof. Therefore, man-made diversion recharge han a significa- tion meaning that can prevent not only fresh water resource from drying up and water quality deterioration but also environmental surrounding more worse, and can make human living environment better who are in the desert.

2 RESEARCH METHOD

2.1 Test introduction of diversion recharge

The Sibaishu Haize Lake has run dry, its terrain is higher east and lower west on the whole. According to the eighteen test boring materials, the depth of embedment of underground water is from 1 to 2.2 m in the higher terrain zone, and it is from 0.5 to 0.6 m in the lower. Seepage pit is located in the northern corner of Sibaishu Haize lower zone, where the lowest water level is only 0.255 m.

Recharge water is underground water extracted by digging a well on account of no surface water and wells that can be utilized. The water pumping well named ZK1 lies in the north- west of seepage pit, whose depth is 30 m, two observation well are seated on the line that links pumping well and reference point, which are named G1, G2 and also 10 m deep, water bearing bed belongs to fine sand layer and its underground water depth is 3 m.

Recharge test lasted for 367.6 hours, the time of pumping is 5 hours, the way of pumping water is continuous steady flow, its flow rate is 8.62 m³/h and the amount of pumping water for 5 hours account to 43.12 m³. Water-collecting area is 319 m² during the test.

Based on the the ZK1 pore (Figures 2~4), the water level of two observation well is stable after pumping for 80 minutes. The water level of G1 observation pore nearly resumed after we stop extracting water for 26 minutes, and the water level of G2 totally resumed after we stop extracting water for 126 minutes. The parameters are calculated in the Table 1 in the light of the test data, in which permeability coefficient K of water bearing bed is 6.8 m/d and the radius R of influence of the experiment is 74 m.

The curve of positive surge and water-level depression of ZK1 pore during the pumping test and after it (Figures 5 and 6) show the process that water-level of seepage pit ascended

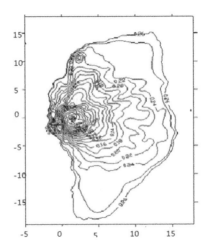

Figure 1. Seepage pit terrain isoline of recharge test in Sibaishu Haize Lake.

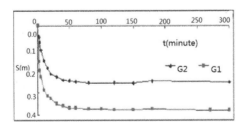

Figure 2. Observation pore drawdown of pumping test.

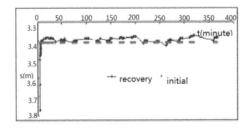

Figure 3. G1 pore's water-level recovery of pumping test.

154

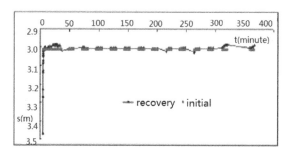

Figure 4. G2 pore's water-level recovery of pumping test.

Table 1. ZK 1 pore's data and result of pumping test in Sibaishu Haize Lake.

Pumping flow rate Q (m³/d)	Aquifer thickness H (m)	G1 distance r1 (m)	G2 distance r2 (m)	G1 drawdown s1 (m)	G2 drawdown s2 (m)	Osmotic coefficient K (m/d)	Radius of influence R (m)
207	27	9	18	0.38	0.255	6.8	74

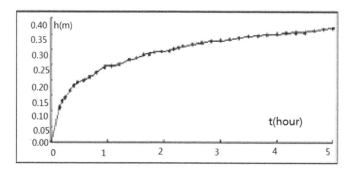

Figure 5. The curve of positive surge of seepage depression of pit during pumping.

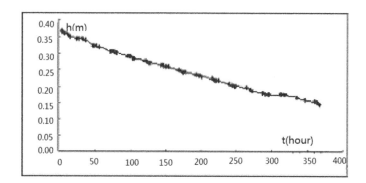

Figure 6. The curve of water level seepage pit after ending pumping.

rapidly at the first, climbing speed is getting slow with time going, it is the reply that the later increased water injection rate manifests mainly the area of seepage pit raising and water layer adding slowly. The water-level depression of seepage pit totally occurs more stable and has the trend from quick to slow, but the whole course has no obvious change and it reflects the balance of recharge and loss. When the stronger rainfall happened on the day of 24, it recharged

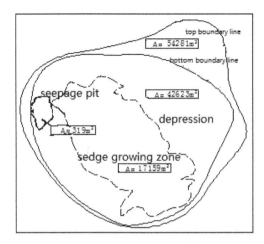

Figure 7.   The shape of Sibaishu Haize and the location of seepage pit.

into seepage pit and made positive surge vast scale. Infiltration intensity is 0.31 m/d, during pumping water, it begins at 0.02 m/d, and it end stably at 0.01 m/d, after stopping pumping. The early infiltration intensity is larger than the late, possibly because quite a few of initial water of seepage pit makes dry soil horizon wet, it is obvious that infiltration intensity after stability have good representation on simulating diversion recharge.

2.2   *Double-ring infiltration test*

In order that the infiltration parameters varying on the surface land in Sibaishu Haize Lake can be obtained, 10 spots are located symmetrically in the depression and water seepage time lasts for 10 hours.

On the synthesis of infiltration test's results above introduced, it is reliable that the calculated value of the last infiltration intensity is taken as infiltration parameters of diversion recharge of Sibaishu Haize Lake.

2.3   *Diversion recharge analysis*

Sibaishu Haize Lake is like a circular depression, high in the east and low in the west, mean depth is 1~2 m. According to Figure 7, the total area covered by the depression is 54,281 m², the area of the depression's bottom is 42.623 m². The infiltration intensity of the depression is 0.01 m/d that is the result of seepage recharge test. The infiltration capacity is 426 m³/d after the depression is entirely submerged, and it is 160,000 m³ every year.

The average annual precipitation of Jingbian county is 396.1 mm and its average annual evaporation degree is 1,835 mm, the conversion coefficient for evaporation is selected as 0.58, the pure evaporation capacity amount yearly is about 30,000 m³ in the submerged annually zone of Sibaishu Haize Lake.

The amount of annual balanced water inflow is the sum of maximum infiltration capacity and pure evaporation capacity, it is 190,000 m³. If the minimum diversion recharge is intended to be 200,000 m³/a, it is possible to ensure the annal water resource of Sibaishu Haize Lake.

3   NUMERICAL SIMULATION AND THE EFFECTIVE ANALYSIS OF DIVERSION RECHARGE

The underground numerical model is established that includes two water source areas named by Sibaishu and Lianyouchang and the reservoir lying in the Haize and Yangjiawan, into which the minimum intake of Sibaishu Haize Lake is added as an infection well. The model can predict underground water-level variation tendency of future 30 years and caused by Sibaishu Haize diversion recharge under the condition of current extraction situation.

The forecast outcome indicates that the whole tendency has no obvious variation of ground water level in the model zone, a closed funnel takes shape in the center of Sibaishu Haize Lake, ground water level of the northern blown-sand region increases a little. Water-level increasing degree is more than 1 m in Sihaizehaize Lake and it is only 0.4~0.6 m.

4 ENVIRONMENTAL EFFECT ANALYSIS OF DIVERSION RECHARGE

Taking all factors into consideration that are the quality, quantity, development and utilization of surface water resource in Jingbian county, the water resource amount of multi-year mean available reservoir is 29,080,000 m^3/a, of which ninety percent fraction is 19,220,000 m^3/a. Usually in the program of water resource development and utilization, it is planned in ninety percent fraction, so the water resource in normal years more than ninety percent fraction can be used as the water recharge resource. In view of some factors such as the high-flow, par-flow and low-flow interchanged and recharge reliable degree increased, water inflow is planned in fifty percent fraction reduction. On the basis, the water resource amount extracted from recharge is 4,930,000 m^3/a.

The simulation prediction makes clear that the number of water recharge resource is 200,000 m^3/a, the number of infiltration recharge is 160,000 m^3/a and the area of Haize Lake is increased 42,623 m^2. The amount of water diversion resource extracted from the reservoirs is 4,930,000 m^3/a and the amount of water recharge resource is 2,900,000 m^3/a, the recharge area of Haize Lake is 1 km^2. Therefore the diversion recharge is an efficient path of improving the whole zone's environmental and ecological effect.

5 CONCLUSION

The environmental influence of Sibaishu Haize Lake is analyzed compositely through field experiment after the diversion recharge. The study shows:

The reliable infiltration coefficient of diversion recharge is obtained by field test, which is 0.01 m/d, and the minimum diversion recharge is intended to be 200,000 m^3/a to ensure the annal water resource of Sibaishu Haize Lake.

The groundwater numerical model is established and its result makes known that the whole tendency has no obvious variation of ground water level in the model zone, ground water level of the northern blown-sand region increases a little. Water-level increasing degree is more than 1 m in Sihaizehaize Lake and it is only 0.4~0.6 m.

For the other 25 Haize similar to Sibaishu Haize, it is no doubt that he diversion recharge is an efficient path of improving the whole zone's environmental and ecological effect, on the two aspects of pumping and recharge of groundwater developing benignly, fulfilling the resume of oasis ecology and making human settlement better.

REFERENCES

Wang Wei, Ma Si-jin, Guo Hong-jun. 2003. Optimal Estimation on Parameters in 3D Hydrogeological Numerical Model of Sibaishu Water Source place [J]. *Ground Water* 25(3):141–146.

Wang Wei, Ma Si-jin, Chen Jian-min. 2008. The study of conservation zone of the Sibaishu Water Source place [J]. *Yellow River* 30(3):48–49.

Zhu Si-yuan, Tian Jun-cang, Li Quandong. 2008. Current Situation and Development Trend of Research on Groundwater Reservoir [J]. *Water-Saving Irrigation* 8(4):23–27.

Hydraulic Engineering – Xie (Ed.)
© 2013 Taylor & Francis Group, London, ISBN 978-1-138-00043-8

Numerical model of the spiral flow in the 90° elbow

Yamei Lan
College of Engineering Science & Technology, Shanghai Ocean University, Shanghai, China

Wenhua Guo
Estuarne & Coastal Science Research Center, Shanghai, China

Qiuhong Song & Yongguo Li
College of Engineering Science & Technology, Shanghai Ocean University, Shanghai, China

ABSTRACT: By CFD software Fluent, the spiral flow was simulated in the 90° elbow in the paper. Using the finite volume method, 6 kinds of spiral flows were studied which formed by various angles of tangential inlet velocity. The formation, development and attenuation of spiral flows in the elbow were investigated in detail. The results show that tangential velocity becomes the largest when the angle of inlet is 60°, which is good for removing the impurities deposited in a pipeline. Also, it is shown that the elbow has a good effect on maintaining the flow. Meanwhile, the decay and the development of a spiral flow are different as the chang of angles. Results show that the angle should not be too large. According to a variety of data, the tangential inflow angle of 60° is an appropriate point of view, it can produce a higher strength and a durable spiral flow.

1   INTRODUCTION

Spiral flow is in nature a common flow form. Tsunamis, cyclones and sudden change of Flow rate with tide and ebb are spiral flow. Studying the mechanism of spiral flow is good for promoting industrial and economic development, and has important significance for predicting and preventing of natural disaster. Because the spiral flow contains more energy, which plays an important role in the flow, it has been the subject of scientific workers.

In general, the spiral flow is produced in three ways: the inlet flow by tangential angle, installating the guide vane and rotating the pipe. Since 1980, K.Q. Zhang has been engaged in the spiral flow for sediment, particularly committed to the vortex tube. Such aspects were studied as the diversion ratio, sand interception rate, carrying capacity, vortex flow structure, sediment mechanism and application conditions. X.F. Wang studied on spiral flow in detail. X.H. Sun investigated the force characteristics and particle suspension mechanism. X.E. Zhang, X.H. Sun and Y.M. Lan test the flow resistance by experiment. On the other hand, through numerical simulation, characteristics and distribution laws of the velocity and pressure are studied, which provided strong basis for the further study of transportation by spiral flow. Using the finite element method, H.M. Zhang simulated the flow field around the drainage device and calculated the velocity field and pressure. Zhang Yu studied the characteristics of sediment transportation by spiral flow. Y.Lin studied the rotary actuator structure, resistance loss and rotation efficiency. In a word, the problem of the sedimentation in the elbow is involved little. So in the paper, the inlet flow by tangential angle is used to simulate the spiral flow in the 90° elbow.

2 NUMERICAL MODEL

2.1 *Governing equations*

Governing equations include the momentum equations, the continuous equation, and $k-\varepsilon$ turbulence model. They can be described as follows.

$$\frac{\partial(\rho u)}{\partial t} + \mathrm{div}(\rho u\mathbf{u}) = \mathrm{div}\left[(\mu+\mu_t)\mathrm{grad}u\right] - \frac{\partial p}{\partial x}$$
$$+\frac{\partial}{\partial x}\left[(\mu+\mu_t)\frac{\partial u}{\partial x}\right] + \frac{\partial}{\partial y}\left[(\mu+\mu_t)\frac{\partial v}{\partial x}\right] + \frac{\partial}{\partial z}\left[(\mu+\mu_t)\frac{\partial w}{\partial x}\right] \quad (1)$$

$$\frac{\partial(\rho v)}{\partial t} + \mathrm{div}(\rho v\mathbf{u}) = \mathrm{div}\,\mathrm{div}\left[(\mu+\mu_t)\mathrm{grad}v\right] - \frac{\partial p}{\partial y}$$
$$+\frac{\partial}{\partial x}\left[(\mu+\mu_t)\frac{\partial u}{\partial y}\right] + \frac{\partial}{\partial y}\left[(\mu+\mu_t)\frac{\partial v}{\partial y}\right] + \frac{\partial}{\partial z}\left[(\mu+\mu_t)\frac{\partial w}{\partial y}\right] \quad (2)$$

$$\frac{\partial(\rho w)}{\partial t} + \mathrm{div}(\rho w\mathbf{u}) = \mathrm{div}\left[(\mu+\mu_t)\mathrm{grad}w\right] - \frac{\partial p}{\partial z} + g$$
$$+\frac{\partial}{\partial x}\left[(\mu+\mu_t)\frac{\partial u}{\partial z}\right] + \frac{\partial}{\partial y}\left[(\mu+\mu_t)\frac{\partial v}{\partial z}\right] + \frac{\partial}{\partial z}\left[(\mu+\mu_t)\frac{\partial w}{\partial z}\right] \quad (3)$$

$$\frac{\partial\rho}{\partial t} + \mathrm{div}(\rho\mathbf{u}) = 0 \quad (4)$$

$$\frac{\partial(\rho k)}{\partial t} + \mathrm{div}(\rho k\mathbf{u}) = \mathrm{div}\left[\left(\mu+\frac{\mu_t}{\sigma_k}\right)\mathrm{grad}k\right] + G_k - \rho\varepsilon \quad (5)$$

$$\frac{\partial(\rho\varepsilon)}{\partial t} + \mathrm{div}(\rho\varepsilon\mathbf{u}) = \mathrm{div}\left[\left(\mu+\frac{\mu_t}{\sigma_\varepsilon}\right)\mathrm{grad}\varepsilon\right] + \frac{C_{1\varepsilon}\varepsilon}{k}G_k - C_{2\varepsilon}\rho\frac{\varepsilon^2}{k} \quad (6)$$

$$\mu_t = \rho C_\mu \frac{k^2}{\varepsilon} \quad (7)$$

where ρ is for the fluid density; t is for time; u,v and w are the corresponding velocity scalar in x, y, and z directions, respectively; $\mathbf{u}$ in bold type refers to the velocity vector; μ is for the viscosity coefficient of the fluid; μ_t is for the eddy viscosity coefficient. The values of C_μ, $C_{1\varepsilon}$, $C_{2\varepsilon}$, σ_k, σ_ε are 0.09, 1.44, 1.92, 1.0, 1.3, respectively; G_k is for the production item of k, owing to the gradient of mean velocity, which is expressed as equation (8). So the formulas (1)~(8) are the complete governing equations for the flow.

$$G_k = \mu_t\left\{2\left[\left(\frac{\partial u}{\partial x}\right)^2 + \left(\frac{\partial v}{\partial y}\right)^2 + \left(\frac{\partial w}{\partial z}\right)^2\right]\right.$$
$$\left. + \left(\frac{\partial u}{\partial y}+\frac{\partial v}{\partial x}\right)^2 + \left(\frac{\partial u}{\partial z}+\frac{\partial w}{\partial x}\right)^2 + \left(\frac{\partial v}{\partial z}+\frac{\partial w}{\partial y}\right)^2\right\} \quad (8)$$

2.2 *The geometry model of the 90° elbow*

In the pre-treatment software **GAMBIT**, the geometry model of the 90° elbow is established. At first, in descarte coordinates, 3d elbow is built, $d = 0.02$ m, $r = 0.1$ m, $\theta = 90°$, $l = 0.1$ m,

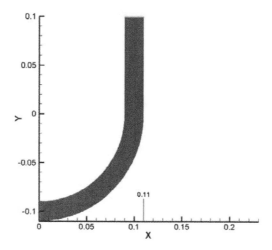

Figure 1.   Geometry size of the 90° elbow.

where d, r, θ and l denote diameter of the pipe, arc radius of the elbow, corresponding central angle and length of the straight section, respectively. The detailed geometry size of the 90° elbow is shown in Figure 1. The quality of the grid has an important effect on accuracy and efficiency of numerical results. Structured mesh, that is hexahedral one, is used in the paper. In dividing mesh, the entrance surface is divided into the boundary layer, and the method of grid refinement is used on the wall surface. At last, the interval size mesh on the surface is 0.015 m, and the one on the volumn is 0.05 m.

2.3   Boundary conditions

The sketch of the 90° elbow is shown in Figure 1. Non-slipping boundary conditions are applied on the inner wall of the elbow. The boundary condition of the entrance of the pipe is velocity-inlet, which is equivalent to the installation of a generator at the entrance. So the velocities in three directions, that is x, y, z, can be directly given. In the paper, 6 kinds of spiral flows are studied which formed by various angles of tangential inlet velocity. For the convenience of simulation, the velocities in x, y directions remain constant, that is $u = 1$ m/s and $v = 1$ m/s. While the velocity in z direction depends on the one in y direction, that is v. The relationship between v and w are as follows.

$$w = v \cdot \tan\theta \tag{9}$$

6 kinds of entrance angles are used in the paper, that is $\theta = 20°$, 30°, 40°, 50°, 60°, 70°, respectively. For example, when $\theta = 20°$, $w = 1 \cdot \tan 20° = 0.364$ m/s.

3   NUMERICAL RESULTS

3.1   The effect of the entrance angle

The velocities of the spiral flow can be divided into radial, tangential and axial velocity. Previous studies declare that compared to the tangential and axial velocity, the absolute value of the radial one is very small, which is close to zero, can be neglected. While the magnitude of the tangential velocity is much more than the axial one. Moreover, the tangential velocity plays a big role in transporting the sediment. In conclusion, the tangential velocity of the elbow is mainly discussed in the paper. Figure 2 shows the relationship between the tangential velocity on the outlet and the entrance angle. The horizontal coordinate is along the diameter

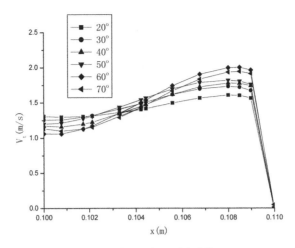

Figure 2. The tangential velocity on the outlet surface with different entrance angle.

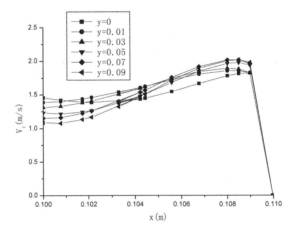

Figure 3. The tangential velocity on the outlet surface with different entrance angle.

of the outlet. While $x = 0.100$ m refers to the center and $x = 0.110$ m refers to inner wall of the outlet. From Figure 2, the tangential velocity becomes the largest when the angle of inlet is 60°, which is good for removing the impurities deposited in a pipeline. Meanwhile, the tangential velocity will decrease if the entrance angle continues to increase. For example the tangential velocity when $\theta = 70°$ is larger than that of 60°, and the spiral strength will be reduced accordingly. On the other hand, the tangential velocity distribution along the diameter direction presents from low to high. The peak appears arround the inner wall of the pipe.

3.2 The effect of the 90° elbow

In the paper, the formation of the spiral flow is not only owing to the tangential inlet, but also to the 90° elbow. For further analysis, one of the 6 different entrance angles is selelced, that is $\theta = 60°$. In this condition, the effect of the spiral flow is the best than others. Figure 3 shows the relationship between between the tangential velocity on the different section with rear and straight pipe. The horizontal coordinate is along the diameter of the outlet, which is same to that of Figure 2. While $x = 0.100$ m refers to the center and $x = 0.110$ m refers to inner wall of the outlet. From Figure 3, the tangential velocity does not decrease rapidly in the black of the straight pipe. On the contrary, the tangential velocity increases while the section is further back along the inner wall of the pipe. It is shown that 90° elbow can not only accelerate the formation of spiral flow, but also effectively maintain the flow.

4 CONCLUSIONS

The spiral flow was simulated in the 90° elbow in the paper. Using the finite volume method, 6 kinds of spiral flows were studied which formed by various angles of tangential inlet velocity. The results show that tangential velocity becomes the largest when the angle of inlet is 60°, which is good for removing the impurities deposited in a pipeline. Also, it is shown that the elbow has a good effect on maintaining the flow. Meanwhile, the decay and the development of a spiral flow are different as the chang of angles. At the straight section of the pipe, the spiral flow has been significantly attenuated as the tangential angle gets 70°. This shows that the angle should not be too large. Results show that the tangential inflow angle of 50°–60° is an appropriate point of view, it can produce a higher strength and a durable spiral flow. On the other hand, the tangential velocity distribution along the diameter direction presents from low to high. The peak appears arround the inner wall of the pipe, which is favorable to transport the sediment.

ACKNOWLEDGEMENTS

This work is financially supported by the Key Project of the Shanghai Science and Technology Commission (No. 11160501000) and Special Project of National Ocean Bureau (SHME2011GD01).

REFERENCES

Guo, Y.H. & Liu, C. 2007. Numerical Simulation of Fluid Flow and Heat Transfer in the Spiral Flow in Pipe. *Thermal Engineering* 36: 27–31.
Lin, Y., Lei, P. & Li, S. 2011. Generation Mechanism of Pipe Spiral Flow and Design of Guiding-Vane Type Local Spiral Flow Generator, *Mining Machinery* 39: 54–57.
Lv, P. 2006. The Research on Testing Characteristics of Spiral Folw in Cricle Pipe. *Shanxi Hydrotechnics*, 3: 25–27.
Yan, Y.X., Li, W. & Zhang, J. 2006. Study on the Experiment of Spiral Energy-counteract Dingus With the Side-diagonal Inflow. *Sci-Tech Information Development & Economy* 16: 163–165.
Zhang, L.L. & Zhang, Y. 2011. The Section Flow Field and Flow Characteristics of the Pipe Spiral Flow. *China Rural Water and Hydropower* 9: 48–50.
Zhang, X.D. & Sun, X.H. 2005. Experimental Study on Measurement for Maitaining the Spiral Flow in Pipe. *Shanxi Water Resources*, 1: 74–76.
Zhang, X.E., Sun, X.H. & Huo, D.M. 2001. The Numerical Simulation of Spiral Flow Field with Continuous Guiding Vane. *Journal of Taiyuan University of Technology* 32: 252–254.
Zhang, Y., Zhang, X.E. & Peng, L.S. 2005. Experimental Study on Sediment Transportation of Spiral Flow in Pipe. *Journal of Sediment Research* 2: 34–38.
Zhu, Y., Wang, S.L. & Shi, X.J. 2008. Numerical Simulation of the Spiral Flow in the 90° Bend Pipe Based on Phoenics. *China Petroleum Machinery* 36: 19–22.

Hydraulic Engineering – Xie (Ed.)
© 2013 Taylor & Francis Group, London, ISBN 978-1-138-00043-8

Drainage system vulnerability analysis of Beijing under the rainstorm

Wang Yan
Institute of Geographic Sciences and Natural Resources Research, CAS, Beijing, China
Graduate University of Chinese Academy of Sciences, Beijing, China

Fang Chuanglin
Institute of Geographic Sciences and Natural Resources Research, CAS, Beijing, China

ABSTRACT: Recently, waterlogging happened in several cities in our country which is caused by rainstorm and brought too much trouble to the daily life of local residents. Beijing city has experienced a rainstorm in July 2012, many streets had been submerged, a lot of cars were destroyed, and the disaster made huge economic losses and many casualties. Urban drainage system is an important part of urban Infrastructural facilities, as well as key engineering for preventing waterlogging. As a result, study on problem existed in the urban drainage system construction has become more and more urgent. This paper collected 96 sample points of accumulating water positions and its water depth in Beijing city, and obtained the coordinates of the sample points by the Google Earth, then established the spatial database of the Stormwater point. Based on the vulnerability study of the urban drainage system, the paper is supposed to be useful for the improvement of the drainage system and make some suggestions for the response to the rainstorm disaster.

1   INTRODUCTION

1.1   Background

Accompanied by the rapid urbanization, many cities in china are constructed by leaps and bounds, the appearance of the cities changes rapidly, but the underground facilities are lagged behind, especially the planning and construction of the drainage system. The integrity of urban storm water systems is poor, no one is responsible for overall supervision and it has no long-term planning. The drainage system is the general term of infrastructure for the treatment of municipal sewage, industrial wastewater and rainwater, it maintains the city's metabolism and circulation, which is the important symbol of measuring the modernization level and the precondition of sustainable development society. The drainage system is not only a livelihood projects, but also lifeline engineering.

1.2   Present situation

Domestic and foreign scholars launched a multi-level multi-angle study on rainstorm waterlogging issues. In the field of urban storm disaster research, scholars proposed formula for the calculation of the storm intensity and storm water amount (Zhou et al. 2012; Wang et al. 2012; Ma et al. 2012; Zhou et al. 2011), and select the case to explain the application of the formula (Zhao et al. 2012). On urban storm with urban waterlogging, the scholars try to develop urban rainstorm waterlogging forecasting and early warning system (Luo et al. 2012; Yin et al. 2011), and take Chengdu, Shanghai for example to expand the rainstorm warning systems research. In the field of urban drainage systems, Scholars had done some research on the construction of drainage systems, measures of drainage systems dealing with rainstorm (Yang et al. 2010; Fang et al. 2009). In addition, scholars had study on improvement measures of urban storm waterlogging, including the utilization of urban underground space, the improvement of drainage systems, the application of low-impact development techniques, etc (Tan et al. 2012; Zhu et al. 2011; Li et al. 2011; Li et al. 2011). Currently, there are study on urban drainage systems from a vulnerability perspective. The urban waterlogging caused by storm reflect the vulnerability of the urban drainage system, based distribution of rainstorm water sections in July 2012, this paper analyses the vulnerability of the drainage system in Beijing rainstorm contexts, and propose corresponding countermeasures and suggestions.

2 STUDY AREA AND DATA

2.1 *Study area*

Beijing, as the capital of China, is located in the northern end of the North China Plain, southeast of Tianjin connected; the rest is surrounded by Hebei Province. In late 2011, the total population of a million people, to achieve GDP 47.1564 trillion yuan. The most active areas of Beijing's economic development and urban construction, including 6 formed zone: Dongcheng District, Xicheng, Chaoyang District, Haidian District, Fengtai District Shijingshan District (Fig. 1). The Beijing urban drainage system including urban river, municipal sewer lines, drainage pumping station, and in accordance with the the Drainage nature can be divided into the stormwater drainage system and sewage drainage system. Stormwater drainage system in accordance with the city Tonghuihe, Liangshuihe, Qinghe, Bahe rivers Department is divided into four watersheds, which consists of 27 main canal and over 270 rain water mains responsible for urban rainwater collection and troubleshooting tasks, the urban flood control drainage of the main body. The research data from the NetEase (http://news.163.com) Beijing has been torrential rains thematic reports, from which the 63 water points in Beijing urban location and hydrocephalus level. Google Earth to get the latitude and longitude coordinates of each sample point, volume built coordinates tools to create Beijing stormwater point spatial database using ArcGIS software.

2.2 *Data*

The research data from the Net Ease (http://news.163.com) Beijing has been torrential rains thematic reports, from which the 63 water points in Beijing urban location and

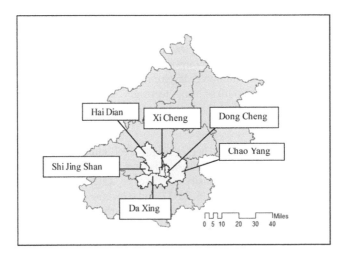

Figure 1. Study area.

hydrocephalus level. Google Earth to get the latitude and longitude coordinates of each sample point, volume built coordinates tools to create Beijing storm water point spatial database using ArcGIS software.

3 EVALUATION METHOD

3.1 *Drainage system vulnerability concept*

The vulnerability is a measure of the system sensitivity to disturbance resistance indicators. Drainage system vulnerability refers to the urban drainage system sensitivity and the lack of capacity to respond to external disturbances (storm) so that the change in the structure and function of the system is prone to a property. It is derived from within the system, a property inherent to this property manifested only when the system is subjected to disturbance. The internal features of the system is the system vulnerability generated, the direct cause of the disturbance and interaction between its vulnerability to zoom in or out, the system vulnerability drivers of change occurred, but this use of the driving factors affect the vulnerability of the system of internal features of the system is changed, and eventually through the system to face disturbance sensitivity and response capacity to reflect (Li et al. 2008).

3.2 *Evaluation method*

In this paper, GIS technology, space database based on the Beijing storm water point, to calculate space some of the characteristics of the water points, found that the spatial distribution of the stagnant water area characteristics at the same time clear which areas of the drainage system is fragile, for urban renewal and The drainage system planning to provide a basis for decision making.

4 CONCLUSION AND DISCUSSION

4.1 *Conclusion*

Drainage vulnerability region centered on the ring road and Bridge District. Loop area affected is more serious, is the number of water points and water points depth, more than in other regions; 96 water points, 51 water points for the bridge area, more than half of the total number of water points. The loop become exposed to serious water concentration. The stagnant water is the most serious of Beijing, Hong Kong and Macao Expressway Beijing 17.5 km Nangang depression railway bridge, stagnant water sections of about 900 meters, with an average depth of 4 m, maximum depth of 6 meters. Stagnant water under the bridge, more than 20 million cubic meters, water lead to some of the vehicles were flooded, Beijing, Hong Kong and Macao high-speed Changyang section of the road traffic disruption. In the East Second Ring Road the Guangqumen bridge, the maximum water depth of 2 meters. Drainage system in the loop and loop around the region more vulnerable to serious loss of life of the people. Therefore, from the point of view of urban planning, the planning and construction of the loop to see him the pros and cons; further enhance its drainage capacity needs has been opened to traffic loop loop for new planning and construction in the building early is necessary to ensure that the drainage capacity. In addition, in the urban development in the future, for the planning and construction of the loop, we must be cautious.

Haidian District drainage system is more frangible than other district. Is evident from the distribution of water points and water points in Haidian District, presents the central tendency. 96 water points in Beijing, Haidian District, more than 30, accounting for about a third, to become the hardest hit by heavy rain in Beijing. Rainfall from the southeast to the northwest showing an increasing trend, Haidian District, located in the northwest of the city, the rainfall is too large; another party, Haidian District, Beijing Urban Construction earlier, its infrastructure building are quite outdated. Recent years, despite the construction

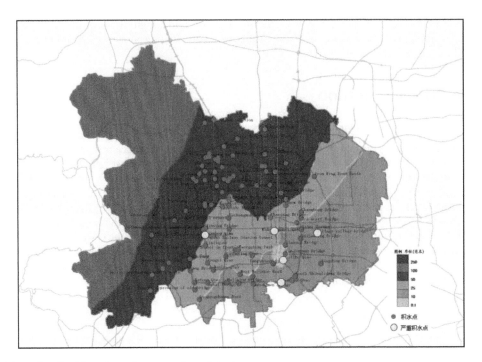

Figure 2. Stacking chart of water and precipitation.

of the city's ground in the growing urban underground infrastructure systems and not with the development of the city's updated, resulting in the urban drainage system is unable to meet the ever-accelerating pace of urbanization, and the serious consequences caused by the rainstorm waterlogging.

Drainage facilities need to be further strengthened and enhanced. In fact, the Beijing storm, presents a problem, also give the city a wake-up call. City flood capacity problems, not just in the capital. Judging from the present situation, waterlogging is a common and serious with each passing day. On one hand, in recent years, our extreme severe weather emergency, multiple, some places appeared many large scale strong rainfall, flood disaster increased range, to the flood control and waterlogging prevention stress; on the other hand, as the city with the rapid urbanization, city area of rapid expansion, heavy and light underground problem exists City, rainstorm and flood control in flood control system become a weakness.

4.2 *Discussion*

Some water conservancy scholars believe that, in many cities, waterlogging standard flood control standards are not unified, the waterlogging standard below the flood protection standards. Reduction of drainage standards, but also with the process of urbanization is almost synchronized. Urban expansion process of the construction of a new town in some big cities as well as small and medium-sized cities, some builders seem quick success, lack of overall planning, investment in water conservancy facilities are also inadequate. Sometimes, more focus on the image of the surface of the city, and not placed enough emphasis on the lining project. Many cities are a case of heavy rain on the people "see the sea" reason. No doubt, include waterlogging system construction, including, on the one hand, to increase investment, on the other hand, properly handle the relationship between ground engineering and underground engineering, immediate and long-term interests of the. Each torrential rains are waterlogging engineering inspection the waterlogging system as a basic project, you need to stand up to the necessary test. In this paper, the urban water points in the rainstorm context data, analysis vulnerability of the urban drainage system, a reflection of the existing problems in the planning and construction of urban drainage systems. In this study, can be learned from the reality the real problem of the urban drainage system exists to provide basis for decision making, planning and improvement of the drainage system for urban planning. Should get firsthand disaster loss data by several post-disaster field surveys, combined with previous research, to build a variety of land use types of economic losses vulnerability equation, at the same time learn from foreign countries against storm surge floods population casualties—drowned depth relationship research results, a more comprehensive study of the vulnerability of urban drainage systems.

REFERENCES

Downing, TE. 1991b. Vulnerability to Hunger and Coping with Climate Change in Africa. *Global Environmental Change* 5(1): 365–438.

Fang Guoliang, Xie Yiyang. 2009. The urban storm flood warning system research in Shanghai. *Atmospheric Research and Application* 6(02): 32–41.

He Minglei. 2012. Control of urban storm waterlogging disasters based on low-impact development techniques. *Construction and Design* (05): 135–137.

Li He, Zhang Pingyu & Cheng Yeqing. 2008.The concept of vulnerability and its evaluation method. *Progress in Geography* 27(02): 18–25.

Li Xia, Yan Xuemei. 2012. Municipal drainage system to deal with stormwater measures. *Chinese high-tech enterprises* (13): 56–58.

Liu Tao. 2011. The problems and countermeasures of the construction of drainage system in urbanization process—Taking Beijing as an example. *Economist* (10): 20,23.

Luo Linan, Li Hongquan & Zhang Xiliang. 2012. Urban storm the waterlogging warning forecast system development in Huzhou. *Zhejiang Meteorological* 33(01): 31–35.

Ma Fengbi. 2009. Beijing urban flood control and drainage system problems and countermeasures. *The Beijing Waterworks* (05): 6–8.

Ma Junhua, Li Jingfei. 2012. Solution to the Problem on Urban Drainage System Overflow in Rainy Season with Storm Water Management Model (SWMM). *Water purification technology* 31(03): 10–15,19.

Tan Qiong, Zhang Jianpin & Shi Zhenbao. 2012. Improvement measures of storm water drainage system in Shanghai. *Water Supply and Sewerage* (03): 35–38.

Wang Zongmin, Zhang Jie & Zhao Hongling. 2012. Selection of the Interpolations of Rainfall for Calculating the Water Logging Disasters Caused by the Urban Storms. *Yellow River* 34(08): 24–26.

Yang Dong. 2010. The Chengdu city rainstorm waterlogging forecasting and early warning systems research and development of GIS-based. *University of Electronic Science and Technology.*

Yin Zhane, Yin Jie & Xu Shiyuan. 2011. The city rainstorm waterlogging scenario simulation and disaster risk assessment (in English). *Journal of Geographical Sciences* 21(02): 274–284.

Zhang Ling, Chen Xiaohong & Wang Zhaoli. 2011. Vulnerability of the flood control system of fuzzy the maximum entropy diagnosis. *Systems engineering theory and practice* 31(08): 1006–1607.

Zhao Zicheng, Yu Huaqian. 2012. City dump heavy rain drainage network nodes at the stagnant water quantity study. *Urban Roads Bridges & Flood* (02): 57–58,80.

Zhou Yuwen, Weng Yaoyao & Li Ji. 2012. Urban storm intensity formula deducing system development. *China Water and Wastewater* 28(02): 25–28.

Zhou Yuwen, Weng Yaoyao & Zhang Xiaoxin. 2011. The application of annual maximum method for deducing the urban storm strength formula. *Water Supply and Sewerage* 37(10): 40–44.

Zhou Yuwen, Yao Shuanglong & Weng Yaoyao. 2012. Urban storm intensity formula data sampling a new method. *China Water and Wastewater* 28(06): 9–12.

Zhu Zheng, Zheng Bohong & He Qingyun. 2011. The impact of urban storm disasters and Countermeasures—A Case Study of Changsha. *Journal of Natural Disasters* 20(03): 105–112.

Hydraulic Engineering – Xie (Ed.)
© 2013 Taylor & Francis Group, London, ISBN 978-1-138-00043-8

Study on the reservoirs ecological operation in Shule River irrigation area

Ai Xueshan
State Key Laboratory of Water Resources and Hydropower Engineering Science, Wuhan University, Wuhan, China

Samuel Sandoval-Solis
University of California, Davis, Department of Land, Air and Water Resources, Davis, California, USA

Gao Zhiyun
Hubei Urban Construction Vocational and Technological College, Wuhan, Hubei, China

ABSTRACT: According to the idea of reservoir ecological operation, this paper describes an integrated human-environmental reservoir operation model including three reservoirs in Shule river irrigational area. The model maximize irrigation assuring rate and economic incomes as objects and subject to the demand for irrigation, industry and drinking water systems, hydropower, environment and transport of sediments from the reservoirs. The platform used is c++ by using Object Oriented Programming. Under the condition of a natural runoff, the result of the model shows that the model is effective and practical.

1 INTRODUCTION

Reservoir operation can be flexible, the difference in the economic benefits may be subtle at different scheduling in some ways, but changing the hydrology of the river may cause significant impact on the environment. Thus, we should design reservoir operations that support a certain level of economic benefits and maintain the environmental health of the riparian ecosystem (Fang, 1984). Integrated water resources management consider the operation of water systems for different purposes (flood control, power generation, water supply, transportation, etc.), taking into account the impact of these economic activities on the environment and providing suitable water quantity and quality for the riparian ecosystem, so that it can remain in an adequate ecological state (Wang, 2006). Reservoir operations for anthropogenic and environmental purpose are becoming more important nowadays.

2 THE GENERAL SITUATION OF RESERCH AREA

2.1 *The base information*

Shule River situates at the western end of the Hexi Corridor, the area of river basin is about 41,300 km², the average annual runoff is 1.031 billion m³, the average annual precipitation is less than 50 mm, and the capacity of evaporation up to 3,200 mm. It is a typical mainland arid desert climate (Ma, 2007). Shule River irrigation area lays in the middle reaches of the Shule River region, located in the Hexi Corridor in Yumen and Anxi city, Gansu Province, China.

The Shule irrigation area is the largest agricultural cultivation base of Gansu Province. It has 300 km long from east to west, 100 km width from north to south (Liu, 2006); it plays an important supporting role to the local socio-economic development in the region (Ma, 2006).

The irrigation area of Changma is 33,343 ha, that of Twin-Towers is 23,071 ha and Huahai is 8,727 ha. Of the three reservoirs, Changma and Twin-Towers reservoirs lie on the Shule Mainstream, and Red Gold Gap reservoir (RGG reservoir) lies on the mainstream of adjacent Oil river valley, shown on the Figure 1 (Dong, 1993).

2.2　Water demands

2.2.1　Irrigation
Agricultural Irrigation is the largest water use in the basin; a reliable water supply on-time is needed for this category. The irrigated area for each crop is considered fixed and known. Another consideration is a canal utilization factor of 0.62 and the irrigation water process.

2.2.2　Industrial and municipal use
The mining area of Gansu province is the major industrial user in this region, long-term planning for water is 80 million m³ per year.

2.2.3　Hydropower
There are 16 small hydropower stations (powerhouse, main and west canal) with 9.8 MW total installed capacities in Changma reservoir, the design hydropower capacity is 250 million kWh. These stations generate electricity based on the irrigation schedule; the quantity and variation of flow (in the channels) have a direct impact on the efficiency and generated electrical energy. So providing an adequate steady flow, as much as possible, can maximize the efficiency of the hydropower stations.

2.2.4　Environment
The amount of water dedicated to protect the environmental is 150 million m³ of the total annual average runoff of the basin, which is 1.032 billion m³.

2.2.5　Sediment control
During the flood season, from July 1 to July 31, Changma reservoir is operated at dead water level to flush out the sediments (mostly silt) accumulated in the reservoir.

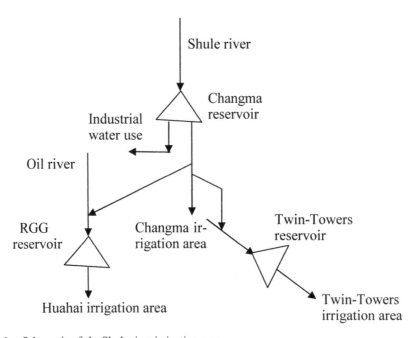

Figure 1.　Schematic of the Shule river irrigation area.

3 THE ECOLOGICAL OPERATION MODEL OF SHULE RIVER IRRIGATION AREA

3.1 Objective functions

The main task of reservoirs is to improve the irrigation guaranteed rate, and keep the diversions of Changma reservoir as steady as possible (for hydropower), while meeting the environmental flow requirements. Therefore, the objective functions can be defined as follows:

1. The highest irrigation guarantee rate

$$\max \sum_{i=1}^{I} \lambda_i p_i \tag{1}$$

where

$$p_i = K_i/T, \quad K_i = \sum_{t=1}^{T} k_{i,t}, \quad k_{i,t} = \begin{cases} 1, qout_i(t) \geq qx_i(t) \\ 0, otherwise \end{cases}$$

λ_i is the weight of irrigation guarantee rate of the ith reservoir; p_i is the irrigation guarantee; T is the number of total time span; $qout_i(t)$ is the runoff from reservoir of the ith reservoir the tth time span, $10^4 m^3$; $qx_i(t)$ is water demand of the ith reservoir the tth time span, $10^4 m^3$; I is the total number of irrigation reservoir.

2. The largest power generation efficiency of the hydropower station:
   The outflow of Changma reservoir should be the maximum of the minimum discharge flow.

$$\max_m \{\min_{t_m} O_i(t_m)\} \quad t_m = t_{m1} \sim t_{mk}, m = 1 \sim M$$

where m is the number of the dead water level, M is the max number of the dead water level, t_m is the variable of time span of the mth dead water level to the next dead water level of the reservoir, t_{m1}, t_{mk} is the mth time span number of the initiation and the termination from the empty to the next empty of the reservoir, $O_i(t_m)$ is the t_m runoff of the reservoir of the ith reservoir the mth from the empty to the next empty of the reservoir in $10^4 m^3$.

3.2 Restrictive conditions

1. Water balance (continuity equation)

$$V_i(t+1) = V_i(t) + Qin_i(t) - O_i(t) \quad i = 1,2,3, \quad t = 1,2, \ldots\ldots T \tag{2}$$

where $V_i(t)$, $V_i(t+1)$ is the begin and the end capacity of the tth time span the ith reservoir in $10^4 m^3$; $Qin_i(t)$ is the inflow water of the tth time span the ith reservoir in $10^4 m^3$, $O_i(t)$ is the outflow water of the tth time span the ith reservoir in $10^4 m^3$.

2. Capacity constraints

$$0 \leq V_i(t) \leq Vn_i \quad i = 1,2,3, \quad t = 1,2, \ldots\ldots, T+1. \tag{3}$$

where, $V_i(t) = 0$ $i = 1,2$ $t = 1,2,3,4, T+1$ (on July and the end of June Changma and RGG reservoirs running on the dead water level), Vn_i is the available storage of the ith reservoir in $10^4 m^3$.

3. Downstream environmental water constraints

$$Wst_i \geq WST_i \tag{4}$$

where Wst_i is the diverted water for the environment from the ith reservoir in $10^4 m^3$, WST_i is the environmental water requirement of the ith reservoir in $10^4 m^3$.

4. The cycle of environmental water constraints:

Environmental flows should be released in accordance with the sequence of natural multi-year average inflow water supply, while meeting the minimum environmental water constraints. The downstream environmental water constraint is an overall indicator, when there is an increase in the released water at a certain period to meet the minimum requirements for the environment, this water should be subtracted from the abundance period.

• To calculate the ecological water-sharing ratio during each time of:

$$k_i(t) = QA_i(t) / \sum_{t=1}^{T} QA_i(t) \qquad (5)$$

where $QA_i(t)$ is the reservoir storage of the t-th period, the i-th reservoir in the calculative representation year in $10^4 m^3$, $k_i(t)$ is the environmental section coefficient of water sharing of the tth time span the ith reservoir.

• Prorated ecological water demand in various periods

This function balance the reservoir releases for environmental purposes in two ways. First, if the reservoir releases for environmental purposes in a period are below the minimum environmental water requirements, then this function supply water according to the lowest environmental water requirements. Second, if the reservoir releases for environmental purposes in a period are above the minimum environmental water requirements, this function reduce the water releases of the corresponding period, and maintain the total environmental water requirements unchanged.

$$q_{st,i}(t) = WST_i \times k_i(t)$$

$$\left.\begin{array}{l} \max\limits_{t} q_{st,i}(t) = \max\limits_{t} q_{st,i}(t) - (\underline{q_{st_i}}(t) - q_{st,i}(t)) \\ q_{st,i}(t) = \underline{q_{st_i}}(t) \end{array}\right\} \quad if \quad q_{st,i}(t) < \underline{q_{st_i}}(t) \qquad (6)$$

where, $q_{st,i}(t)$ is the distributed environmental water demand of the tth time span the ith reservoir in $10^4 m^3 \cdot q_{st,i}(t)$ is the minimize environmental water demand of the tth time span the ith reservoir in $10^4 m^3$, $\max q_{st,i}(t)$ is the time span of the maximize environmental water demand of the ith reservoir in $10^4 m^3$.

5. Non-negative constraints: All variables should be non-negative.

4 MODEL SOLUTION

4.1 Calculation of the water demand process

1. Calculate downstream process of the RGG and the Twin-Towers reservoirs:
The water demand of each reservoir is calculated by the following formula

$$qx_i(t) = q_{st,i}(t) + q_{gg,i}(t), \quad i = 2,3 \qquad (7)$$

where $qx_i(t)$ is the downstream water demand of the tth time span the ith reservoir in $10^4 m^3$, $q_{gg,i}(t)$ is the downstream irrigation water demand of the tth time span the ith reservoir in $10^4 m^3$.

2. Calibrate water demand for RGG and Twin-Towers from Changma reservoir:
In accordance with the principle of continuity and the strategy of streamline operations, this consideration calculates the water transfer processes from the outside.

174

$$\left.\begin{array}{l} qd_i(t)= qx_i(t)-V_i(t)-Qin_i(t) \\ V_i(t+1)= 0 \end{array}\right\} \quad while \quad qx_i(t)-V_i(t)-Qin_i(t) > 0$$

$$\left.\begin{array}{l} qd_i(t)= 0 \\ V_i(t+1)= V_i(t)+ Qin_i(t)- qx_i(t) \end{array}\right\} \quad others \qquad i = 2,3 \tag{8}$$

where $qd_i(t)$ is supplementary water demand of the tth time span the ith reservoir in $10^4 m^3$.

3. Calculation on the downstream water need process of Changma reservoir

$$qx_i(t)= q_{st,i}(t)+ q_{gg,i}(t)+ q_{gy,i}(t)+ \sum_{j=2}^{3} qd_{j,i}(t), \quad i = 1 \tag{9}$$

where $q_{gy,i}(t)$ is industry water demand of the tth time span the ith reservoir in $10^4 m^3$.

4.2 *Identify the process of water supply*

1. Calculation process of uniform releases from Changma reservoir:
   First, releases from the Changma reservoir are calculated in accordance with the simplify operation strategy.
   Second, evaluate whether there is an abundant water period or not. If there is an abundant water period, then allocate the abundant water to the period before when water is scarce and re-calculate the level process of the reservoir. After several iterations, it is possible to obtain a relatively uniform reservoir releases.
   Finally, distribute water to all users according to the water demand process of downstream of Changma reservoir. If there is an excess of water beyond the channel water capacity, then distribute only the channel capacity and the rest of the water spill into the lower reaches.
2. Calculate uniform releases from RGG and the Twin-Towers reservoirs:
   After water distribution of Changma reservoir, a water allocation system has been defined to RGG and Twin-Towers reservoirs. This allocation depends on to the operation of Changma, these reservoirs will release water as uniform as possible.
3. Calculate the irrigation guaranteed rate and interactive adjustment:
   According to the reservoirs' uniform releases process, calculate the irrigation guaranteed rate. If necessary, adjustment the outflows of the reservoirs, the proportion of RGG and Twin-Towers reservoirs, or other key coefficient parameters manually, and then recalculated.

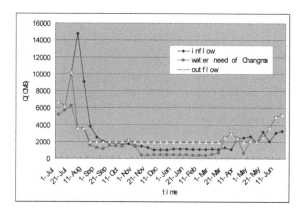

Figure 2. The flow of Changma reservoir.

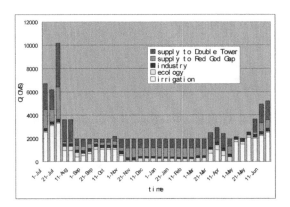

Figure 3.   The water use of Changma outflow.

5   RESULTS AND ANALYSIS

According to the previous approach, an optimization model was built using Visual C++ plat-form. It was developed the corresponding constraints and functions under the idea of object-oriented programming. When the allocation ratio between RGG and Twin-Towers reservoirs was 50% of the water supply, compared to the overflow water from Changma reservoir, the storage of Changma are shown in Figures 2 and 3.

6   CONCLUSION

The scarcity of water resources in western China is particularly challenging, how to distribute the limited water resources on scientific and rational basis has become an important research and practical topic. In addition, there is a need for scientific guidance to support engineer-ing measures, integrated water resources management will be the main strategy used to solve competing water resources systems. The three reservoirs in the Shule river irrigation area provide a case of study where integrated water resources management can be implemented, through the joint operation of these three reservoirs. A simulation-optimization model was developed in Visual C++ platform using adopted object-oriented programming. The model is a holistic integration between optimization and simulation techniques. This model can support the decision making process for the Irrigation area, water regulations can be derived from this optimization-simulation techniques to address decision makers concerns. The model plays a greater role in arid and semi-arid areas with limited water resources.

REFERENCES

Dong Feng. 1993. Shule River Basin water resources simulation and optimization scheduling and the scope of the project demonstrated. Gansu water conservancy and hydropower technology 9:6–13.
Fang Ziyun. 1984. Tan Peilun. the preliminary study on the reservoir operation in order to improve the ecological environment. Yangtze River 6:65–67.
Liu Qiang. 2006. Research on the Shule River Irrigation District Information System. China rural water and hydropower 8:18–19.
Ma Dehai. 2007. Experiment of Salt Leaching Ration of Newly Reclaimed Salinity Land in the Shule River Irrigation District. China rural water and hydropower 7:22–24,27.
Ma Dehai. 2006. Information Construction on Shule River irrigation area of Gansu Province. Water Resources Development Research 3:44–47.
Wang Kai. 2006. Wang Shucheng talked about the ecological scheduling in China Institute of Water Resources at the 2006 academic year. China Water Resources News/2006/November/14/001 version.

Hydraulic Engineering – Xie (Ed.)
© *2013 Taylor & Francis Group, London, ISBN 978-1-138-00043-8*

A simple approach for tunnel face stability in soft ground

Lv Gao-Feng
Beijing MTR Construction Administration Corp., Beijing, P.R. China

Jia Cang-Qin, Huang Qi-Wu & Liu Jiao
School of Engineering and Technology, China University of Geosciences, Beijing, P.R. China

ABSTRACT: Tunnel construction processes more and more frequently involve full-face excavations. Therefore engineers have to analyze the tunnel face stability, and to design adequate counter- measurements. This paper describes a new kinematical approach, called upper bound method, enabling to quickly estimate face support pressure. This method can be used to estimate a tunnel face safety factor, and thus to easily manage the tunneling construction. Critical pressures determined for failure model are compared with published analytical solutions and laboratory test results.

1   INTRODUCTION

Tunneling has been widely used in the construction of subway and sewage systems in China so as to avoid contributing to the congestion of traffic in streets. The stability of tunnel face is one of the most critical components that should be secured for the successful tunneling. It is especially true for the tunneling in urban environment and even more when large diameters are contemplated, where the excessive settlement and ground deformation may lead to catastrophic and costly consequences.

Many researchers and engineers have successfully presented various theoretical and empirical/experimental methods to evaluate the tunnel face stability and to assess the required face support pressure. The analytical approaches may be used to assess the face support pressure, but they do not provide any information about surface settlement and face deformation characteristics. Currently, only a three-dimensional numerical analysis is in a position to provide complete information on face stability, required face support pressure and ground deformation and subsidence. However, an attempt to evaluate the required face support pressure using a series of numerical simulations was not made yet due to the restriction on available time and resources. In this paper, using a three-dimensional failure mode and kinematical theorem technique, the face stability was evaluated and equations that evaluate the face support pressure necessary to avoid excessive deformation of the ground near the tunnel heading were developed.

During tunnel construction, soil is removed from the tunnel face. The soil layer in front and above the tunnel face exerts active earth pressure. The presence of infrastructures or surcharge also contributes as additional earth pressure. For the tunnel alignment below the groundwater table, water pressure is the another significant component of pressure acting at the tunnel face (in Fig. 1).

For stability, the layers of soil at the tunnel face should have sufficient strength to balance these forces. In many projects, tunnels will encounter several layers of loose soil or weathered rock. The face may not be strong enough to bear such pressures or may be unstable. Therefore, the soil mass in front of the cutterhead can collapse which would then result in excessive settlement at the surface. Support pressure needs to be built up at the face of tunnel, to counterbalance the pressure generated by the soil, water and overlying infrastructures.

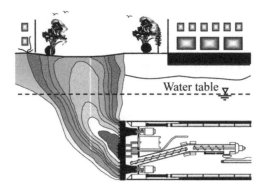

Figure 1. Failure mechanism at tunnel face.

This pressure is known as support pressure. Sometimes, even with stable geology, face support pressure needs to be built up in order to prevent the inflow of water into the tunnel. A decrease in the groundwater level may result in consolidation and thereby surface settlement.

In cases of mechanised tunnelling, support mediums will be used to build the required face support pressure. Common support mediums used are bentonite slurry, earth paste, and compressed air. Choosing a support medium depends on various factors, a few of which are properties of soil and the type of TBM used. There are some adverse effects to applying excessive support pressure as it may lead to surface heave and ground distortion. Inadequate support pressure may cause ground settlement. Therefore, an adequate range of face support pressure is needed to stabilize the face, which in turn will minimize settlement, avoid ground collapse, prevent ground heaving and allow for reliable advance of the TBM (EPB).

2 FAILURE MECHANISM

2.1 *Combined falling and sliding block system at tunnel face*

This collapse model is formed by an upper block bearing over a lower wedge block that slides towards tunnel face as shown in Figure 2.

The upper block mechanism derives from a two dimensional upper bound solution presented by Atkinson & Potts (1977). Their solution is based on a work rate calculation of a falling wedge with upper planes inclined by an angle of ϕ with vertical, where ϕ is the soil dilatancy angle. Hence the work done by friction is null. This solution of the theory of plasticity implies an associated flow rule $\varphi = \phi$, where ϕ is the soil friction angle.

This solution was modified to allow for:

a. Three dimensional blocks. The upper planes delimitating this block are at an angle v with vertical planes that contain their lower edges.
b. Cohesive forces acting on the block upper surfaces. The work done by cohesion is given by $C \times \cos(v) \times \delta$ where C is the resultant cohesive force acting on that plane, and δ is the virtual vertical displacement.
c. Water pressure resultant force, acting on the block. This force may be evaluated by studying a flow net around the tunnel face and integrating the resulting pore pressure acting on the block upper surfaces.

The top edge of the upper block may be perpendicular or parallel to the tunnel axis, depending on the ratio between its width (B) and depth (A or A′) as shown in Figure 3.

For the condition:

$$A \leq B \times Tan\left(\frac{\pi}{4} - \frac{\phi}{2}\right) \tag{1}$$

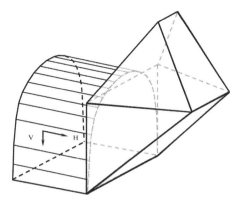

Figure 2.   Combined block mechanism at tunnel face.

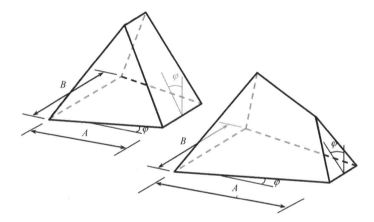

Figure 3.   Typical upper block shapes $\varphi = \phi > 0$.

The block weight is given by

$$W_t = \frac{\gamma}{12Tan\phi} \times \left[3A^2B - A^3 \times \left(2Tan\phi + Tan\left(\frac{\pi}{4} + \frac{\phi}{2} \right) \right) \right] \qquad (2)$$

where γ is the soil natural unit weight and $\varphi = \phi > 0$ is the soil dilatancy angle.
    For the condition:

$$\frac{B}{2Tan\phi} \geq A' \geq B \times Tan\left(\frac{\pi}{4} - \frac{\phi}{2} \right) \qquad (3)$$

the upper block weight is given by:

$$W_t = \frac{\gamma}{12Tan\phi} \times \left[\begin{array}{c} \frac{1}{2}B^3 \times Tan\left(\frac{\pi}{4} - \frac{\phi}{2} \right) \times Cot\phi \\ -\frac{1}{2}(B - 2A'Tan\phi)^3 \times Tan\left(\frac{\pi}{4} + \frac{\phi}{2} \right) \times Cot\phi \end{array} \right] \qquad (4)$$

It is always necessary to ensure that the upper block does not intercept ground surface, otherwise the above formulae are no longer valid.

179

The vertical component of the cohesive force is given by:

$$C_t = C \times \cos(\nu) = \frac{c \times A}{Tan\phi} \times [B - A Tan\phi] \tag{5}$$

valid for both block shapes (A or A'), $\varphi = \phi > 0$, and

$$A(A') \leq \frac{B}{2Tan\phi} \tag{6}$$

For $\varphi = \phi = 0$

$$C_t = c \times 2 \times (A + B) \times H \tag{7}$$

The work rate calculation of the falling upper block holds the same equation as in the corresponding limit equilibrium analysis:

$$F_t = (W_t - C_t + U_t) \tag{8}$$

where $F_t \geq 0$ is the supporting force acting on the upper block base, and U_t is the vertical component of the resultant water thrust acting on the block.

A work rate calculation is performed for the lower wedge block, which also holds the same equation as in the corresponding limit equilibrium analysis. The geometry of this block is shown in Figure 4. The lateral faces of this block are at an angle $\varphi = \phi$ with tunnel main axis. The block sliding base is inclined at an angle α with vertical.

The forces acting on the block are:

- weight W_f,
- reaction to the upper block supporting force $F_t \geq 0$,
- forces resulting from cohesion c acting on the sliding base and lateral surfaces,
- forces resulting from water pressure acting on the sliding base, U_b, and on lateral surfaces, U_l, respectively,
- resultant thrust acting on the tunnel face, E.

The weight of the lower wedge block is given by:

$$W_f = \frac{1}{6} \times \gamma \times H^2 \times Tan\alpha \times [3B - 2H \times Tan\alpha \times Tan\phi] \tag{9}$$

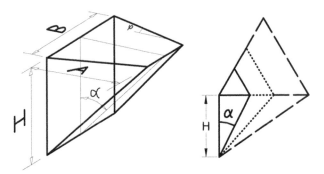

Figure 4.   Lower wedge block geometry and influence of angle α on its depth and on upper block height.

The resultant thrust E is given by:

$$E = \frac{\left(F_t + W_f\right)}{Tan\left(\alpha + \phi\right)} + \frac{U_b \times Sin\phi}{Sin\left(\alpha + \phi\right)} + \left(2U_l \times Sin\phi\right)$$
$$- c \times H \times \frac{B \times Cos\phi + H \times Sin\alpha \times \left(Cos\phi - \dfrac{Sin\phi}{Cos\alpha}\right)}{Cos\alpha \times Sin\left(\alpha + \phi\right)} \qquad (10)$$

2.2 Combined upper and lower wedge block mechanism

The above derived equations were implemented in an electronic calculation sheet. The equivalent required support pressure at tunnel face, p_{eq}, can then be evaluated by dividing the maximum thrust by the lower wedge face area:

$$p_{eq} = \frac{E}{\left(H \times B\right)} \qquad (11)$$

The resultant thrust E is then calculated for α angles varying in 1° steps, which leads to the maximum thrust value, as shown in the graph of Figure 5.

2.3 Limitations

a. The combined block mechanism could be considered as a quasi-upper bound solution, since the virtual displacement of the upper block is vertical, and that of the lower wedge block has a horizontal component. The work done at the contact face between the two blocks is neglected. Therefore this mechanism should be merely considered as a solution that satisfies equilibrium in a limited way.
b. A series of soil modeling limitations apply to the above analysis, including:
   – the models are only applicable to a continuum medium, as opposed to a jointed medium.
   – the combined block mechanism is based on planar surfaces, whereas observed failure surfaces tend to be curved. Also the lower wedge block face is rectangular, therefore not geometrically matching circular tunnel faces.
c. A consequence of the previous considerations is that it cannot be ensured in any of the above solutions that collapse load will always be approached from a safe or unsafe side (i.e. the solutions presented in this paper are not lower or upper bound solutions).

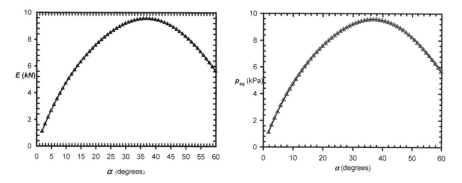

Figure 5.   Graphical example of resulting thrust and support pressure acting on the tunnel face, as function of angle α.

181

Table 1. Critical pressures for Mohr Coulomb material.

		Critical pressure/kPa			
		Lower bound	Upper bound		Present
C/D	γ (kN/m³)	Leca & dormieux	Leca & dormieux	Centrifuge tests	analysis
1.0	15.3	29	2	6	4.5
1.0	16.1	29	3	3	5.5
2.0	15.3	46	2	4	4.5
2.0	16.1	44	3	4	5.5

Table 2. Comparison between the present results and those of the centrifuge tests.

C/D	γ (kN/m³)	σ_t (kPa) (centrifuge tests)	p_{eq} (kPa) (present analysis)
2	15.3	4.4	3.5
2	16.1	4.1	4.6
4	26.47	16.6	17
6	26.47	20.3	19.5

3 COMPARISON WITH PUBLISHED MODELS AND LABORATORY TEST RESULTS

3.1 Circular tunnel heading in cohesive and frictional Mohr Coulomb material

Leca & Dormieux (1990) presented upper and lower bound solutions applicable to cohesive and frictional material and compared their solution with laboratory centrifuge tests based on a rigid cylinder with an exposed face, carried out in Nantes, France.

Their comparative table is reproduced in Table 1, to where results from the combined block system were added. Two materials were tested: $\gamma = 15.3$ kN/m³, $c' = 2.3$ kPa, $\phi' = 35.2°$, and $\gamma = 16.1$ kN/m³, $c' = 1.1$ kPa, $\phi' = 38.3°$, for two conditions of cover to diameter ratio, as indicated in Table 1. It is interesting to note that support pressure in both solutions are relatively close to each other, and excepting the first case, also close to centrifuge test results.

It should be noted that the effects of groundwater percolation forces were not simulated in these tests, as well as in the other published laboratory tests found by the Author.

3.2 Circular tunnel heading in a purely frictional material

Centrifuge tests have been carried out in Nantes (France) to study the face stability of tunnels in case of collapse. Two soil conditions were examined including a loose sand and a dense sand. Shear tests on these soils has shown that $c = 2.3$ kPa, $\varphi = 35.2°$ for the loose sand, and $c = 1.1$ kPa, $\varphi = 38.3°$ for the dense sand.

The results obtained by Chambon and Corte (1989) are presented in Table 2 and compared to those given by the present analysis. Note that the σ_t given by the present analysis is calculated from direct maximization of this pressure. As we can see, there is good agreement between experimental and theoretical results.

4 PROPOSAL OF A SIMPLE METHOD TO EVALUATE THE FACE STABILITY

Mihalis et al. (2001) suggested the use of the Tunnel Stability Factor (TSF) for the estimation of the stability behavior of underground openings in weak soil conditions. The tunnel stability factor combines all the above factors and it is possible to consist a significant parameter for the initial estimation of tunnel cross-sections. The inclusion of size (equivalent diameter D) of the underground openings in TSF, results from the practical experience that in similar

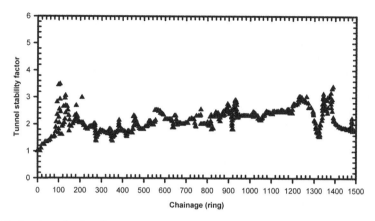

Figure 6.   Evaluation of safety factor of tunnel face stability.

geotechnical conditions and at the same depth, tunnels of different size exhibit modes of deformational behavior of different scale and degree of criticality.

In order to simplify the evaluation procedure, Figure 6 shows schematic drawing of the site conditions based on the assumptions including actual shield machine. It can also observed from Figure 6 that the shield tunnel has constructed safely during the tunnel advance maintaining Fs = 3, except for the start and end of the tunnel. It, therefore, could be said that the tunnel face stability is generally assured during tunnel excavation. It, however, is assumed that the start of the tunnel had a lower tunnel face stability compared to other sections because of the process of installing and settling the shield machine.

This result is a qualitatively analyzed case study of the tunnelling situation using theoretical approach through field data of the completed shield tunnel. Since the case study ignored site conditions during tunnelling and only used the measured tunnel face pressure, comprehensive analysis of detailed site conditions and ground behavior measurements are required additionally.

5   CONCLUSION

The analytical stability models presented in this paper are intended to enable a quick preliminary assessment of tunnel face stability conditions, in three dimensions and considering the effects of water pressure. Comparisons with published analytical solutions, and few divulged laboratory tests, indicate reasonably good agreement. However no tests were found where the effect of groundwater was simulated, and most tests were modeled in plane strain conditions.

This paper presents a study of the tunnel face behaviour during shield tunnel construction, concentrating on the study of the short-term stability of the tunnel face in soft ground. The design concept and the strong influence of construction procedures are demonstrated and evaluated based on the predicted and measured tunnel face behaviour during shield tunnelling. An important feature of these case studies is to investigate the influence of tunnel face behaviour due to soil type, tunnel size, and construction procedure.

According to the results presented in the paper, the future programme to be developed technically is suggested in order to provide a practical knowledge for soft ground tunnelling in China.

REFERENCES

Atkinson, J.H., Potts, D.M. 1977. Stability of a shallow circular tunnel in cohesionless soil. *Geotechnique* 27(2): 203–215.

Chambon, P., Corté, J.F. 1994. Shallow tunnels in cohesionless soil: Stability of tunnel face. *Journal of Geotechnical Engineering* 120(7):1148–1165.

Davis, E.H., Gunn, M.J., Mair, R.J., Seneviratne, H.N. 1980. The stability of shallow tunnels and underground openings in cohesive material. *Geotechnique* 30(4): 397–416.

Leca, E., Dormieux, L. 1990. Upper and Lower bound Solutions for the Face Stability of Shallow Circular Tunnels in Frictional Material. *Geotechnique* 40(4): 581–606.

Mühlhaus, H.B. 1985. Lower bound solutions for circular tunnels in two and three dimensions. *Rock Mechanics and Rock Engineering* 37–52.

Mashimo, H., Suzuki, M. 1998. Stability conditions of tunnel face in sandy ground. *Centrifuge '98*, Balkema, Rotterdam.

Mihalis I.K., Kavvadas M.J. & Anagnostopoulos A.G. 2001. Tunnel Stability Factor-A new parameter for weak rock tunnelling. *Proc. 15th Inter. Conference on Soil Mechanics and Geotechnical Engineering*, 2, Istanbul, Turkey, 1403–1406.

Sozio L.E. Analytical Stability Models for Tunnels in Soil. In Kwast E.A. (ed.), *Geotechnical Aspects of Underground Construction in Soft Ground; Proceedings of the 5th International Symposium TC28*. Amsterdam, 15–17 June 2005, Netherlands: Taylor & Francis.

Hydraulic Engineering – Xie (Ed.)
© 2013 Taylor & Francis Group, London, ISBN 978-1-138-00043-8

Novel computational implementations for stability analysis of rock masses

Jia Cang-Qin, Huang Qi-Wu, Wang Gui-He & Chen Bo
School of Engineering and Technology, China University of Geosciences, Beijing, China

ABSTRACT: By exploring the nature of the analogy between optimum trusses and optimum layouts of discontinuities, a novel numerical analysis method for rock masses is proposed in this paper. The procedure is used to determine the critical layout of discontinuities and associated upper bound limit analysis for stability problems. Potential discontinuities which interlink nodes laid out across the problem domain are permitted to crossover one another, giving a much wider search space than when such discontinuities are located only at the edges of finite elements of fixed topology. Highly efficient SOCP (second-order cone programming) solvers can be employed when certain popular failure criteria are specified (e.g. Hoek-Brown and Mohr-Coulomb). Stress/velocity singularities are automatically identified and visual interpretation of the output is straightforward. Several numerical examples including rock slope are studied by the new method, and the results are very close to those calculated by using analytical method and FEM.

1 INTRODUCTION

The progress of numerical limit analysis strongly relies on the development of both discretisation methods and mathematical programming techniques. Over the past four decades, many numerical solution methods for limit analysis problems have been developed, and in parallel significant progress has been made in developing powerful numerical analysis and optimisation techniques.

Computational limit analysis has become a powerful tool for analyzing the stability of problems in soil mechanics, and a huge amount of work has been done in the field over last few decades. Lysmer (1970) originally proposed a numerical procedure using finite elements and linear programming to compute lower bound limit loads in soil mechanics. In the paper, the soil mass was discretised into a number of 3-noded triangular elements, the nodal stresses being the unknowns, and in contrast to standard finite element formulation each node was unique to a particular element. This meant that more than one node could share the same coordinate, a key feature of discontinuous finite element methods.

Together with the development of finite element technology and mathematical programming algorithms, limit analysis techniques for geotechnical problems have been developed by many researchers over the last two decades. Significant work in the field has been carried out by Sloan and his collaborators, i.e. Sloan (1988, 1989); Yu & Sloan (1994); Lyamin & Sloan (2002) and Zhao *et al.* (2007). In Sloan(1989), an upper bound analysis was carried out using constant strain triangles and the finite meshes were arranged in a specific manner to avoid volumetric locking problems. Yu *et al.* (1994) used linear strain triangles with straight edges. In this paper, the shearing direction between elements was specified in advance in order to avoid locking problems. More recently, Makrodimopoulos & Martin (2006) developed simplex strain elements for use in combination with second-order cone programming algorithms to apply to upper bound limit analysis problems. Two and three dimensional bearing capacity of footings in sand problem have recently been investigated by Lyamin *et al.* (2007). Unfortunately, the solutions

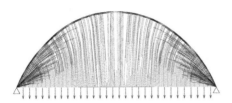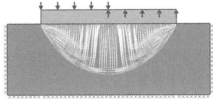

Figure 1.   Optimum layout of truss bars compared to the arrangement of slip-lines for geo-structures.

obtained using finite element limit analysis are often highly sensitive to the geometry of the original finite element mesh, particularly in the region of stress or velocity singularities. And also in order to improve the efficiency it have to consider the identifying the discontinuities which form at failure.

As an alternative, a truly discontinuous model called Discontinuity Topology Optimization (DTO) was proposed. In fact, a successful discontinuous limit analysis procedure is able to identify the critical arrangement of discontinuities in a problem from a wide, preferably near infinite, number of possibilities.

The problem is thus similar to the problem of identifying the optimum layout of grid-like structures. Figure 1 illustrates the analogy between a truss layout optimization and a strip footing bearing capacity problem in geo-mechanics. The DTO procedure can overcome both the volumetric locking and stress/velocity singularity limitations of finite element limit analysis. Apart from application in the determination of critical yield line patterns in concrete slabs, and identification of critical slip-line patterns in metal forming problems, the DTO numerical analysis procedure is potentially applicable to geo-mechanics problems, such as stability analysis.

2   ANALOGY BETWEEN STRUCTURAL OPTIMIZATION AND NUMERICAL LIMIT ANALYSIS

It is known that the optimum layout of bars in 'Michell' trusses mirrors the arrangement of slip-line discontinuities in plane strain metal plasticity problems; the geometry of both turn out to be Hencky-Prandtl nets, forming orthogonal curvilinear coordinate systems.

Figure 1 provides a visual comparison of the optimum layout of bars needed to carry a spatially fixed uniformly distributed load between pinned supports together with the slip-lines at failure identified when modelling a rotating foundation resting on a clay soil. Note that the lines in Figure 1(a) represent truss bars, whereas the lines in Figure 1(b) represent slip-lines.

Note that all these numerical solutions contain a dense set of inter-node connections, representing truss bars or slip-lines. In order to show the solutions clearly, connections with near-zero bar force or slip displacement must be filtered out. This can occasionally result in apparently anomalous solutions, characterized by missing connection lines. However, this is simply because some inter-connecting lines have been filtered out. The same layout optimization technique used for truss optimization problems can be applied to limit analysis problems. Smith and Gilbert (2007) demonstrated this, referring to the new method for solving limit analysis problems as discontinuity layout optimization.

2.1   *Mathematical formulation*

Smith and Gilbert (2007) define the kinematic DLO formulation for the plane strain analysis of a quasi-statically loaded, perfectly plastic cohesive body, discretized using m nodal connections (slip–line discontinuities), n nodes and a single load case as:

$$\min E = \mathbf{g}^T\mathbf{d}$$

$$\text{Subject to:}\ \begin{matrix}\mathbf{Bd} = \mathbf{u}\\ \mathbf{d} \geq \mathbf{0}\end{matrix} \tag{1}$$

where: E is the total internal energy dissipated due to shearing along the discontinuities;

$$\mathbf{d}^T = \{s_1^+, s_1^-, s_2^+, s_2^-, \ldots, s_m^-\}$$

s_i^+, s_i^- are the relative shear displacement jumps between blocks along discontinuity i ($i = 1 \ldots m$), which are the LP variables;

$$\mathbf{g}^T = \{c_1 l_1, c_1 l_1, c_2 l_2, c_2 l_2, \ldots, c_m l_m\}$$

l_i and c_i are respectively the length and associated cohesive shear strength of discontinuities i;

B is a suitable ($2n \times 2m$) compatibility matrix;

$$\mathbf{u}^T = \{u_1^x, u_1^y, u_2^x, u_2^y, \ldots, u_n^y\}$$

u_j^x and u_j^y are the x and y components of the (virtual) displacement imposed at node j ($j = 1 \ldots n$).

The classical primal LP formulation is concerned with finding the minimum volume of a loaded truss structure with prescribed boundary conditions:

$$\min V = \mathbf{q}^T\mathbf{c}$$

$$\text{Subject to:}\ \begin{matrix}\mathbf{Bq} = \mathbf{f}\\ \mathbf{q} \geq \mathbf{0}\end{matrix} \tag{2}$$

where: V is the total volume of the structure

$$\mathbf{c}^T = \{l_1/\sigma_1^+, -l_1/\sigma_1^-, l_2/\sigma_2^+, -l_2/\sigma_2^-, \ldots, -l_m/\sigma_2^-\}$$

$$\mathbf{q}^T = \{q_1^+, -q_1^-, q_2^+, -q_2^-, \ldots, -q_m^-\}$$

B is a suitable ($2n \times 2m$) quilibrium matrix;

$$\mathbf{f}^T = \{f_1^x, f_1^y, f_2^x, f_2^y, \ldots, f_n^y\}$$

where l_i, q_i^+, q_i^-, σ_i^+, σ_i^- represent respectively the length and tensile and compressive member forces and stresses in member i, and finally, f_j^x, f_j^y are x and y load components applied to node j. Using the formulation of Dorn et al. (1964) the linear programming problem variables are clearly the member tension and compression forces q_i^+, q_i^- in **q**, and an LP solver can be used to minimize the volume of the structure, while satisfying equilibrium constraints at all nodes.

The dual formulation can be derived from the primal using duality principles; the dual problem turns out to involve maximizing the virtual work done by the applied loads, subject to members having prescribed virtual strain limits:

$$\max W = \mathbf{f}^T\mathbf{u}$$

$$\text{Subject to: } \mathbf{B}^T\mathbf{u} \leq \mathbf{c} \tag{3}$$

Table 1. Analogy between limit state analysis and design topology optimization problem.

	Limit analysis	Truss topology optimization
Objective function	Minimum work, corresponds to maximum collapse factor as in equation	Minimum weight or cost, corresponds to maximum virtual work as in equation
Variables	Discontinuity displacements	Internal forces in truss bars equilibrium
Constraints	Compatibility	Equilibrium
Results	Failure mechanism (hodograph)	Optimum layout of truss members (Maxwell force diagram)

where $u^T = \{u_1^x, u_1^y, u_2^x, u_2^y, \ldots, u_n^y\}$ and u_j^x, u_j^y are the x and y direction virtual nodal displacement components, i.e. the objective is to maximize the virtual work done by the given applied loads, subject to limits on the virtual member strains. In this case the variables are the nodal displacements u_j.

Comparison of equation (1) and (2) or the formulations it is shows that two problems are closely related and key features of the analogy are summarized in Table 1. In DTO the problem involves using optimization to identify the critical layout of discontinuities (rather than truss bars), together with the associated upper bound load factor (rather than structural volume). This is achieved by minimizing the total internal energy dissipation, constrained by compatibility conditions (analogous to minimizing the volume of a truss structure subject to equilibrium constraints). Note that for convenience the terms energy dissipation and displacement are used here as shorthand for rate of energy dissipation and displacement rate, respectively.

2.2 *Discontinuity topology optimization (DTO)*

The DTO procedure can be posed in the form of a mathematical optimization problem, as follows,

$$\min \lambda f_L^T d = -f_D^T d + g^T p$$

$$\text{Subject to: } \begin{array}{l} Bd = 0 \\ Np - d = 0 \\ f_L^T d = 1 \\ p \geq 0 \end{array} \tag{4}$$

where λ is the unknown load factor at collapse; f_d and f_L are vectors containing respectively dead and live loads at discontinuities; d contains displacements along the discontinuities, $d^T = \{s_1, n_1, s_2, n_2, \ldots, n_m\}$, where s_i and n_i are the relative shear and normal displacements between blocks at discontinuities i; p is a vector of plastic multipliers, and g contains the corresponding dissipation coefficients. B is a suitable compatibility matrix containing direction cosines, and N is a suitable flow matrix. A key feature of the formulation is that compatibility is enforced at the nodes. The discontinuity displacement in d and the plastic multipliers in p are the variables in the optimization problem, which can be solved using linear programming or second order cone programming when a Mohr-Coulomb failure criterion is used. The solutions obtained will represent upper bounds on the true plastic collapse load, with accuracy being controlled by the number of nodes used to discretise the problem.

3 ALGORITHM

The ground structure is the union of all potential members, first proposed by Dorn et al. For a problem comprising n joints the ground structure will consist of $m = n(n - 1)/2$ potential members (including overlaps). The optimization solver will search for the best combination of these members to obtain the overall lowest structural volume. This is equivalent to searching for the optimum structure from a total of *2m* potential structures. Obviously, most of them are unstable or insufficient to carry the load, but checking all possible solutions by direct enumeration is clearly not an achievable goal. In contrast, using classical gradient based methods such problems can be handled easily.

The main problem with the traditional ground structure method is the large number of potential members, and hence problem variables, which are involved, making the problem very computationally expensive to solve. For a long time it was widely—but wrongly assumed that the only way to achieve a provably optimum solution was to solve the full problem from the outset, as there 'were no rigorous methods of introducing new members during optimization procedure'. This has changed recently, as it has been found to be possible to solve a smaller problem containing fewer potential members initially, and then adding other members iteratively, using a so called 'adaptive member adding' procedure.

In the adaptive member adding technique a minimally connected, initial base structure, is used in the first iteration; this structure should at least:

1. connect all external loads sufficiently to supports in order to allow a stable structure to be identified by the solver;
2. fix all Degree of Freedom (DOF) in order to ensure nodal displacements in the dual problem can all be computed. If one or more DOFs were not constrained in the initial structure then the displacements will be unknown at that iteration, and the 'adaptive member adding' algorithm will add members to constrain these DOFs before other more optimum members in the next iteration. If the members were not sufficient to constrain all DOFs then other iterations will be needed until all DOFs are constrained. This will not lead to premature termination of the algorithm, but additional iterations will be required to calculate all displacements.

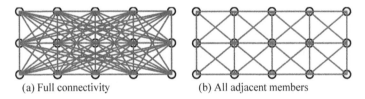

(a) Full connectivity   (b) All adjacent members

Figure 2. Base structures with different connectivities in 2D.

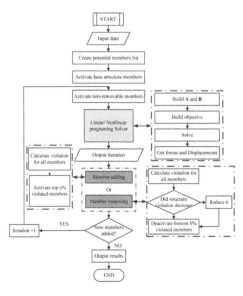

Figure 3. Adaptive member refinement flowchart.

189

For uniform nodal grids, a base structure could include only connections to adjacent nodes, i.e. short members (without overlap) in both Cartesian directions together with diagonals, giving a total number of members in 2D of approx. $4n$, where n is the number of nodes as shown in Figure 2(b).

One of the main advantages of using the adaptive member adding procedure is making the number of members in the base structure m a multiplier of n, instead of being a multiplier of n^2 or n^3 in 2D and 3D respectively.

The base structure determines the size of the problem in the first iteration, and can influence the number of iterations needed before a final solution is reached. Gilbert and Tyas (2003) mentioned using different initial nodal connectivity strategies. It was mentioned that when a base structure with inadequate base connectivity is used, several iterations might needlessly be required at the beginning of the procedure to constrain DOFs (if these were inadequately constrained at the outset).

4 VALIDATION AND APPLICATION

To verify the accuracy of the solutions obtainable using the procedure, it will now be initially applied to bearing capacity problems from the literature. And then the procedure will be applied to more complex problems involving slopes with non-convex geometries. To facilitate the efficient analysis the following strategy has been adopted: (a) discontinuities that cross over the boundary of any region are disallowed; (b) increased numbers of nodes are positioned along all boundaries, helping to compensate for the absence of such disallowed discontinuities.

4.1 *Load capacity of shallow foundations on Mohr-Coulomb materials*

The new procedure was used to determine the ultimate bearing capacity of soil using low nodal density (the target number of node, 250). A symmetry plane was defined along one vertical edge and a rigid footing placed adjacent to this along top surface of the soil. Results are present in Table 2. From Table 2 it is evident that reasonable approximations of the bearing pressure were obtained even when a low nodal density was used. It is also clear that, as well as permitting larger problems to be solved in the first place so as to reduce memory and time requirements, the adaptive nodal connection procedure also permits solutions to small problems to be obtained more quickly. At last, as expected it is found that the accuracy of the solution degrades with increasing angle of friction. This is partly a consequence of the larger extent of the failure mechanism in a soil with high angle of friction, which means that the important zone below the footing becomes increasingly sparsely populated with nodes, when the nodal density is uniform throughout the problem domain.

4.2 *Slope stability*

Figure 5 illustrates sample 3D slope stability solutions using DTO that can be used to investigate the non-convex geometry stability problem. The stability of the overhang subject to its self-weight is analysed. And the target number of nodal density is medium, 500.

The procedure has used to model other types of blocky problems, the weathered rock mass shown in Figure 6, where a foundation is to be placed at the top left of a slope, close to the location of an existing tunnel. When the domain geometry is concave inter connecting

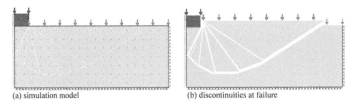

(a) simulation model   (b) discontinuities at failure

Figure 4.   Simple vertically loaded footing analysis.

Table 2. Results for bearing capacity on soil with or without surcharge problem (undrained footings).

Simulation	Benchmark	Results by DTO	Discrepancy on collapse load (%)
Undrained footing (weightless soil)	5.14	5.2	1.16
Undrained footing with surcharge 1 kPa (weightless soil)	6.14	6.2	0.97
Undrained footing	5.14	5.2	1.16
Undrained footing with surcharge 1 kPa	6.14	6.2	0.97

Table 3. Results for bearing capacity on weightless soil with surcharge 1 kPa (drained footings).

Simulation	Benchmark	Results by DTO	Discrepancy on collapse load (%)
$\phi' = 20$	6.4	6.52	1.84
$\phi' = 30$	18.4	18.99	3.12
$\phi' = 40$	64.2	67.75	5.25

Table 4. The parameters used in numerical analysis.

Type	Shear strength c' (kPa)	Angle of shearing resistance ϕ' (degrees)	Shear strength c_u (kPa)	Unit weight γ (kN/m^3)
Layer 1	0	30	50	20
Layer 2	2	18	30	19
Layer 3	5	25	150	22

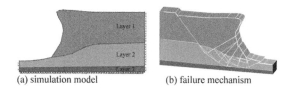

(a) simulation model   (b) failure mechanism

Figure 5. Overhang stability analysis (FOS = 0.998).

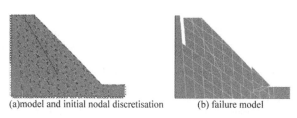

(a)model and initial nodal discretisation   (b) failure model

Figure 6. Overhang stability analysis (FOS = 6.148).

Table 5. The parameters used in numerical analysis.

Type	Shear strength c' (kPa)	Angle of shearing resistance ϕ' (degrees)	Shear strength c_u (kPa)	Unit weight γ (kN/m^3)
Rock	50	45	1000	24
Fissure*	25	26.5	500	24
Weak fissure*	15	16.7	300	24

*Mohr-Coulomb with no-tension cutoff.

all nodes within a concave domain with potential connections can be memory and CPU expensive. This is because checks need to be made to ensure each connection lies entirely within the domain. It is wise that the concave part can be split into several convex parts.

5 CONCLUSION

The real novelty of the DTO procedure lies in the expression of the limit analysis problem formulation entirely in terms of layout of discontinuities, rather than in terms of arrangement of elements. Using the adaptive refinement scheme recently developed for discontinuity topology optimization problems can be extended to enable internode connections to be removed as well as added. Provably optimum solutions can be obtained in a reasonable time on a standard PC even for large problem.

Motivated by the need of simple, yet robust, approaches to solve this problem for practical, often large-scale structures, we attempt to take advantage of the increased availability of advanced and powerful algorithm.

This paper has been primarily concerned with solving the upper bound method. Useful extensions of the present work, made possible by the positive conclusions reached regarding the adaptive member adding or removel. From the computational viewpoint, it would be worthwhile to carry out more extensive testing of the algorithm on similar and other problem types, and to investigate use of the more efficient DTO formulation, coupled with some robust and efficient search strategy.

ACKNOWLEDGEMENTS

This work was financially supported by the National Natural Science Foundation (40902085), and the Fundamental Research Funds for the Central Universities (2010ZY39, 2011YYL039).

REFERENCES

Dorn, W.S. et al. 1964. Automatic design of optimal structures. *J. de Mechanique* 3(1): 25–52.
Gilbert, M. & Tyas, A. 2003. Layout optimization of large-scale pin-jointed frames. *Engineering Computations* 20(8): 1044–1064.
Lyamin, A. et al. 2007. Two and three dimensional bearing capacity of footings in sand. *Geotechnique* 57(8): 647–662.
Lyamin, A.V. & Sloan, S.W. 2002. Upper bound limit analysis using linear finite elements and nonlinear programming. *International Journal for Numerical and Analytical Methods in Geomechanics* 26(2): 181–216.
Lysmer, J. 1970. Limit analysis of plane problems in soil mechanics. *Journal of the Soil Mechanics and Foundations Division* 96(4): 1311–1334.
Makrodimopoulos, A. & Martin, C.M. 2006. Upper bound limit analysis using simplex strain elements and second-order cone programming. *International Journal for Numerical and Analytical Methods in Geomechanics* 31(6): 835–865.
Sloan, S.W. 1988. Lower bound limit analysis using finite elements and linear programming. *International Journal for Numerical and Analytical Methods in Geomechanics* 12, 61–67.
Sloan, S.W. 1989. Upper bound limit analysis using finite elements and linear programming. *International Journal for Numerical and Analytical Methods in Geomechanics* 13, 263–282.
Smith, C.C. & Gilbert, M. 2007. Application of Discontinuity Layout Optimization to Plane Plasticity Problems. *Proceedings of the Royal Society A: Mathematical, Physical and Engineering Sciences* 463 (2086): 2461–2484.
Yu, H.S. & Sloan, S.W. 1994. Limit analysis of anisotropic soils using finite elements and linear programming. *Mechanics Research Communications* 21(6): 545–554.
Zhao, J.D. et al. 2007. Limit theorems for gradient-dependent elastoplastic geomaterials. *International Journal of Solids and Structures* 44(2): 480–506.

The 2nd SREE workshop on environment and safety engineering

Hydraulic Engineering – Xie (Ed.)
© 2013 Taylor & Francis Group, London, ISBN 978-1-138-00043-8

Study on smoke movement with semi-transverse smoke extraction in subway tunnel fire

Yanyan Lu, Fang Liu & Gang Li
Faculty of Urban Construction and Environmental Engineering, Chongqing University, Chongqing, China

Shujiang Liao
Chongqing Public Security Fire Brigade, Chongqing, China

ABSTRACT: The smoke movement with the semi-transverse smoke extraction in subway tunnel fire has been simulated by Fire Dynamics Simulation software (FDS). The comparisons of the smoke flow and temperature field with the different semi-transverse smoke extraction schemes have been conducted with the different opening smoke vent numbers, smoke vent distances and smoke vent areas. It has been shown from the study that smoke temperature and concentration on the height of 1.8 meters were within the standard scope with the semi-transverse smoke extraction, which was safety to evacuate people; it was a good method for the smoke extraction to open the smoke vents near the fire and the smoke vent area had little effect on smoke extraction.

1 INTRODUCTION

With the economic development and accelerated urbanization, the urban population increases constantly. Traffic blocking has become an urgent urban problem to be solved. Subway, for its large traffic volume, fast, punctual, safety, low energy consumption, easy traffic, and comfort, has been widely used around the world. However, due to it is buried deep underground, closed-environment, poor ventilation, and during the construction and operation of subway, along with fire, storm, flood, earthquake explosion and other disasters, the tunnel fire accident is the most serious (Xie, Q. & Chen, H. 2004, Niu, M.P. 2007, Zhao, C.J. 2007, Hu, L.H. 2006). Toxic and high temperature smoke is the main factor to threaten the human lives, so that the smoke control in tunnel fire is the most critical thing (Alarie, Y. 2002, Han, X. & Cui, L.M. 2002). Natural and mechanical smoke exhaust are common ways to exhaust smoke, and mechanical smoke exhaust includes longitudinal smoke exhaust, semi-transverse smoke exhaust and transverse smoke exhaust, the natural smoke extraction effect is poor, the longitudinal smoke exhaust effect is good, and the semi-transverse and transverse smoke exhaust have better effects, while the transverse smoke exhaust has high engineering cost (Huang, Y.D. et al. 2012, Wang, B.B. 2011, Yang, H. et al. 2009, Roh, J.S. et al. 2009, Park, W.H. et al. 2006). Currently, there are few studies of the semi-transverse smoke exhaust, but how to set up a smoke vent of semi-transverse smoke exhaust system, the specification has not yet provided, so the study on the setup of the smoke outlet and the distance between the exhaust ports with the semi-transverse smoke exhaust is especially significant (Yi, L. et al. 2011, Li, Y.Z. et al. 2009, Zhang, Y.C. et al. 2009, Xu, L. & Zhang, X. 2008, Liang, Y. 2007). In this study, Fire Dynamics Software (FDS) has been employed to simulate the smoke flow situation in subway tunnel under semi-transverse ventilation mode. The comparison of the smoke extraction effect with the semi-transverse smoke extraction schemes has been conducted with the different opening smoke vent numbers, distances and areas.

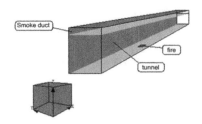

Figure 1. Physical model of tunnel.

Table 1. Numerical simulation cases.

Case	Fire power (MW)	Fire location	Smoke exhaust volume (m³/s)	Opening smoke vent situation		
				Number	Distance (m)	Single area (m²)
1	10	Center of tunnel	60	5	10	4
2	10	Center of tunnel	60	5	15	4
3	10	Center of tunnel	60	7	10	4
4	10	Center of tunnel	60	7	15	4
5	10	Center of tunnel	60	9	10	4
6	10	Center of tunnel	60	9	15	4
7	10	Center of tunnel	60	5	10	6

2 NUMERICAL SIMULATION PROGRAMS

2.1 Physical model of tunnel

A subway tunnel in this study was a prototype tunnel in Chongqing. It was simplified into a rectangular section as a physical model, which was 150 m length and with the section size of 4.8 m × 7.0 m. There was a smoke duct at the top of the tunnel whose section size was 4.8 m × 1.6 m. A smoke exhaust fan was set on the right side of the smoke duct. The smoke extraction rate of tunnel section is more than 2 m/s and less than 11 m/s in case of single hole in national code "Subway Design Specifications" (GB50157-2003) of our country. According to the requirements above, smoke extraction rate has been taken as 2.3 m/s and the volume of smoke extraction has been calculated as 60 m³/s in this study. Irrespective of the impact of the outdoor environment, the left longitudinal wind speed of the tunnel was 0 m/s. The fire size was 2.0 m × 2.0 m, the fire power was 10 MW, which was in the center of the model tunnel. As shown in Figure 1.

Fire Dynamics Simulation software FDS5.0 has been used to do simulation, the grid number of three directions including length, width, and height for the tunnel were 750, 24 and 27 respectively, while for the smoke duct were 375, 12 and 4 respectively.

2.2 Numerical simulation cases

In order to obtain the best smoke extraction scheme from different smoke vent setting programs, 7 cases have been simulated as shown in Table 1. Among all the simulations, single smoke vent areas were 4 m² and 6 m², respectively, and one smoke vent was located just above the fire, while the others were distributed symmetrically on both sides of the smoke vent. The smoke vent distances were 10 m and 15 m, respectively.

3 ANALYSIS OF THE SIMULATION RESULTS

3.1 Analysis of the steady smoke flow

According to the study, Figure 2 has shown that some smoke flew into the smoke ducts, the others were still in the tunnel when the smoke got stable. The smoke spread to the tunnel

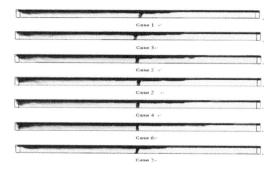

Figure 2.   Smoke spread distance under every case.

entrance toward the left side of the fire, and to the rightmost smoke vent toward the right side of the fire. Because the exhaust fan was set on the right side of the smoke duct, the closer the distance between the smoke vent and the exhaust fan, the greater the speed of smoke exhaust, which was more conducive to extract the smoke from the exhaust port into the smoke duct. The reason why the smoke couldn't be completely discharged into the smoke duct was that the smoke exhaust rate of the smoke vent was not great enough.

By comparing the first 6 cases, the effect of the opening smoke vents number and smoke vent distance on smoke spread distance and smoke exhaust effect could be obtained. When the smoke exhaust opening number and the smoke vent distance were less, the smoke spread distance was shorter and smoke layer was thinner. Comparing the simulations of case 1 with case 7, the results were almost same, which has shown that the smoke vent area had little impact on the smoke extraction effect. The smoke extraction effects of case 1 and 7 were best, which were most beneficial for evacuation and rescue. This was because when a fire occurred, due to the buoyancy effect, the hot smoke would first rise up to the bottom of the smoke duct clapboard above the fire, then flew to the both ends of the tunnel. The smoke extraction rates of smoke vents had pumping effect on the hot smoke, while the smoke extraction rate of the smoke vent was the largest on the far right of the fire and the pumping effect on the smoke was also the largest. So the smoke generally spread to the far right smoke vent toward the right side of the fire. It has indicated that it could achieve satisfactory result by opening the smoke vents near the fire, while it was not conducive to exhaust smoke with too many opening smoke vents.

3.2   *Analysis of the temperature field*

The temperature distribution of the center longitudinal profile of the tunnel below smoke duct clapboard and on the height of 1.8 m under various cases has been shown in Figure 3. It has indicated that the smoke temperature was the highest in the fire place and it was relatively low in other positions. This was because the mixing of the smoke with cold air above the fire was less, while the other locations was more. Below the smoke duct, the smoke temperature on the left side of the fire was higher than that on the right side of the fire. The reason was that the smoke on the left of the fire was relatively more compared to the right side, and the smoke layer was rather thicker. It has also shown that in addition to the place of fire, the smoke temperature below the smoke duct was higher than that on the height of 1.8 m, which has illustrated that the temperature above the smoke layer was higher than below.

Comparing the temperature of the center longitudinal profile of the tunnel below smoke duct clapboard under each case shown in Figure 3a and 3b, it could be found that opening smoke vents program had little impact on the smoke temperature on the left side below the smoke duct clapboard. From the results of the first 6 cases, it has shown that the effect of smoke vent number and smoke vent distance on the smoke temperature on the right side below smoke duct clapboard was obvious. Figure 3b has shown that the temperature difference between case 1 and 7 was small and almost zero, which has illustrated that smoke

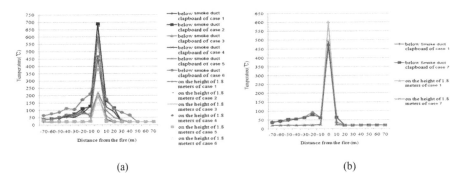

(a)   (b)

Figure 3.   Temperature distribution of the center longitudinal profile of the tunnel under every case.

vent area had little effect on smoke temperature below smoke duct clapboard. The smoke temperatures on the right side of the fire under case 1 and 7 were lower, which has shown that the two cases had better smoke extraction effect.

The temperature distribution of the center longitudinal profile of the tunnel on the height of 1.8 m under variety of cases in Figure 3a and 3b has shown that the smoke temperature distributes symmetrically on both sides of the fire in all simulation cases, in addition, the temperatures were all about 20°C. On the one hand, it has indicated that the opening smoke vent number, distance and area had little effect on the smoke temperature on the height of 1.8 m. On the other hand, it has proved that semi-transverse smoke extraction was a good smoke exhaust method for evacuation and fire rescue.

4   CONCLUSIONS

Fire Dynamics Simulation software (FDS) has been employed to simulate the subway tunnel fire smoke movement under the semi-transverse smoke extraction. According to the comparisons of the smoke flow and temperature field with different opening smoke vent numbers, smoke vent distances and smoke vent areas, the following conclusions could be obtained:

1. It has been shown that smoke temperature on the height of 1.8 meters under the semi-transverse smoke extraction was low, which was good for evacuation and fire rescue.
2. It was a good method for the smoke extraction to open smoke vents near the fire in the semi-transverse smoke extraction schemes, and the smoke vent area had little effect on smoke extraction effective. But more researches were needed.

In this study, the simulation fire was in the center of the tunnel, while the actual fire location was random, in other words, in the actual project smoke vents were set according to the characteristics of tunnels. It was a good method for the smoke extraction to open smoke vents near the fire instead of opening all the smoke vents, but the best distance between the vent and fire was not discussed. The tunnel also had a certain slope, which also had a certain impact on the semi-transverse smoke extraction, so these issues need to be further studied.

ACKNOWLEDGEMENT

The authors would like to thank Ministry of Science and Technology, Chongqing, China for their financial support to this work through the following projects: "Research on Subway Tunnel Smoke Control" (Project No. CSTC2009BB6193).

REFERENCES

Alarie, Y. 2002. Toxicity of Fire Smoke. *Critical Reviews in Toxicology* 32(4):259–289.

Han, X. & Cui, L.M. 2002. Brief Introduction to Tunnel Fire Test Research at Home and Abroad. *Chinese Journal of Underground Space and Engineering* 32(4):259–289.

Hu, L.H. 2006. Study on thermophysical property of tunnel fire smoke spread. Hefei: Doctoral Dissertation of University of Science and Technology of China.

Huang, Y.D., Li, C. & Chang, N.K. 2012. A numerical analysis of the ventilation performance for different ventilation strategies in a subway tunnel. *Journal of Hydrodynamics* 24(2):193–201.

Li, Y.Z. & Yi, L. 2009. Numerical Study on Effect of Configuration of Exhaust Inlets on Semi-horizontal Extraction in Tunnel Fire. *China Public Security* (2):77–80.

Liang, Y. 2007. Numerical simulation of Cross-river tunnel fire under combined ventilation. Chengdu: Master Dissertation of Southwest Jiaotong University.

Niu, M.P. 2007. Study on subway tunnel fire smoke flow characteristic. Beijing: Master Dissertation of Beijing University of Technology.

Park, W.H., Kim, D.H. & Chang, H.C. 2006. Numerical predictions of smoke movement in a subway station under ventilation. *Tunnelling and Underground Space Technology* 21(3–4):304.

Roh, J.S., Ryou, H.S. & Park, W.H. 2009. CFD simulation and assessment of life safety in a subway train. *Tunnelling and Underground Space Technology* 24(4):447–453.

Wang, B.B. 2011. Comparative Research on FLUENT and FDS's Numerical Simulation of smoke Spread in Subway Platform Fire. *Procedia Engineering* (26):1065–1075.

Xie, Q. & Chen, H. 2004. Few Words on Distinguishing Features of Fire in Metro Line and Construction of Fireproof Projects. *Modern Urban Transit* (2):38–40.

Xu, L. & Zhang, X. 2008. Effect of Smoke outlet characteristics on smoke control for highway tunnel with Central Smoke extraction Systems. *Heating Ventilation and Air Conditioning* 38(3):76–79.

Yang, H., Jia, L. & Yang, L.X. 2009. Numerical analysis of tunnel thermal plume control using longitudinal ventilation. *Fire Safety Journal* 44(8):1067–1077.

Yi, L., Li, Y.Z. & Xu, Z.S. 2011. Experimental Study on Effect of Semi-horizontal Extraction on Tunnel Fire. *Journal of Disaster Prevention and Mitigation Engineering* 31(1):85–90.

Zhang, Y.C., He, C. & Zeng, Y.H. 2009. Study on Extra Long Highway Tunnel under Semi-transverse Smoke Extraction. *Highway Tunnel* (1):55–58.

Zhao, C.J. 2007. Study on subway tunnel fire smoke flow and ventilation type of Guangzhou. Guangzhou: Master Dissertation of Guangzhou University.

Hydraulic Engineering – Xie (Ed.)
© *2013 Taylor & Francis Group, London, ISBN 978-1-138-00043-8*

Risk analysis on extreme precipitation events over China based on peaks over threshold model

Xiaotian Gu
Key Laboratory of Environmental Change and Natural Disaster, Beijing Normal University, Beijing, China

Ning Li
Key Laboratory of Environmental Change and Natural Disaster, Beijing Normal University, Beijing, China
State Key Laboratory of Earth Surface Processes and Resource Ecology, Beijing Normal University, Beijing, China

ABSTRACT: Study on change of weather and climate extremes has become a hot-spot issue in recently climate change research. Based on the daily precipitation data from 752 stations during 1951 to 2010, this article uses the model of Peaks Over Threshold (POT) and Generalized Pareto Distribution (GPD) to analysis the thresholds of extreme precipitation event. According to the characteristics of the thickness of the tail that from POT model, and combining with the scale parameters, shape parameters and extreme precipitation return period that are reached by MCMC estimation methods, we can get the geographical spatial distribution of extreme precipitation risk intensity. So we must pay more attention to the disaster caused by precipitation in coastal areas.

1   INTRODUCTION

In recent years, the frequency and strength of extreme weather and climate events have changed a lot, it has a big impact for people's lives and economy and society. Extreme weather and climate events are one kind of minor probability events, and extreme precipitation event is a rare event. Due to the instability and complexity of the extreme precipitation event, different areas of the world will experience different problems from extreme precipitation event.

As the importance of climate change, the research about the extreme precipitation event has become the focus of attention. In the early 1900s, Fisher-Tippet pointed out that when the sample size is more enough, the asymptotic distribution of extreme values can obey one of three kinds limiting distributions (Fisher RA & Tippett LHC 1928). So, on the study of extreme precipitation, many scholars use distribution fitting methods to analysis the extreme precipitation. For example, Panmao Zhai et al, put forward the definition method of extreme precipitation extremum (Zhai Panmao & Pan Xiaohua 2003), Zhihong Jiang et al, (Jiang et al, 2007) used Gumbel extreme value distribution to fit the regional extreme precipitation based on Markov Chain Monte Carlo (MCMC) method, normal, quasi normality and skewness Gamma distribution models are used to fit the precipitation data in different periods by Cheng Bingyan et al, (Cheng et al, 2003).

In 1982, Embrechet (Embrechets & Paul, 1999) had analyzed the tail thickness of probability distribution, several common distribution types of tail situation are divided into three classes by them: thin tail, tail and thick tail. Through a large number of researches, scholars thought that the thick tail problem about extreme precipitation events has a lot of guidance for the future risk researches. Zhuo Zhi et al, (Zhuo et al, 2011) discussed the relative advantage of the POT model in fitting catastrophe thick tail risk in theory; Xie Qiang

(Xie Qiang, 2009) applied the **POT** model to fit the heavy rain loss data, the research showed that it is more reasonable to estimate the catastrophe risk loss distribution. But there have been few studies to examine the thick tail risk about extreme precipitation event. So, in this paper, combined with PBdH theorem, we choose generalized Pareto distribution to analysis the extreme precipitation event, and make use of GPD probability distribution parameters to study the thick tail risk problem about extreme precipitation event.

2 MATERIAL AND METHODS

2.1 Data source

The meteorological data sets used in this paper were daily precipitation of 752 stations in 1951–2010 in China, the material is provided by the National Climatic Data Center of China Meteorological Administration (CMA). Through the simple data quality control, 624 sites are chosen in this paper finally (Fig. 1).

2.2 Research methods

Generalized Pareto Distribution (GPD) is one of the theoretic distributions that are described the threshold value, it can screen the extreme values from the given sample time series to establish the extreme probability distribution through the given threshold. Compared with the other extreme value distributions (General Extreme Value Distribution (GEV) and Gumbel Distribition), the fitting effect of GPD model is better (Jiang et al, 2009).

The GPD distribution function over a threshold can be written as:

$$F(x) = 1 - \left[1 - k\left(\frac{x-\varepsilon}{\alpha}\right)\right]^{1/k}, \left(k \neq 0, \varepsilon \leq x \leq \frac{\alpha}{k}\right)$$ (1)

And the corresponding distribution density function (PDF) is:

$$f(x) = \left(\frac{1}{\alpha}\right)\left[1 - k\left(\frac{x-\varepsilon}{\alpha}\right)\right]^{1/k-1}$$ (2)

Among the formula, ε is the threshold value, α is the scale parameter, k is the shape parameter, x is the stochastic variable.

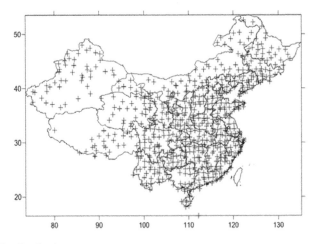

Figure 1. The site distribution.

202

In this article, the daily precipitation material is used to fitting by GPD model, so the threshold value is given according to certain weather conditions. Generally speaking, the choice of threshold value must accordance with the local climate conditions, in addition, the value of threshold will produce certain effect for the result of the research. We hope that the value of threshold is high enough, so that we can ensure that the extreme conform to GPD distribution that are chosen before, but the higher the threshold chosen, the less the sample got, then it may influence the accuracy of parameter estimation. Because in the condition of the smaller sample size, the reliability of maximum likelihood estimation method is not strong, parameter estimation results is instability, so in this paper, we use Markov Chain Monte Carlo (MCMC) method for Generalized Pareto Distribution parameter estimation.

3 PARAMETER ESTIMATION

3.1 *Thick tail detection*

In order to be able to determine the tail distribution of the precipitation in each station is thick tail type visually, so we draw histogram and Q-Q plot according to the precipitation, as shown in Figure 2. From the graphs, we can see that the rainfall distribution has obvious non-normality in Figure 2a, and compared with the optimal shape, the shape of all the precipitation samples looks more concave deviation, so Figure 2 shows that daily precipitation distribution has thick tail type.

3.2 *The threshold value and parameter estimation*

Smith proposed maximum likelihood method to estimate parameters of GPD distribution, and in the world of finance, the risk value based on the theory of extreme value is usually estimated by the maximum likelihood method. But Gui Wenlin et al, (Gui et al, 2010) thought that maximum likelihood estimation is effective only in large sample conditions. So take the size of samples into account, we choose MCMC parameter estimation method to analysis. Table 1 gives the parameters that are estimated by using different number samples, we found that, parameters that are estimated by MCMC method are still relatively stable with small samples.

In the meteorological study, 99 or 95 percentile value point is often chosen as the extreme threshold value. By analyzing the precipitation data in 624 sites, and combining with the requirements of POT model, the daily precipitation sample (more than 0 mm) in 1951–2010 is in ascending sort, and then we take the 31st value as the critical value, so the previous 30 data is chosen as the sample that we study, because 30 sample size relatively belongs to small sample, so this paper take MCMC estimation method to estimate parameters.

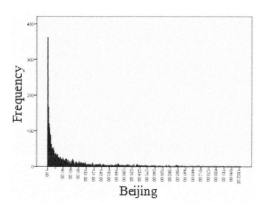
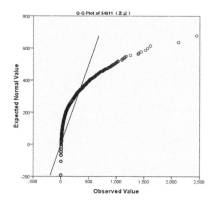

Figure 2.   Daily precipitation histogram (a) and daily precipitation Q-Q plot (b).

4 RESULTS ANALYSIS

4.1 *Spatial distribution of fitting parameters*

There is some relevant relationship between the tail thickness of GPD distribution and the shape parameter. The tail thickness of distribution belongs to "thick tail", and with the shape parameter increasing, the tail is becoming thick. The scale parameter is standard deviation linear function, so the bigger the scale parameter, the more unstable the extreme precipitation. The spatial distribution of parameters has great significance on the analysis of extreme precipitation event and the disaster risk caused by extreme precipitation.

Figure 3 is the spatial distribution of the GPD distribution scale parameter and shape parameter. From the analysis of Figure 3a, we can see that scale parameter distribution from the southeast to the northwest have obvious decrease trend in China, and the scale parameter in China's coastal areas is almost more than 20, so it shows that the extreme precipitation is not stable in these areas, the disaster risk caused by the precipitation is relatively big, that is means that we need to pay more attention; And in the most regions of Sichuan and Gansu, Tibet, Qinghai, Xinjiang and Inner Mongolia, the scale parameter is mostly under 10, and it shows that the extreme precipitation in these areas is relatively stable, the precipitation does not have a large mutation change; From the analysis of the shape parameter spatial distribution (Figure. 3b), compared with the distribution of scale parameter, the distribution of shape parameter has no obvious regional characteristics. If the absolute value of shape parameter is above 0.2, it means that the possibility of extreme precipitation events occurring is big.

4.2 *Spatial distribution of return period extreme*

Because of the differences of Regional climate characteristics, so different areas have different threshold values for extreme precipitation, and this leads directly to the adaptability to extreme precipitation in different areas is different, then, it is important for the future environmental security. So, using the relationship of return period and generalized Pareto distribution function, we can get different quantile that corresponding with return period, (Ding YuGuo):

$$X_t = \beta + \frac{\alpha}{k}\left[1 - \left(\frac{n}{M}T\right)^{-k}\right] \tag{3}$$

N for excess number, *M* for sample, *k* for critical value point.

By using this formula, the spatial distribution of extreme precipitation is given in Figure 4 (return period is 50 years and 100 years). Compared the two pictures (Figure 4a and b), there are similarities in the spatial distribution of the extreme precipitation value change in

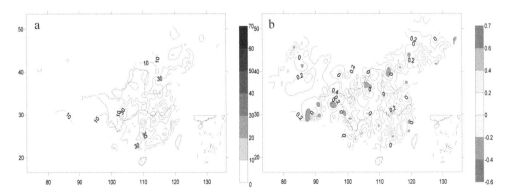

Figure 3.   Spatial distribution of scale parameters (a) and shape parameters (b) in China.

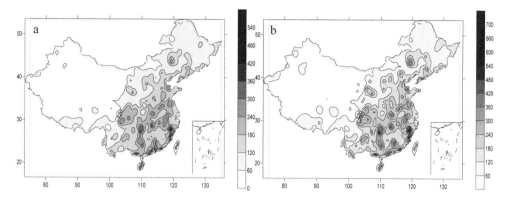

Figure 4. Spatial distribution of daily extreme precipitation in China.

50 years and 100 years. There has a tendency of increase from northwest to southeast coastal areas, strong precipitation distribution characteristics in different return period cases is corresponding with the scale parameter distribution that is shown in Figure 3a.

5 CONCLUSION AND DISCUSSION

This paper analyses the daily extreme precipitation over China by using POT model, and estimates the scale and shape parameter of GPD distribution by using MCMC method, then we can get the spatial distribution of the scale, shape parameter and extreme strong precipitation value under different return period. Based on analyzing the characteristics of extreme precipitation, we can understand China's disaster risk strength distribution. Then according to the analysis of the results, we can catch on to the degree of the environmental safety.

1. The size of GPD scale parameter and the stability of extreme precipitation are in high correlation, and scale parameter is decreasing from the southeast coastal to northwest inland. So that extreme precipitation in China's inland areas is fairly constant, and in coastal areas the contrary is the case.
2. Under Different return period conditions, China's strong precipitation extreme value distribution and the scale parameter spatial distribution are in high correlation, there is a increasing tendency from northwest to southeast coastal areas.

In conclusion, we must pay more attention to the disaster caused by precipitation in coastal areas.

ACKNOWLEDGMENT

This work was supported by "973 Program" of China (no. 2012CB955402), NSFC project (no. 41171401), National project (no. 105560GK), (2012DFG20170) and (no. 20100003110019). Cordial thanks should give to the editors and two anonymous reviewers for their constructive suggestions and comments that improved the manuscript.

REFERENCES

Cheng Bingyan, Ding Yuguo, Wang Fang. A diagnosis Method of the Extreme Features of Weather and Climate in Time Based on Non-Normal Distribution [J]. Chinese Journal of Atmospheric Sciences, 2003, 27(5):920–928.
Embrechets, Paul. Extreme Value Theory in Finance and Insurance. Department of Mathematics, 1999, ETH Zurich, Switzerland.

Fisher RA, Tippett LHC. Limiting forms of the frequency distribution of the largest or smallest member of a sample [J]. Proceedings of the Cambrigde Philosophical Society, 1928, 24(2):180–190.

Jiang Zhihong, Ding Yuguo, Cai Ming. Monte Carlo experiments on the sensitivity of future extreme rainfall to climate warming [J]. Acta Meteorologica Sinica, 2007, 67(2):272–279.

Jiang Zhihong, Ding Yuguo, Zhu Lianfang, et al. Extreme Precipitation Experimentation over Eastern China Based on Generalized Pareto Distribution [J]. Plateau Meteorology, 2009, 28(3):573–580.

Wenlin Gui, Zongyi Wang, Zhaozhou Han. The Choice of GPD Estimation and Risk Measure in POT Model [J]. Mathematics in Practice and Theory, 2010, 40(5):66–77.

Xie Qiang. The Application of POT Model in Loss Data: Based on MCMC Method [J]. Statistics & Information Forum, 2009, 25(2):84–87.

Zhai Panmao, Pan Xiaohua. Change in Extreme Temperature and Precipitation over Northern China During the Second Half of the 20th Century [J]. Acta Geographica Sinica, 2003, 58:1–10.

Zhuo Zhi, Wang Weizhe. Catastrophe Risk Thick Tail Distribution: POT Model and Application [J]. Insurance Studies, 2011, 8:13–19.

Hydraulic Engineering – Xie (Ed.)
© 2013 Taylor & Francis Group, London, ISBN 978-1-138-00043-8

Application of an anaerobic unit—combined constructed wetland system for swine wastewater treatment

Yin Chu, You-Hua Ma, Peng Zhu
School of Resources and Environment, Anhui Agricultural University, Hefei, China

Zhi-Qiang Hu
College of Mathematics and Information, Taizhou Teachers College, Taizhou, China

ABSTRACT: Animal wastewater has been a major contributor to water pollution in China. A two-stage anaerobic unit—combined constructed wetland system was applied to treat wastewater from a small confined piggery with 50 pigs. The combined constructed wetland unit included a Surface Flow (SF) cell followed by two series of Sub-Surface Flow (SSF) cells which were in horizontal baffling layout. The system demonstrated high treatment efficiency for organics, nitrogen and phosphorus: with the average removing rates being 96%, 97%, 75%, 82% and 88% for COD_{cr}, BOD_5, TN, NH_4–N and TP, respectively. The treated water can well meet the discharge standard of pollutants for livestock and poultry breeding (GB18596-2001) for organics and nitrogen. For phosphorus about 70% of the effluent samples could meet the criteria of TP. This system was proved to be cost effective and easy in maintenance and suitable for swine wastewater treatment in China.

1 INTRODUCTION

With the development of stock raising animal wastewater has been a major contributor to water pollution in China (Wang 2006). Swine wastewater is one of the most important sources of pollution because of the relatively large amount of pork consumption. Pre-survey research showed that effluent from piggery farms has been linked to downstream eutrophication of surface water system of Chaolake, which is the fifth largest fresh lake of China and is undergoing serious eutrophicaiton. The distribution of these confined piggery farms is characterized by large amount and small scale. One part of the piggery waste is applied to nearby field. The rest is usually disposed with very simple or without any treatment. Piggery waste control is considered important in pollution prevention of Chao lake and its water system according to the '12th-five-year plan of water pollution control of Chao Lake Basin' (2010).

Anaerobic and biogas facilities have been advocated by the local government to be applied in large scaled piggery farms (>500 pigs) by offering about a half to two-thirds of the construction investment. But the rest of construction cost and latter operation and maintaining cost are still too much for most of the farmers. So even though some of the farmers did build the facilities they would not pay to maintain it. What's more, wastewater treated only in this way cannot meet the emission criteria. It is also the case for very small-scaled anaerobic and/or biogas facilities which are always buried under the ground. For these piggery farms, practical wastewater treatment facilities in this region should be effective, low cost in construction and maintaining and easy to manage. The use of constructed wetlands for the treatment of wastewater is an approach that has received increasing attention over the last 30 years due to their innate ability to deal with a wide range of highly variable influent types (Harrington et al. 2012) and to its low cost and easy management (Kandasamy & Vigneswaran 2008, Zhang et al. 2009). This paper presented a specifically designed treatment system and its first application on the treatment of water from a small confined piggery was investigated.

2 MATERIAL AND METHODS

2.1 *The pilot treatment system*

A pilot on-site wastewater treatment system (Fig. 1), namely two-stage anaerobic unit—combined constructed wetland system, was designed and constructed near a small pig farm with 50–60 live pigs. In this pig farm dry pig feces were removed for manure compost. The wastewater mainly came from pigpen rinsing. The average concentration of COD_{cr}, BOD_5, TN and TP of the wastewater were about 2800 mg/L, 1100 mg/L, 650 mg/L and 65 mg/L respectively. The anaerobic unit was the primary treatment part of the system, which served both as a treatment unit and as a storage tank. To greatly reduce the high organic matter in the swine wastewater, two series of anaerobic tanks (AI and AII in Figure 1) were included. In the inlet of the first anaerobic tank grilles were set so that large suspended solids can be filtered. In each tank separating wall with holes in the middle-lower part was constructed to avoid short circuiting. The bottom of the tank was conical.

The Constructed Wetland (CW) unit was made up of a Surface Flow (SF) cell (SF in Figure 1) followed by two series of horizontal Sub-Surface Flow (hSSF) cells (hSSF1 and hSSF2 in Figure 1), which were separated and arranged in a horizontal baffled way. The cells comprised different inflow and outflow structure according to their position in the system. The water inlet of SF was connected with the outlet of pretreatment unit. Between SF and hSSF1 was an overflow weir with controlled height. Between hSSF1 and hSSF2 were over-flow weirs in the top and drainage holes in the upper and middle part. The outlet of hSSF2 was a drainage hole. The bottom of each cell keeped a slope of 1%. Gravel, sand and soil were used as the padding material. In the inlet and outlet parts of each cell only gravel with the size of 2–5 cm were stuffed. *Eichhornia crassipes*, *Typha* and *Vetiver* were planted in SF, hSSF1 and hSSF2 respectively.

2.2 *Size and running parameters*

The maximum treatment capacity was set to 1.5 m³/d based on the discharge of the farm. So the inflow and outflow was about 1.04 L/min. The volume of the two anaerobic tanks were 10 m³ (2.5 m * 2 m * 2 m) and 8 m³ (2 m * 2 m * 2 m) respectively. The total hydrologic retention time for the two anaerobic tanks was about 12 days.

The area of the combined CWs was calculated by the following formula (Kandasamy & Vigneswaran, 2008):

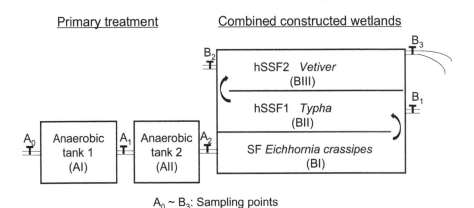

$A_0 \sim B_3$: Sampling points

Figure 1. Layout of the pilot swine wastewater treatment system.

$$A_{cw} = Q_i \cdot A_T \tag{1}$$

where Q_i = Influent discharge to wetland (m³/day); A_T = Area required to treat the influent to a suitable level given the time period (m²).

The value range of A_T varies depending primarily upon the type and concentration of the target pollutant and the treatment efficiency of the wetland. To guarantee treatment result, here it was set to 24 m², which was a little larger than proposed range: 10 m² < A_T < 20 m² (Kandasamy & Vigneswaran 2008). The total surface area was 36 m². Total hydrologic retention time for the combined CWs was set to 4 days. The total retention time of the whole system was 16 days.

2.3 *Sampling and water analysis*

The raw water and outlets of each cell (Fig. 1) were sampled two times a month. Monitoring indicators included COD_{cr}, BOD_5, TN, NH_4–N and TP. COD_{cr} and BOD_5 were analyzed with dichromate method and dilution and seeding method respectively. Alkaline potassium persulfate digestion-UV spectrophotometric method for TN, Nseeler's reagent colorimetric method for NH_4–N and ammonium molybdate spectrophotometric method for TP were applied.

Besides, the removing rate was calculated by the difference of inflow and outflow concentration divided by inflow concentration.

3 RESULTS AND DISCUSSION

3.1 *Treatment efficiency*

The general treatment efficiency of the two-stage anaerobic unit—combined constructed wetland system during the 7-month running was expressed in removing rate and average concentrations of the five indicators of treated effluent (Table 1). The system showed high removing rate especially for the organic matter represented in COD_{cr} and BOD_5, which were higher than 95%. Total nitrogen had the lowest removing rate among all the indicators. According to the emission standards for livestock and poultry breeding, for which the COD_{cr}, BOD_5, NH_4–N and TP are 400, 150, 80 and 8.0 mg/L respectively, discharge from the system well met the requirement except for TP. About 30% of the effluent samples could not meet the criteria of TP and the maximum, average and minimum effluent concentration were 11.2, 8.1 and 7.2 mg/L respectively.

3.2 *Removal dynamics*

Since temporal variation during the running period was small, both effluent concentration and removing rate were represented in averaged values (Table 2) to study the removal dynamics of different pollutant along the system.

The swine wastewater was rather high in organic matter loading with the average concentrations of COD_{cr} and BOD_5 reaching 2858 and 1120 mg/L respectively (Table 2).

Table 1. Treatment efficiency of the system.

Indicators	Concentration (mg/L)	Removing rate (%)
COD_{cr}	119.6	95.8
BOD_5	33.5	97
TN	165.5	74.8
NH_4–N	64.3	81.7
TP	8.1	88.4

Table 2. Removal dynamics along the system: the organic matter.

Parts of the system	COD$_{cr}$			BOD$_5$		
	Inlet (mg/L)	Outlet (mg/L)	Removing rate (%)	Inlet (mg/L)	Outlet (mg/L)	Removing rate (%)
AI	2857.8	1883.0	34.1	1119.6	788.9	29.4
AII	1883.0	771.5	58.8	788.9	287.5	63.5
BI	771.5	405.2	47.5	287.5	131.5	54.2
BII	405.2	216.8	46.6	131.5	67.2	48.9
BIII	216.8	119.6	44.8	67.2	33.5	50.1
AI+AII	2857.8	771.5	73.0	1119.6	287.5	74.3
BI+BII+BIII	771.5	119.6	84.5	287.5	33.5	88.4
Whole system	2857.8	119.6	95.8	1119.6	33.5	97.0

This treatment system demonstrated important removing rate for the organic matter: 95.8% and 97% of the COD$_{cr}$ and BOD$_5$ were respectively removed, of which 73% and 74.3% were removed by the anaerobic tanks. The AI tank showed higher removing rate than the AII tank, especially for BOD$_5$. Complex organic matter in wastewater was first broken down into simpler organic matter mainly in the AI tank and then the latter will be further removed through the second stage anaerobic treatment in AII. Separating the anaerobic processes in two stages may help keeping specific conditions in each tank. The following CW cells were important in further removing organic matter and demonstrated good separated removing rates.

The removal dynamics of nitrogen indicators (Table 3) was quite different from that of organics. The anaerobic tanks showed low (less than 15% for TN) or even negative removing rate (−11.5% for NH$_4$−N). The organic forms of nitrogen (e.g. protein) were first hydrolyzed under anaerobic conditions into peptide and amino acid, the latter forms of nitrogen might be further degraded into amine and ammonia. That was why the concentration of NH$_4$−N was increased in the anaerobic tanks, demonstrating negative removing rates. The nitrogen was mainly removed in the following CW cells through ammonification, nitrification and denitrification processes under aerobic or anoxic conditions and through plant uptaking (Meers et al. 2005).

Concentrations of TP also increased in the two anaerobic tanks (Table 4), which was due to phosphorus release from Phosphorus Accumulating Organisms (PAOs) under anaerobic conditions. TP was removed by the following CW cells. Although the system demonstrated rather high removing rate, the effluent concentration fluctuated around the standard value (8.0 mg/L). To further lower the TP concentration in the effluent measures like prolonging the CW cells and trying special padding material can be applied (Reed et al. 1992).

Above tables (Tables 2–4) showed that the two anaerobic tanks were important in removing organics but seemed no use for nitrogen and phosphorus removal. For nitrogen the benefits of the anaerobic parts were that organic nitrogen were broken down and ready for further reduction. The constructed wetland cells demonstrated stable treatment effect for the three types of pollutants.

3.3 Cost and maintaining

The construction cost of this system was about 1000$. The main maintenance work was planting and simple plant management. No pumping facilities were needed in this case due to the nature declining on-site terrain. In other cases where pumping is needed, less than 5 ¢ of power cost is needed for one stage pumping either before entering the AI or BI unit. Comparing with other treatment systems, this two-stage anaerobic unit—combined constructed wetland system is cost effective and easy for maintenance.

Table 3. Removal dynamics along the system: the nitrogen.

Parts of the system	TN			NH_4-N		
	Inlet (mg/L)	Outlet (mg/L)	Removing rate (%)	Inlet (mg/L)	Outlet (mg/L)	Removing rate (%)
AI	657.3	627.7	4.5	351.4	359.6	−2.4
AII	627.7	561.2	10.6	359.6	391.7	−8.9
BI	561.2	353.6	36.9	391.7	194.7	50.2
BII	353.6	236.5	33.1	194.7	107.1	44.8
BIII	236.5	165.5	30.1	107.1	64.3	39.8
AI+AII	657.3	561.2	14.6	351.4	391.7	−11.5
BI+BII+BIII	561.2	165.5	70.5	391.7	64.3	83.6
Whole system	657.3	65.5	74.8	351.4	64.3	81.7

Table 4. Removal dynamics along the system: the phosphorus.

Parts of the system	TP		
	Inlet (mg/L)	Outlet (mg/L)	Removing rate (%)
AI	69.7	70.9	−1.6
AII	70.9	74.1	−4.6
BI	74.1	38.9	47.5
BII	38.9	18.2	53.5
BIII	18.2	8.1	57.5
AI+AII	69.7	74.1	−6.2
BI+BII+BIII	74.1	8.1	89.0
Whole system	69.7	8.1	88.4

4 CONCLUSIONS

The application of this two-stage anaerobic unit—combined constructed wetland system on the treatment of wastewater from a small piggery farm demonstrated high treatment efficiencies for organics, nitrogen and phosphorus: with the average removing rates being 96%, 97%, 75%, 82% and 88% for COD_{cr}, BOD_5, TN, NH_4-N and TP respectively. The treated water can well meet the discharge standard of pollutants for livestock and poultry breeding (GB18596-2001) for organics, nitrogen. For phosphorus about 70% of the effluent samples could meet the criteria of TP.

The removal dynamics along the system was quite different. The anaerobic tanks played an important role in organics removing, where more than 70% of the organic load was removed. The treatment effect by anaerobic processes was not good in terms of removing rate for nitrogen. But complex organic nitrogen was broken down and ready for further reduction. The constructed wetland cells demonstrated stable treatment effect for the three kinds of pollutants. Phosphorus was removed in the constructed wetland cells. To further lower the TP concentration in the effluent measures like prolonging the CW cells and trying special padding material can be applied.

This system was proved to be cost effective both in construction and in running. The management and maintenance are simple and convenient. The system can be also suitable for the treatment of sewage from small and medium-sized towns and rural area.

ACKNOWLEDGMENTS

This study was supported by Project of Non-profit Research Foundation for Agriculture from the Ministry of Agriculture, China (No. 201003014).

REFERENCES

Harrington C, Scholz M, Culleton N & Lawlor PG. 2012. The use of Integrated Constructed Wetlands (ICW) for the treatment of separated swine wastewaters. Hydrobiologia, 692:111–119.

Kandasamy J & Vigneswaran S. 2008. Constructed Wetlands. New York: Nova Science Publishers, Inc.

Meers E, Rousseau DPL, et al. 2005. Tertiary treatment of the liquid fraction of pig manure with Phragmites australis. Water, Air, & Soil Pollution, 160:15–26.

Reed SC, Brown DS, Reed C & Brown S. 1992. Constructed Wetland Design: The First Generation Constructed first wetland generation design. Water Environment Research, 64:776–781.

Wang XY. 2006. Management of agricultural nonpoint source pollution in China: current status and challenges. Water Science and Technology, 53:1–9.

Zhang D, Gersberg RM & Keat TS. 2009. Constructed wetlands in China. Ecological Engineering, 35:1367–1378.

Hydraulic Engineering – Xie (Ed.)
© 2013 Taylor & Francis Group, London, ISBN 978-1-138-00043-8

Risk assessment methods study of the breakage asphalt runway pavement based on entropy

Mingjie Li & Rong Shi

Civil Aviation Flight University of China, Guanghan, Sichuan, China

ABSTRACT: In order to reduce the subjectivity of the indexes weights calculation in the breakage risk assessment of the asphalt concrete runway pavement of the airport, this paper took the breakage of asphalt concrete runway pavement, surface damage, evenness and skid resistance performance four as the assessment indexes, and established the indexes grades according to the influence degree on the evaluation result. Then, it used the entropy theory to calculate the proximity of the scores given by the efforts. After that, it used the fuzzy comprehensive evaluation method to evaluate the risk of damage on the airport asphalt concrete pavement, and reflected the safety state by numerical value. It will support the airport managers and workers with academic method on airport runway pavement safety assessment and maintenance and repair work.

1   INTRODUCTION

Runway is the most important infrastructure in the aerodrome. Its performance usually decays by the long-term interaction of aircraft loading and natural factors. Therefore, finding out the potential safety hazards timely and assessing pavement performance accurately have become the prime tasks of airport management and maintenance departments.

The U.S. Army Construction Engineering Research Institute studied the airport pavement evaluation system in 1970s, raising pavement condition index PCI (Pavement Condition Index) and a set of related investigations, calculations, analysis and evaluation methods, developed the world's first set of Airport Pavement Management system, named PAVER (now upgraded to Micro PAVER). Since then, the United States and some other countries take Micro PAVER as airport pavement management systems as the support software, such as the O'Hare, Midway, Albany Airport and Hong Kong International Airport and other. At the same time, Japan and Europe also developed their own standards on pavement evaluation and management.

Since the 1990s, many Chinese exports have carried out the research in evaluation and management of airport pavement. The CAAC promulgated the <Technical standards for airfield area of civil airports> in 2000, but it mainly emphasized pavement maintenance, lacking systematic and advanced pavement evaluation theory. Tong Ji University learned from the developed countries, carried out a large number of airport pavement evaluation research, established the shanghai airport pavement management system (SHAPMS) in 2002. <Technical specifications of aerodrome pavement evaluation and management> was issued by CAAC in 2009, took the airport pavement evaluation with guidance value.

But, overall, the airport pavement evaluation system in China still lacks of advanced tools and methods. Although some airports had done some research in asphalt concrete pavement evaluation, but they didn't assess and analyze the survey data scientifically. The information almost basing on engineering experience lacked of continuity and quantitative data support.

This paper introduced entropy into the fuzzy comprehensive evaluation of pavement performance, and amended the index weights impacted by the experts' subjectivity seriously.

The method is especially suitable for the calculation and correction of the objective indicators weights when the scores with large differences, and eliminates human factors in evaluation results, makes the evaluation results with more reality.

2 FUZZY COMPREHENSIVE EVALUATION MODEL BASED ON ENTROPY

2.1 Establishment of the factor sets

The evaluation index sets are the basis of the evaluation object, which are composed of a collection of various influencing factors, denoted as $U = (U_1, U_2, ..., U_n)$.

2.2 Comment set

It could be formed by several evaluation levels because different evaluation indicator has different value. We can evaluate each U_i comprehensively, then $V = (V_1, V_2, ..., V_n)$.

2.3 Establishment of entropy weights

To determine the role in the overall evaluation of evaluation index, we should judge the value of each index, meaning to determine its weight. The value of entropy weight is determined by how much information the evaluators passed to us. The calculation result of entropy value is much smaller when the scores given by evaluators with larger differences. But the entropy weight of the index will be much bigger meaning that it impacts on the evaluation result seriously[9], calculating as follows:

If the number of evaluators are m, the number of indexes are n, x^*_j is the value given by evaluator i to index j, x^*_j is the most biggest value to j. The value of x^*_j is various to the index' characteristic. When the index is a profitability indicator, the bigger the better, but when it is a loss one, the smaller the better. Therefore, the proximity of x_{ij} and x^*_j can be expressed as

$$d_{ij} = \begin{cases} x_{ij}/x^*_j & \text{profitability indicator} \\ x^*_j/x_{ij} & \text{loss indicator} \end{cases} \tag{1}$$

If $d_j = \sum_{i=1}^{m} d_{ij}$, the entropy value of j is:

$$e_j = -\frac{1}{\ln m} \sum_{i=1}^{m} \frac{d_{ij}}{d_j} \ln \frac{d_{ij}}{d_j} \tag{2}$$

Therefore the objective weight of j is:

$$\omega_j = \frac{1 - e_j}{n - \sum_{j=1}^{n} e_j} \tag{3}$$

It is known that $0 \le \omega_j \le 1, \sum_{j=1}^{n} \omega_j = 1$.

2.4 Establishment of the judgment matrix

R_n is the evaluation result of the indicator n, r_{nm} means the membership of evaluation index n to Evaluation level m, which reflects the fuzzy relations between the evaluation factors and evaluation levels. The number of elements in line is determined by the number of indexes sets, and the comment sets decide the number of elements in the column.

$$R = \begin{bmatrix} R_1 \\ R_2 \\ \cdots \\ R_n \end{bmatrix} = \begin{pmatrix} r_{11} & r_{12} & \cdots & r_{1m} \\ r_{21} & r_{22} & \cdots & r_{2m} \\ \cdots & \cdots & \cdots & \cdots \\ r_{n1} & r_{n2} & \cdots & r_{nm} \end{pmatrix} \tag{4}$$

2.5 *Fuzzy comprehensive evaluation*

The fuzzy synthetic evaluation of U can be expressed as:

$$B = \omega R = (\omega_1, \omega_2, \ldots, \omega_n) \begin{pmatrix} r_{i1} & r_{i2} & \cdots & r_{im} \\ r_{i1} & r_{i2} & \cdots & r_{im} \\ \cdots & \cdots & \cdots & \cdots \\ r_{n1} & r_{n2} & \cdots & r_{nm} \end{pmatrix} = (b_1, b_2, \ldots, b_m) \tag{5}$$

Finally, the evaluation object result could be obtained according to the principle of maximum membership.

3 PERFORMANCE EVALUATION OF AIRPORT PAVEMENT

3.1 *Establishment of evaluation objects sets, indexes sets and comment sets*

Step 1: Evaluation objects sets
A case, studying in asphalt concrete runway pavement, evaluates its running security in the event of damage.

Step 2: Factor sets
According to the survey, the breakage type of airport asphalt concrete pavement could be divided into four categories, summarized as in Table 1.

Therefore, $U = \{$surface breakage, surface damage, surface evenness, surface skid resistance$\} = (u_1, u_2, u_3, u_4)$.

Step 3: Comment set
According to the indicators' characters, the comment set can be divided into four grades, such as {High risk, Moderate risk, Low risk, Safe}.

Table 1. Breakage type of airport asphalt concrete pavement.

Number	Surface breakage type	Name
I	Surface breakage	Crack
		Block splitting
		Longitudinal and transverse crack
		Reflective crack
		Sliding crack
II	Surface damage	Loose aged
		Repair damaged
		Corrosion of oil
III	Surface evenness	Subsidence
		Uplift
		Pit
		Washboard
		Pushing
		Pumping
IV	Surface skid resistance	Polished
		Oil pan

215

3.2 Establishment of evaluation matrix

Firstly, a set of evaluation staff (such as airport airfield area management staff, technicians, operators, and so, from 20 to 50 persons), carry out the grade evaluation in the specific situation of the airport according to their work experience. Secondly, we conduct a statistical survey on the evaluation results. An airport asphalt concrete pavement damage statistical survey is shown in Table 2.

The evaluation matrix can be shown as

$$R = \begin{pmatrix} 0.25 & 0.40 & 0.30 & 0.05 \\ 0.20 & 0.30 & 0.45 & 0.05 \\ 0.10 & 0.15 & 0.25 & 0.50 \\ 0.30 & 0.30 & 0.30 & 0.10 \end{pmatrix}$$

3.3 Calculation of entropy weights

This paper consulted ten experts on the pavement evaluation, and required the rating value of road risk evaluation indexes. The rating value is divided into four grades, namely [100, 75], (75, 50], (50, 25], (25, 0], meaning stronger impact, moderate impact, slight impact and little impact, the higher the score is, the stronger the impact is. Experts' scores are shown in Table 3 as shown. The entropy value and indicators' weights are listed in Table 4.

3.4 Fuzzy comprehensive evaluation

Step 1: Calculations of evaluation results

$$B = A \cdot R = (0.368\ 0.222\ 0.124\ 0.296) \begin{pmatrix} 0.25 & 0.40 & 0.30 & 0.05 \\ 0.20 & 0.30 & 0.45 & 0.05 \\ 0.10 & 0.15 & 0.25 & 0.50 \\ 0.30 & 0.30 & 0.30 & 0.10 \end{pmatrix} = (0.236\ 0.318\ 0.327\ 0.119)$$

Table 2. Statistical survey of pavement damage.

Number	Factors	High risk	Moderate risk	Low risk	Safe
1	u_1	5	8	6	1
2	u_2	4	6	9	1
3	u_3	2	3	5	10
4	u_4	6	6	6	2

Table 3. Mark of risk factors given by experts.

Expert	1	2	3	4	5	6	7	8	9	10
u_1	85	95	65	90	70	95	70	60	80	75
u_2	70	75	90	60	65	75	80	85	70	85
u_3	55	50	50	60	55	65	60	50	55	50
u_4	80	75	80	85	70	60	90	95	65	70

Table 4. Calculating table of entropy weights.

Indicator	e_j	ω_j
u_1	0.995053	0.358
u_2	0.995053	0.222
u_3	0.998275	0.124
u_4	0.995909	0.296

216

Step 2: Improvement of evaluation results

This paper introduced weighted vector into evaluating according to the characteristics of the risk factors because the membership could not reflect the evaluation results visually. Setting the weighted vector $\mu = (1\ 0.75\ 0.5\ 0.25)$, it could turn the evaluation results into a comparable composite value through weighting calculating by weighted vector. Then, it divided the security state into four intervals [1, 0.75], (0.75, 0.5], (0.5, 0.25], (0.25, 0], corresponding to quite normal, normal, low-risk state and risk state.

$$N = \mu \cdot B'^T = (1 \quad 0.75 \quad 0.5 \quad 0.25) \quad (0.236 \quad 0.318 \quad 0.327 \quad 0.119)^T = 0.668$$

In this case, its pavement is in the state of normal. It is need to check the pavement regularly to prevent deterioration of the surface.

4 CONCLUSION

Practice has proved, the weights calculation of evaluation indexes basing on the entropy theory can effectively overcome the limitations that single evaluation method on the determination of indexes weights, and especially suitable for the calculation and correction of the objective indicators weights when the scores with large differences. The evaluation method is simple and easy, with strong practicability.

ACKNOWLEDGEMENT

This work are supported by youth fund of civil aviation flight university of China (Q2011–45) and research fund of civil aviation flight university of China (J2011–07).

REFERENCES

China Airport Construction Group Corporation of CAAC. 2000. *Maintenance manual for airfield area of civil airports*. Beijing: Civil Aviation Administration of China.

Jian-ming Ling, Yue-feng Zheng, Wei-ming Jin, 2001. On Airport Pavement Evaluation System. *Journal of Traffic and Transportation Engineering* 1(1):29–33.

Jie Zhan, Hao Wang, 2009. The Application of Fuzzy Analysis based on Entropy in Construction Risk Assessment. *Journal of Henan Institute of Engineering* 1 (21):29–32.

Study on Shanghai Hong Qiao airport flexible runway damage security risk identification and treatment principles 2006.

Tongji University. Technical specifications of aerodrome pavement evaluation and management (MH/T5024-2009), Civil Aviation Administration of China.

Yu-xiong JI, Sheng-nan Kan, Qi Feng, Li-jun Sun, 2004. Evaluation Models of Airport Pavement Management. *Journal of Tongji University(Natural Science)* 32(6):731–735.

Yu-zhong Yang, Li-yun Wu, Jian-chun Cong, 2009. Fuzzy synthetic evaluation on safety of coalmine transportation systems based on entropy weight [J]. *Journal of Harb in Institute of Technology* 41(4):257–259.

Zheng-wen Xie, Fan-yu Kong, 2007. Fuzzy comprehensive evaluation method for mines based on entropy technology. *Journal of China Jiliang University* 18 (1):79–82.

Hydraulic Engineering – Xie (Ed.)
© 2013 Taylor & Francis Group, London, ISBN 978-1-138-00043-8

Correlation between Chla concentration and environmental factors in Poyang Lake, China

Mao-Lin Hu
School of Life Sciences and Food Engineering, Nanchang University, Nanchang, Jiangxi, China

Hui-Ming Zhou
Jiangxi Fisheries Research Institute, Nanchang, Jiangxi, China

Fei Li
Jiangxi Entry-Exit Inspection and Quarantine Bureau, Nanchang, Jiangxi, China

ABSTRACT: Spatiotemporal variation of Chlorophyll a (Chla) concentration and its relationship with other water quality factors in the Poyang Lake outlet were analyzed during the period of 2007 to 2008. The results indicated that the Chla concentration was 4.46 ± 1.53 mg·m^{-3} in 2007 and 4.60 ± 1.67 mg·m^{-3} in 2008. Additionally, the Chla concentration reached a peak in August or September while bottomed out in January or December. And its values were significantly higher in the left sampling site than that in the middle and right sampling sites. Moreover, the Chla concentration in Poyang Lake outlet showed significant positive correlation with Z, WT, SD while SS, DO, COD and nutrients were inversely related. But no obvious correlation was found between pH and Chla in Poyang Lake outlet. The study would highlight further research needs for the distribution patterns of chlorophyll a in association with hydrological features in Poyang Lake.

1 INTRODUCTION

Jiangxi province is located in south China, to the south of middle and lower reaches of Yangtze River and lies between 113°34′36″–118°28′36″E and 24°29′14″–30°04′41″N (Fig. 1A). Poyang Lake, the largest freshwater body in China covering about 4,000 km^2 when full, is located in the north of Jiangxi Province. The area immediately surrounding Poyang Lake consists of low-lying alluvial plains prone to flooding. Mountains close to the boundaries of Jiangxi Province surround this region and all the five major rivers in the province flow into the Poyang Lake. The watershed of Poyang Lake covers almost 162,000 km^2, accounting for 96.8% of Jiangxi Province. Drainage for Poyang Lake is a narrow outlet named Hukou, which issues into the Yangtze River and marks the northern border of the province. Headwaters of the rivers in Jiangxi Province are located in the surrounding mountains. The five major rivers in the province flowing into the Poyang Lake are Ganjiang River, Xinjiang River, Fuhe River, Raohe River and Xiuhe River (Fig. 1A). The five major rivers and their tributaries are interlaced with Poyang Lake to form a complete riverine-lacustrine network. Poyang Lake is important nationally and internationally due to its geographical locality, history, and human activity. It was listed in the Global Ecoregion 200 by the World Wildlife Fund (WWF) as a site that should receive priority for conservation efforts (Huang et al. 2011).

Due to the rapid development of industry and agriculture in the last decade, large amounts of nitrogen and phosphorus associated with fertilizers were discharged to oceans, rivers, lakes, reservoirs and other waters, subsequently causing eutrophication (Abreu et al. 2010, Chen et al. 2011, Philippart et al. 2010, Zang et al. 2003). Accelerated eutrophication has

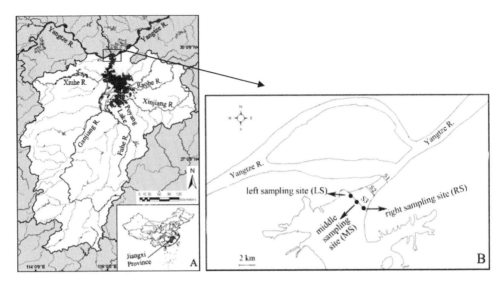

Figure 1. Location map of Poyang Lake in Jiangxi Province of China (A), and location of water sampling stations on Poyang Lake outlet (B).

become one of the world's most serious environmental problems, and associated symptoms, including the growth of attached and planktonic algae, can be identified in about 75% of the global closed waters (Chen et al. 2011, Lau & Lane 2002). Chlorophyll a is an important indicator for the presence of algae, and it is often considered as the dominant factor for assessing eutrophication (Chen et al. 2011, Lau & Lane 2002, Liu et al. 2011). Algal growth is closely related to a variety of environmental water quality parameters such as total nitrogen, total phosphorus, light intensity, water temperature, pH and DO. Changes in the number and composition of algae often relate to these parameters (Abreu et al. 2010, Chen et al. 2011, Correa-Ramirez et al. 2012, Kasai et al. 2010, Lau & Lane 2002, Liu et al. 2011, Philippart et al. 2010, Zang et al. 2003).

Currently, quantitative understanding of the temporal and spatial variability of chlorophyll a concentration in Poyang Lake is far from adequate (Huang et al. 2001, Wang et al. 1999). Accordingly, a better understanding could help scientific researchers, environmental managers and policy designers to know when, where, and why algal blooms occur in Poyang Lake. In order to accomplish a better understanding, this paper concentrates on the spatiotemporal variation of phytoplankton chlorophyll a and its relationship with other water quality parameters in Poyang Lake outlet. The study would highlight further research needs for the distribution patterns of chlorophyll a in association with hydrological features in Poyang Lake.

2 MATERIALS AND METHODS

2.1 Site description

Poyang Lake, the largest freshwater body in China, is located in the south of middle and lower reaches of Yangtze River. It drains the five major rivers in Jiangxi Province, such as Ganjiang River, Xinjiang River, Fuhe River, Raohe River and Xiuhe River, and then flows into Yangtze River through a narrow mouth named the Poyang Lake outlet (Fig. 1A). Three water quality monitoring sections (S1, S2 and S3) were selected on the Poyang Lake outlet (Fig. 1B). And then we chose three sites (shown in Fig. 1B) as the sampling stations on each monitoring section. So a total of nine sites were sampled on the Poyang Lake outlet.

2.2 Water sampling and analysis

Monthly sampling was carried out from January 2007 to December 2008. Integrated water samples were taken using a 5 L hydrophore. The sampling depth was 1.0 m beneath the water surface. Water samples at the nine sites were collected and refrigerated in a cold and dark container and processed several hours later in the laboratory. Physical parameters such as Water Temperature (WT), Secchi-Depth (SD) and pH were examined by the multi-meter YSI6600 (Beijing SYTF Science and Development Company, Beiyuan Road, Chaoyang District, Beijing, China) in the field. Chemical analyses of water samples in the laboratory included Dissolved Oxygen (DO), Chemical Oxygen Demand (COD), Suspended Solids (SS) and nutrient concentrations such as Total Nitrogen (TN), ammonium nitrogen (NH_4^+-N), nitrate nitrogen (NO_3^--N), nitrite nitrogen (NO_2^--N), Total Phosphorus (TP) and orthophosphate (PO_4^{3-}-P). All the measurements and analyses followed the standard methods, if not state otherwise (SEPA 2002).

Samples for chlorophyll a concentration (Chla) were filtered on GF/C filters (Whatman). The Chla was extracted from the filters by adding 90% acetone (by volume) and then placing the samples in a 0°C freezer for 24 h. The Chla concentration was analyzed spectrophotometrically at 750 nm, 663 nm, 645 nm and 630 nm, with correction for phaeopigments (SEPA 2002). Daily water level (Z) information of Poyang Lake outlet was obtained from the CJSWW (http://www.cjh.com.cn).

2.3 Statistical analysis

All statistical tests were performed using the Statistical Package for the Social Sciences software package (SPSS 13.0). Significances were defined as $P < 0.05$, if not stated otherwise.

3 RESULTS AND DISCUSSION

3.1 Distribution of chlorophyll a concentration

The Chla concentration was 4.46 ± 1.53 mg·m^{-3} (Mean $\pm$ SD) and ranged from 2.12 to 6.89 mg·m^{-3} during 2007. During 2008, the Chla concentration was 4.60 ± 1.67 mg·m^{-3} and ranged from 2.02 to 7.72 mg·m^{-3}. The Chla concentration reached a peak in August or September while bottomed out in January or December (Fig. 2A). The Chla values were significantly higher in the LS (left sampling site) than that in the MS (Middle Sampling site) and that in the RS (Right Sampling site) (Fig. 2B). But no significant differences were found either among the three monitoring sections (S1, S2 and S3). The seasonal distribution of Chla concentration was partially due to the changes of water parameters in Poyang Lake outlet. As suggested by Huang et al. water temperature is related to phytoplankton Chla in Poyang Lake (Huang et al. 2001).

3.2 Correlation between water parameters and chlorophyll a concentration

The pH is a key chemical water indicator. The pH value is governed by the amount of carbon dioxide, which can react with water, as well as carbonate and bicarbonate to form complex

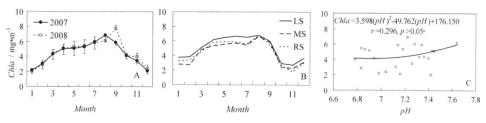

Figure 2. Monthly average (A) and spatial distribution (B) of chlorophyll a (*Chla*), regression curve for chlorophyll a (*Chla*) and *pH* (C) in Poyang Lake outlet.

but reversible carbonate systems (Chen et al. 2011). And its value can rise up to 9 or 10 for eutrophic waters. A high pH may inhibit the photosynthesis of algae (Chen et al. 2011). Regarding natural waters, as the environmental conditions of natural waters are complex, the relationships between pH and Chla are not of consistent nature. Ruan et al. concluded that there was a significant positive linear correlation between pH and Chla. They demonstrated that with the increase of phytoplankton biomass, the consumption of carbon dioxide caused by photosynthesis promoted an increase of the pH value (Ruan et al. 2008). Similarly, Zhang found a significant positive exponential relationship between pH and Chla for the eutrophic Dashahe Reservoir (Zhang 2009). This can be explained by significant photosynthetic activities consuming large amounts of carbon dioxide, which consequently resulted in a significant increase in pH. Therefore, significant positive correlations were found between pH and Chla for the eutrophic water with a mean Chla concentration higher than $10 \, mg \cdot m^{-3}$ (Chen et al. 2011). But no obvious correlation was found between pH and Chla in Poyang Lake outlet. The correlation coefficient r was 0.296 at $p > 0.05$ (Fig. 2C).

Nevertheless, the Chla concentration in Poyang Lake outlet showed significant positive correlation with Z, WT, SD while SS, COD were inversely related (Fig. 3). Similar relationships between these parameters and Chla have been reported in Poyang Lake (Huang et al. 2001). Water level fluctuations in lakes and rivers, especially their extent, frequency and duration, are dominant forces controlling the functioning of these ecosystems (Leira & Cantonati 2008). The water temperature and water level in a lake affect water chemistry (nutrients) and biota (plankton, fish) both directly and indirectly. The effect of warm weather on shallow lakes is particularly strong when it coincides with low water level (Haldna et al. 2008). The water temperature indirectly controls the rate of algal photosynthesis and aquatic respiration through enzymes, leading to changes in the carbon dioxide content due to metabolic activities. Previous studies showed that water temperature can promote algal photosynthesis and aquatic respiration at certain temperature ranges. The solubility of carbon dioxide within water is also affected by temperature. When atmospheric pressure is constant, a rise of water temperature results in the kinetic rate of carbon dioxide to increase. This allows for carbon dioxide to escape easily from the water surface, leading to a lower solubility. As the water temperature decreases, the solubility of carbon dioxide increases at the same time (Chen et al. 2011). The average concentrations of SS changed from $15.94 \, mg \cdot L^{-1}$ in September to $511.15 \, mg \cdot L^{-1}$ in January. The temporal dynamic of SS was driven by difference in the lake-river hydrodynamic. The analyses of the relationship between SS and Chla in the Poyang Lake outlet allowed highlighting the importance of SS as a limiting factor for phytoplankton growth, overcoming its role as a nutrient source in some phases of the hydrological cycle.

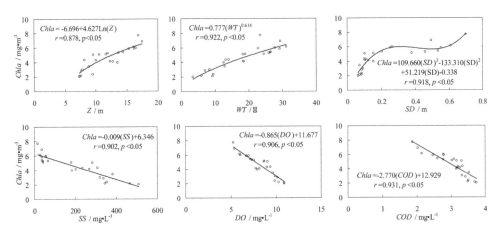

Figure 3. Regression curve for water level (Z), Water Temperature (WT), Secchi-Depth (SD), Suspended Solids (SS), Dissolved Oxygen (DO), Chemical Oxygen Demand (COD) and Chlorophyll a (Chla) in Poyang Lake outlet.

Table 1. Multiple regression models for Chlorophyll a (Chla), including Total Nitrogen (TN), ammonium nitrogen (NH_4^+-N), nitrate nitrogen (NO_3^--N), nitrite nitrogen (NO_2^--N), Total Phosphorus (TP) or orthophosphate (PO_4^{3-}-P).

Parameter/ mg·L^{-1}	Model	R value	P value
TN	Chla = 9.716–2.37 (TN)	0.879	0.05
NH_4^+-N	Chla = 10.022 − 29.813 (NH_4^+-N) + 55.588 (NH_4^+-N)^2 − 44.582(NH_4^+-N)^3	0.959	0.05
NO_3^--N	Chla = 12.536 − 49.992 (NO_3^--N) + 119.372 (NO_3^--N)^2 − 101.831 (NO_3^--N)^3	0.941	0.05
NO_2^--N	Chla = 12.508 − 1119.564 (NO_2^--N) + 52984.901 (NO_2^--N)^2 − 803364 (NO_2^--N)^3	0.867	0.05
TP	Chla = 7.879–32.606 (TP)	0.888	0.05
PO_4^{3-}-P	Chla = 138.414 e$^{-464.823}$ (PO_4^{3-}-P)	0.865	0.05

With respect to natural waters, the significant positive linear and exponential relationships were found between DO, nutrients (including TN, NH_4^+-N, NO_3^--N, NO_2^--N, TP and PO_4^{3-}-P) and Chla (Haldna et al. 2008, Ruan et al. 2008, Zhang 2009). However, this paper shows significant negative correlations ($p < 0.05$) between these parameters and Chla for Poyang lake outlet (Fig. 3, Table 1). This observation contradicts current study findings, and no explanation can currently be found.

4 CONCLUSIONS

The results of the present study indicated that the concentration peaks of chlorophyll a (Chla) typically appeared in summer and autumn while the minima occurred during winter and early spring. The Chla concentration in Poyang Lake outlet showed significant positive correlation with water level (Z), Water Temperature (WT), Secchi-Depth (SD) while Suspended Solids (SS), Chemical Oxygen Demand (COD) were inversely related. No obvious correlation was found between pH and Chla. Similar relationships between these parameters and Chla have been reported in Poyang Lake.

With respect to natural waters, the significant positive linear and exponential relationships were found between DO, Total Nitrogen (TN), ammonium nitrogen (NH_4^+-N), nitrate nitrogen (NO_3^--N), nitrite nitrogen (NO_2^--N), Total Phosphorus (TP), orthophosphate (PO_4^{3-}-P) and Chla. However, this paper shows significant negative correlations ($p < 0.05$) between these parameters and Chla for Poyang lake outlet. This observation contradicts current study findings, and no explanation can currently be found. Therefore, the study would highlight further research needs for the distribution patterns of Chla in association with hydrological features in Poyang Lake.

REFERENCES

Abreu, P.C. Bergesch, M. Proenca, L.A. Garcia, C.A.E. Odebrecht, C. 2010. Short- and long-term chlorophyll a variability in the shallow microtidal Patos Lagoon estuary, southern Brazil. *Estuar. Coasts* 33: 554–569.
Chen, Y.W. Fan, C.X. Teubner, K. Dokulil, M. 2011. Changes of nutrients phytoplankton comparison of relationships between pH, dissolved oxygen and chlorophyll a for aquaculture and non-aquaculture waters. *Water Air Soil Pollut.* 219: 157–174.
Correa-Ramirez, M.A. Hormazabal, S.E. Morales, C.E. 2012. Spatial patterns of annual and interannual surface chlorophyll-a variability in the Peru-Chile Current System. *Prog. Ocean.* 62–65: 8–17.
Haldna, M. Milius, A. Laugaste, R. Kangur, K. 2008. Nutrients and phytoplankton in Lake Peipsi during two periods that differed in water level and temperature. *Hydrobiol.* 599: 3–11.

Huang, L.L. Wu, Z.Q. Li, J.H. 2011. Fish fauna, biogeography and conservation of freshwater fish in Poyang Lake Basin, China. *Environ. Biol. Fish.* doi: 10.1007/s10641-011-9806-2.

Huang, W.J. Chen, J.F. Xu, N. Jiang, T.J. Luo, Y.M. Huang, G.H. Qi, Y.Z. 2001. Analysis of grey models between lake water environmental essential factors and the incidence of chlorophyll-a in Poyang Lake, China. *Acta Hydrobiogica Sinica* 25: 416–419.

Kasai, H. Nakano, Y. Ono, T. Tsuda, A. 2010. Seasonal change of oceanographic conditions and chlorophyll a vertical distribution in the southwestern Okhotsk Sea during the non-iced season. *J. Oceanogr.* 66: 13–26.

Lau, S.S.S. & Lane, S.N. 2002. Biological chemical factors influencing shallow lake eutrophication: a long-term study, Sci. *Total Environ.* 288: 167–181.

Leira, M. & Cantonati, M. 2008. Effects of water-level fluctuations on lakes: an annotated bibliography. *Hydrobiol.* 613: 171–184.

Liu, X.C. Xu, Z.X. Wang, G.Q. 2011. Spatiotemporal variation of Chlorophyll a and its relationship with other water quality factors in the Tai Lake. *Advanced Materials Research* 183–185: 783–789.

Philippart, C.J.M. Iperen, J.M. Cadée, G.C. Zuur, A.F. 2010. Long-term field observations on seasonality in chlorophyll-a concentrations in a shallow coastal marine ecosystem, the Wadden Sea. *Estuar. Coasts* 33: 286–294.

Ruan, X.H. Shi, X.D. Zhao, Z.H. Ni, L.X. Wu, Y. Jiao, T. 2008. Correlation between chlorophyll a concentration and environmental factors in shallow lakes in plain river network areas of Suzhou. *Journal of Lake Sciences* 20: 556–562.

SEPA. 2002. Standard methods for the examination of water and wastewater. Beijing: China Environmental Science Press.

Wang, J. Liang, Y.L. Xie, Z.C. 1999. Comparative studies on Chlorophyll a content and production of planktonic algae in a large river-connected lake (the Poyang Lake) and neighboring sections of the Changjiang River. *Acta Hydrobiogica Sinica* 23: 40–46.

Zang, C.J. Huang, S.L. Wu, M. Du, S.L. Scholz, M. Gao, F. Lin, C. Guo, Y. Dong, Y. 2003. Changes of nutrients phytoplankton chlorophyll-a in a large shallow lake, Taihu, China: an 8-year investigation. *Hydrobiol.* 506–509: 273–279.

Zhang, W.T. 2009. Analysis on limiting factors of eutrophication in Dashahe Reservoir. *Guang Dong Water Resources and Hydropower* 9: 26–28.

Hydraulic Engineering – Xie (Ed.)
© 2013 Taylor & Francis Group, London, ISBN 978-1-138-00043-8

Vertical distribution of sediment organic phosphorus species and simulated phosphorus release from lake sediments

Liming Dong
Beijing Technology and Business University, Beijing, P.R. China

Zhifeng Yang, Xinhui Liu & Guannan Liu
Beijing Normal University, Beijing, P.R. China

ABSTRACT: Vertical distribution of Organic Phosphorus (OP) species in sediment cores of the Baiyangdian Lake was investigated using a soil OP fractionation scheme. Results of chemical fractionation showed that different OP fractions ranked along the sediment depth in the order: humic acid-P > HCl-OP > $NaHCO_3$-OP > fulvic acid-P > residual OP. Results of simulated P release showed that soluble reactive P (SRP) flux rates under anaerobic condition were much more than those under aerobic condition. The experiments of P release under seasonal temperature cycle condition (5 °C—15 °C—25 °C—15 °C—5 °C) were also simulated. The positive P flux rates were observed under both simulated spring (the first 15 °C) and summer (25 °C) conditions from sediment cores in the shallow and deep sites, indicating P release from sediments.

1 INTRODUCTION

Phosphorus (P) is the limiting macronutrient for primary production in most lakes. Lake sediments can act either as sinks or as sources of P in the form of phosphates (Kim et al. 2003; Søndergaard et al. 2003; Christophoridis and Fytianos 2006). The potential P release from lake sediments depends on the amount of sedimentary P that exists in various chemical forms with a marked difference in mobility and bioavailability. Most of the fractionation work focused on IP species and OP was usually treated as a residue or a refractory fraction. So far, there is only one comprehensive scheme, which was first developed by Bowman and Cole (1978) and later improved by Ivanoff et al. (1998), to fractionate soil OP into three separated distinct fractions: labile OP, moderately labile OP and nonliable OP. Quantification of P release from sediments is usually presented with batch experiments or intact sediments simulation (Jin et al. 2006; Lai and Lam, 2008). Some related environmental factors are also researched widely in simulated experiments of sediment P release. Jenson and Andersen (1992) found that the water temperature alone could explain nearly 70% of the seasonal variation in sediment P release in three shallow eutrophic lakes. He also argued that NO_3^- concentration could reduce P release in winter and early summer, but increase it in late summer. Christophoridis and Fytianos (2006) revealed that reductive conditions and high pH values could increase P release rates from surface lake sediments. In this study, the vertical distribution of sediment organic phosphorus species in Baiyangdian (BYD) Lake was investigated for assessing the accumulation trends of P in the sediments. Laboratory studies were also conducted to determine the effects of redox conditions and seasonal temperature on the release of Soluble Reactive P (SRP) and OP from lake sediment cores.

2 MATERIALS AND METHODS

2.1 *Site description*

The Baiyangdian Lake, which is located in the center of Hebei Province in north China (38°49′N and 116°04′E), is a shallow lake. The lake is frozen from December to the next February. Water depth ranged from 2 to 4 m. The water surface area covered 266 km² when the water level was 8.3 m above mean sea level. The lake area near the inflow entrance from the Fuhe River is often shallow, while the area is often deep near the outflow of the Zhaowangxinqu River. The two representative sampling sites from the deep (4.2 m) and the shallow (2.5 m) area were selected, respectively.

2.2 *Sample collection*

Five-liter water was collected 10–15 cm below water surface at each site in November in 2009. Then the sediment samples of 6 cm × 20 cm (diameter × depth) were collected with a core sampler and were sectioned into 15 parts at intervals of 1 cm for 0–10 cm and 2 cm for 10–20 cm. The sediments for P fractionation were then freeze-dried, sieved through a 200-mesh stainless steel-sieve and stored in darkness for further analysis. The other intact cores with some water were stored in plexiglass tubes (60 mm in diameter, 800 mm in length) and transported in darkness to the lab for P flux determination.

2.3 *OP fractionation*

The OP fractionation procedure was primarily based on the scheme described by Ivanoff et al. (1998). OP in lake sediments was classified into labile, moderately labile and nonlabile fractions. Labile OP (LOP) was initially extracted with 0.5 M $NaHCO_3$ at pH 8.5. Moderately Labile OP (MLOP) was extracted with 1.0 M HCl and then 0.5 M NaOH sequentially. The NaOH extract was acidified with concentrated HCl until pH 0.2 to separate the non-labile fraction (humic acid-P) from the moderately labile fraction (fulvic acid-P). Finally, the highly resistant and Non-Labile OP (NLOP) was determined by ashing the residue from the NaOH extraction at 550 °C for 1 h, followed by dissolution in 1 M H_2SO_4. The P concentrations in all extracts were measured colorimetrically using the molybdenum blue method. TP in all extracts was measured after an aliquot was digested with $K_2S_2O_8 + H_2SO_4$. Each test was performed in triplicate.

2.4 *Core incubation experiments*

For determining the effects of redox condition on P release from sediments, sediment cores were kept in the dark in an incubator at 20 °C. Replicate cores were aerated continuously with room air to maintain an aerobic condition for 15 days. The same replicate cores were incubated anaerobically for the same time by purging N_2 gas into the water column to maintain dissolved oxygen below 1 mg L^{-1}. Replicate cores were incubated at a cycle of 5 °C, 15 °C, 25 °C, 15 °C and 5 °C represented the water temperatures in winter, spring, summer and autumn for 4 days, respectively. They were bubbled for 2 hours every day with pumped air just 2 cm below water surface. Water (15 mL) overlying the sediment surface was collected regularly at 1–2 day intervals, and filtered through 0.45 μm membrane (GF/C), then immediately analyzed for Soluble Reactive Phosphorus (SRP) and TP. OP concentrations were calculated as the difference between TP and SRP. Distilled water was added prior to sampling to replenish the evaporative loss of water, while filtered water of known composition was added after sampling to replace the volume of water removed. P flux across the sediment-water interface was then estimated as the slope of the regression line of cumulative SRP or OP mass in the overlying water column against incubation time, divided by the surface area of sediment core. The flux data were expressed as the average of replicate cores.

226

3 RESULTS AND DISCUSSION

3.1 *OP fractionation in sediment cores*

The concentrations of different OP fractions in the sediment cores are presented in Figure 1. In shallow and deep sites, OP was mainly composed of MLOP (HCl-OP and fulvic acid-P) and NLOP (humic acid-P and residual OP). The rank order of different OP fractions was humic acid-P > HCl-OP > NaHCO$_3$-OP > fulvic acid-P > residual OP. In summary, LOP (NaHCO$_3$-OP) and MLOP contributed to 50.6–74.1% in shallow site and 48.3–65.8% in deep site of total extracted OP respectively, which can be degraded for phytoplankton uptake. It showed that there is a potential risk of P release from sediments in the BYD Lake.

The highest LOP contents of 35.4 and 46.0 mg kg^{-1} were observed at the depth of 6 and 4 cm in the sediment cores from shallow and deep sites respectively. Then the LOP contents decreased along the depth and reached the lowest contents at the depths of 14 and 9 cm. Meanwhile, the MLOP contents of top sediments were the lowest of 52.4 mg kg^{-1} (1 cm) and 41.0 mg kg^{-1} (2 cm) in shallow and deep sites, respectively, then increased and reached the peak value at the depths of 14 and 7 cm. There was an opposite change of the NLOP content along the sediment depth in shallow and deep sites. The great difference of vertical distribution of OP fractions should be attributed to the different sediment environments and the history of sediment OP accumulation in these sites.

3.2 *Phosphorus flux*

Phosphorus flux rates from the intact sediment cores under aerobic and anaerobic conditions for the shallow and deep sites were illustrated in Figure 2. Over the 15-day aerobic incubation, the sum of P flux from sediment cores in the shallow site was 0.56 mg m^2 d^{-1}, which was twice of 0.26 mg m^2 d^{-1} from sediment cores in the deep site. Aerobic conditions which often occur in the sediment-water interface of shallow lakes, are always regarded as preventing P release into overlying water, due to the micro-layer of FeOOH with its high sorption capacity for P in oxic sediments (Hupfer and Lewandowski, 2008). However, under appropriate temperatures and aerobic conditions, the microbial activity is increased for organic matter decomposition and further contributes to the P release from oxic sediments. Under anaerobic condition, SRP flux rates from sediment cores increased sharply to 0.60 and 1.15 mg m^2 d^{-1} for the shallow and deep sites, respectively. It should be attributed to the sorbed P by oxidized ferric complexes releasing together with Fe. Compared to those under aerobic conditions, OP flux

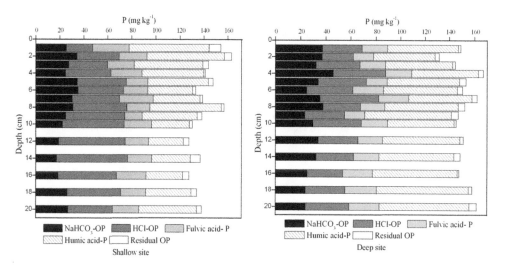

Figure 1. Vertical distribution of organic P species in sediment cores for the shallow and deep sites.

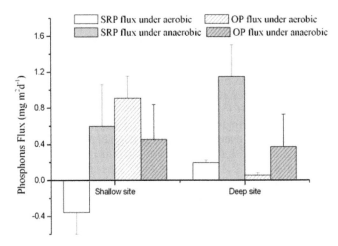

Figure 2. Soluble Reactive Phosphorus (SRP) and Organic Phosphorus (OP) flux rate from the sediment cores under aerobic and anaerobic condition.

rates decreased to 0.46 mg m² d⁻¹ from sediment cores in shallow site, but it increased to 0.37 mg m² d⁻¹ in the deep site under anaerobic conditions. The different compositions of OP compounds together with the different mechanisms of OP release between the sediments for the shallow and deep sites should be the main reason.

It appeared from the Figure 3 that positive P flux rates (the sum of SRP and OP flux rates) of 0.24 and 0.11 mg m² d⁻¹ were observed under simulated water temperature conditions of spring (the first 15 °C) and summer (25 °C) respectively from sediment cores in the shallow site. The sediment cores in deep site also had positive P flux rates of 0.59 and 0.75 mg m² d⁻¹ at these two seasonal temperatures, respectively. It indicated that P release from sediments mainly occurs in spring and summer. Those under other temperature conditions of winter (5 °C) and autumn (the latter 15 °C) from the shallow and deep sites were negative, meaning P retention by sediments. SRP flux rates varied slightly under changed temperature conditions. However, significant changes in OP flux rates were observed at different temperatures from

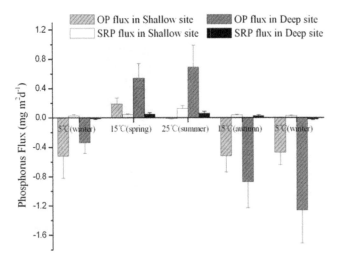

Figure 3. Soluble Reactive Phosphorus (SRP) and Organic Phosphorus (OP) flux rate from the sediment cores under simulated seasonal temperature in water column.

sediment cores. It showed that the temperature is one of the main factors affecting OP release from sediments. However, it seemed to release P from sediments under simulated spring condition of the first 15 °C and to sequestrate P from water column under simulated autumn condition of the latter 15 °C. It should be attributed to the amount of labile OP in sediments and the concentration of P in water column. Although the sediment cores in deep site seemed to have a higher P release than that in shallow site in the simulated experiments, the importance of sediments as P source is further enhanced by the high ratio for sediment surface to water column (Søndergaard et al. 2001). So, its potential influence on water concentration in shallow site is stronger than that in deeper site.

4 CONCLUSIONS

Based on a soil OP fractionation scheme, the vertical distribution of OP species in sediment cores from typical shallow and deep sites in the Baiyangdian Lake in North China was investigated. The rank order of different OP fractions along the sediment depth was humic acid-P > HCl-OP > $NaHCO_3$-OP > fulvic acid-P > residual OP. LOP and MLOP contributed to 50.6–74.1% in shallow site and 48.3–65.8% in deep site of total extracted OP respectively, which can be degraded for phytoplankton uptake. The great differences of vertical distribution of OP fractions were observed in the sediment cores for the shallow and deep sites, due to the different sediment environment and the history of sediment OP accumulation in these sites. Results of simulated P release showed that SRP flux rates from sediment cores increased sharply from –0.35 and 0.20 mg m^2 d^{-1} under aerobic condition to 0.60 and 1.15 mg m^2 d^{-1} under anaerobic condition for the shallow and deep sites, respectively. Compared to those under aerobic conditions, OP flux rates decreased to 0.46 mg m^2 d^{-1} from sediment cores in shallow site, but it increased to 0.37 mg m^2 d^{-1} in the deep site under anaerobic conditions. The positive P flux rates were respectively observed under simulated water temperature in spring (the first 15 °C) and summer (25 °C) from sediment cores in the shallow and deep sites, indicating P release from sediments. However, those under other temperatures in winter (5 °C) and autumn (the latter 15 °C) from the sediments in shallow and deep sites were negative, indicative of P retention by sediments.

REFERENCES

Bowman, R.A. & Cole, C.V. 1978. An exploratory method for fractionation of organic phosphorus from grassland soils. *Soil Science* 125(2):95–101.
Christophoridis, C. & Fytianos, K. 2006. Conditions affecting the release of phosphorus from surface lake sediments. *Journal of Environmental Quality* 35(4):1181–1192.
Hupfer, M. & Lewandowski, J. 2008. Oxygen controls the phosphorus release from lake sediments— a long-lasting paradigm in limnology. *International Review of Hydrobiology* 93(4–5):415–432.
Ivanoff, D.B., Reddy, K.R. & Robinson, S. 1998. Chemical fractionation of organic phosphorus in selected histosols. *Soil Science* 163(1): 36–45.
Jensen, H.S. & Andersen, F. 1992. Importance of temperature, nitrate, and pH for phosphate release from aerobic sediments of four shallow, eutrophic lakes. *Limnology and Oceanography* 37(3):577–589.
Jin, X.C., Wang, S.R., Bu, Q.Y. & Wu, F.C. 2006. Laboratory experiments on phosphorous release from the sediments of 9 lakes in the middle and lower reaches of Yangtze river region, China. *Water, Air, and Soil Pollution* 176(1–4): 233–251.
Kim, L.H., Choi, E. & Stenstrom M.K. 2003. Sediment characteristics, phosphorus types and phosphorus release rates between river and lake sediments. *Chemosphere* 50(1):53–61.
Lai, D.Y. & Lam, K.C. 2008. Phosphorus retention and release by sediments in the eutrophic Mai Po Marshes, Hong Kong. *Marine Pollution Bulletin* 57(6–12):349–356.
Søndergaard, M., Jensen, P.J. & Jeppesen, E. 2001. Retention and internal loading of phosphorus in shallow, eutrophic lakes. *Scientific World Journal* 1(8):427–442.
Søndergaard, M., Jensen, P.J. & Jeppesen, E. 2003. Role of sediment and internal loading of phosphorus in shallow lakes. *Hydrobiologia* 506–509(1–3):135–145.

Hydraulic Engineering – Xie (Ed.)
© *2013 Taylor & Francis Group, London, ISBN 978-1-138-00043-8*

Research on implementation of safety management diagnosis in enterprises

Gong Yunhua
School of Mechanics, Storage & Transportation Engineering, China University of Petroleum, Beijing, China

ABSTRACT: In order to find out defects in safety management and improve safety management in enterprises, enterprise safety management diagnosis was developed. It was essential to define what should be diagnosed to implement safety management diagnosis. So, a safety management Diagnosis Item (DI) system was developed. The item system nearly covered all aspects of enterprise safety management. According to enterprise diagnosis theories, diagnosis procedure was provided, which was composed of diagnosis preparation, initial diagnosis, formal diagnosis and guidance conduction. Four diagnosis methods were proposed, which were document inspection, questionnaire survey, interviews and statistical analysis. There were three safety management diagnosis conclusion patterns in diagnosing, which were comprehensive diagnosis conclusion, item diagnosis conclusion and particular problem analysis. The comprehensive diagnosis conclusion was obtained by Fuzzy Comprehensive Evaluation.

1 INTRODUCTION

Enterprises adopt various methods to manage their safety. Many enterprises are facing problems of how to assess their safety management and find out safety management defects. Lag indicators, such as accident rate, have certain deficiency in evaluating enterprises safety management. Different companies in different industries or of different sizes can not be compared with the same indicators. To solve these problems, enterprise safety management diagnosis was developed to be implemented. Safety management diagnosis was a process with experts going to workplace, using various diagnosis methods to identify major problems, doing quantitative or qualitative analysis to identify the causes of problems, putting forward practical improvement interventions, and guiding the implementation of improvement interventions so as to enhance safety management.

2 LITERATURE REVIEW

Safety management diagnosis was a new concept extended from management diagnosis. To some extent, it was the application of management diagnosis into safety management. However, there were many differences between general management and safety management. The implementation of enterprise safety management diagnosis requests its own theory, method and procedure.

At the moment, there was not any definition of safety management diagnosis. But, there are several definitions related to it. Such as, HSE management system diagnosis (Chen, 2006) and enterprises safety diagnosis. (Zhao, 2008) Fuzzy Comprehensive Evaluation Method was also applied in safety management evaluation in construction industry. (Sun & Liu, 2006)Theories close related to enterprises safety management diagnoses were mainly normal managment consulting theories. There were enterprise culture diagnosis, enterprise strategy diagnosis, human recourses diagnosis, performance management diagnosis, etc. (Tang, 2001)

Diagnosis methods which were usually adopted were observation, onsite survey, experiment, questionnaire, interviews. Diagnosis expert system was also developed, which could make enterprise diagnosis process more efficiently. (Hu, 2003).

3 SAFETY MANAGEMENT DIAGNOSIS IMPLEMENTATION

3.1 Safety management diagnosis item system

Safety management diagnosis should cover all aspects of enterprise safety management, and those aspects can not be overlapped. In this research, the content of safety management was expressed in the form of diagnosis items. All items and their hierarchy constituted a safety management diagnosis system. Literature research and expert consulting were applied into defining safety management item. Items were set based on management theories, safety laws and regulations, relative occupational safety and health management system elements, safety performance indicators, safety assessment indicators, safety culture Indicators etc.

There are many system structures. The most widely used form is the hierarchical structure. The hierarchical structure can not only reflect a hierarchical order, but also facilitate the assessment. Therefore, hierarchical structure was adopted in this study. Safety management item system has 3 first level diagnosis items, 17 second level diagnosis item and 48 third level diagnostic items (Table 1). The comprehensive enterprise safety management diagnosis conclusion was the integration of diagnosis conclusions of each item. Different diagnosis item plays different roles in safety management; therefore, weight value for each diagnosis item should be set. This study used Analytic Hierarchy Process (AHP) to set the weight value of the diagnosis items.

3.2 Safety management diagnosis procedure

Appropriate diagnosis procedure facilitates the implementation of diagnosis and ensures the accuracy of diagnosis results. The procedure of enterprise safety management was divided into four steps, which are diagnosis preparations, initial diagnosis, formal diagnosis and guidance conduction. (Table 2).

3.3 Safety management diagnosis method

In the process of safety management diagnosis, diagnosis consultants need to collect lots of information. Therefore, scientific data collecting methods help them diagnose quickly and accurately grasp the enterprise safety management information. These data collecting methods are called diagnosis methods. Different data collection methods should be adopted according to different characters of information. Four diagnosis methods were adopted, which were document inspection, questionnaire survey, interviews, and statistics analysis.

Lots of information about safety management can be reflected by document inspection. Complete files are an important part of safety management. Through document inspection, safety attitude and risk awareness could be assessed. The procedure of document inspection includes preparation of inspection document list, preparation of main points that should be inspected, doing inspection and filing inspection record.

Doing questionnaire surveys was an important approach to implement safety management diagnosis. Two works are necessary to doing a good survey; one is the design of the questionnaire, the other is the sampling technique.

Doing interview is also a good way to obtain information for diagnosis and enhance communication with the customer. Interview skill is a technique that diagnosis consultants must master. According to interview modes, interviews can be divided into non-structured interviews and structured interviews. In the process of interview, there are a series of information will be generated, which includes interview outline, interview notes, interview log.

Table 1. Enterprises safety management Diagnosis Item (DI) system.

1st level DI	Weight	2nd level DI	Weight	3rd level DI	Weight
Basic safety management	0.2970	Department arrangement	0.2881	Safety department	0.7500
				The structure of safety department	0.2500
		Safety human resources	0.0987	Number of full-time safety managers	0.5338
				Number of registered safety engineers	0.2199
				Qualification and title of full-time safety managers	0.1320
				Working years of full-time safety managers	0.1144
		Safety investment	0.1705	Safety investment collection	0.5000
				Usage of safety investment	0.5000
		Safety plans	0.0877	Safety plan	0.3333
				The implementation of safety plan	0.6667
		Laws and standards	0.0669	Collection and updating of safety laws and standards	0.1667
				Learning of safety laws and standards	0.8333
		Rules and regulations	0.2881	Safety institution	0.0915
				Safety procedure	0.2203
				Safety responsibility system	0.1096
				Implementation of rules and regulations	0.5785
Scientific safety management	0.5396	Management system	0.1894	System setup	0.1047
				System running and audit	0.6370
				System authorization	0.2583
		Safety education	0.10056	Safety education for upper managers	0.2500
				Safety education for safety managers	0.2500
				Safety education for special operation worker	0.2500
				Safety education for other workers	0.2500
		Risk management	0.1894	Major hazard management	0.6370
				Non major hazard management	0.2583
				Occupational health management	0.1047
		Worksite safety management	0.1116	Worksite worker safety management	0.4286
				Worksite machine safety management	0.4286

(*Continued*)

Table 1. (*Continued*)

1st level DI	Weight	2nd level DI	Weight	3rd level DI	Weight
				Worksite condition management	0.1429
		Safety performance	0.0537	Safety performance evaluation procedure	0.5000
				Implementation of safety performance evaluation procedure	0.5000
		Encouragement mechanism	0.0537	Setup of encouragement measures	0.5000
				Implementation of encouragement measures	0.5000
		Emergency management	0.10056	Emergency plan	0.1634
				Emergency assets	0.2969
				Emergency plan exercising	0.5396
		Accident management	0.10056	Accident report and investigation	0.4286
				Accident files and information	0.1429
				Accident lessons learning	0.4286
		Incident management	0.10056	Incident report and investigation	0.4286
				Incident files and information	0.1429
				Incident lessons learning	0.4286
Safety culture	0.1634	Organizational safety culture	0.6667	Safety philosophy	0.4458
				Learning culture	0.2336
				Awareness of safety	0.2336
				Safety environment	0.0870
		Personal safety culture	0.3333	Workers attitude towards safety	0.5000
				Workers safety behavior	0.5000

Some diagnosis items should be analyzed quantitatively. When diagnosing these diagnosis items, statistics analysis should be used.

3.4 *Safety management diagnosis tools*

Any kind of diagnosis must be based on certain criteria. Diagnostic tools provide the criteria for safety management diagnosis. Each of the third level items has its own diagnosis tool. Each of the diagnosis tool this paper proposed contains item name, diagnosis content, item attributes, diagnosis method, diagnosis criteria, diagnosis records, diagnosis conclusions. Four classes of diagnosis methods corresponding to different types of diagnosis methods are provided, which are inspection diagnosis tool, questionnaire diagnosis tool, interviewing diagnosis tool and statistical diagnosis tool.

3.5 *Safety management diagnosis conclusion*

Safety management conclusions could be divided into comprehensive diagnosis conclusion, item diagnosis conclusion and specific problem analysis. Comprehensive diagnosis conclusion

Table 2. Enterprises safety management diagnosis procedure.

Phases	Key steps and job description
Diagnosis preparation	Enterprise application
	Data collection
	Diagnosis group setup
Initial diagnosis	Business briefings
	Company site visiting
	Investigation and analysis
	Plan to develop a formal diagnosis
Formal diagnosis	Concrete investigation
	Application of appropriate diagnosis tools
	Draw diagnosis conclusions
	Intervention development
	Diagnostic report writing
Guidance conduction	Guide the intervention program implementation
	Evaluation of diagnostic results

refers to the overall safety management diagnosis conclusion. Specific item diagnosis is the diagnosis result of a particular item. Specific problem diagnosis was the analysis of specific problem identified during the process of enterprise safety management.

According to research on development of safety management at home and abroad and the process of safety management evolution (Parker, 2006), the enterprise safety management fell into five levels (Table 3). Enterprise safety management diagnosis conclusions provided a specific safety management state of a certain enterprise and indicated the typical characteristics of their safety management. Diagnosis of the first level and second level diagnostic items are deprived from the third level diagnosis items' conclusion. Fuzzy Comprehensive Evaluation was adopted to do comprehensive diagnosis.

Enterprise safety management diagnosis can also provide diagnosis conclusions for items in different levels in the diagnosis item system. The diagnosis conclusions for each individual item also follows into five levels, which are highly immune state, health state, sub-health state, sick state and seriously sick state. Diagnosis conclusion for each individual item is very useful for improving of safety management.

Safety management level is not the only conclusion obtained through safety management diagnosis. What we care more are the specific problems in safety management. Descriptions of

Table 3. Enterprises safety management level.

Safety management status	Description
Highly immune state	Safety has been integrated into the organization and operation of the work, the values of safety has been realized; employees and managers have the attitude of vigilance when they do everything.
Health state	The organization used system to manage safety. Employees and managers truly recognize the importance of safety management and effectiveness of using system safety management; they follow safety procedures spontaneously.
Sub-health state	The organization set up safety management system, but the safety management system was not well implemented. Employees and managers follow those safety procedures proactively. They can't recognize the importance of safety from their heart.
Sick state	The organization search for solutions after accidents or emergencies. Those solutions can only solve problems which happened. Employees and managers only recognize the existence of risk after those incidents.
Seriously sick state	On safety issues, the organization and staff only care about whether they will be caught by relevant departments.

these specific problems include the time when it was found, the phenomenon of the problem, the possible consequences and the measures should be taken. These problems could provide necessary and important information for drawing comprehensive conclusions.

4 CONCLUSIONS

This paper discussed the necessity of safety management diagnosis and how to implement safety management diagnosis. A safety management diagnosis item system was set up. Safety management diagnosis procedure was provided. Safety management diagnosis methods were proposed. And safety management diagnosis conclusion patterns were discussed.

It was admitted that there were some problems that haven't been solved in the process of safety management diagnosis implementation. For example, more diagnosis methods could be selected, and some diagnosis tools were not very easy using. These problems could be solved in the further implementation of safety management diagnosis.

ACKNOWLEGEMENT

This paper was supported by Foundation for Young Teacher in Subjects Less Related to Oil of China University of Petroleum, Beijing.

REFERENCES

Chen L. 2006. *HSE management system improvement of oil well drilling industry*. Beijing: China University of Geosciences (Beijing).

Hu C. 2003. Research on Manufacturing Execution System Modeling and Intelligent Enterprise System: Hang Zhou: Zhen Jiang University.

Parker D. 2006. A framework for understanding the development of organizational safety culture. *Safety science*, 44(6):551.

Sun B., Z. Liu. 2006. Applying Fuzzy Comprehensive Evaluation Method on Safety Management of Building Enterprise. *China Safety Science Journal*, 24(11):125–128.

Tang M. 2001. New discussion on corporate management consultation theory and method. Beijing: Enterprises Management Press.

Zhao X., G. Fu, G. Xing. 2008. Research on Safety Management and its experiment design. *China Safety Science Journal*, 18(7):85–93.

Hydraulic Engineering – Xie (Ed.)
© 2013 Taylor & Francis Group, London, ISBN 978-1-138-00043-8

The study of Hydrogen Peroxide modified activated carbon on the adsorption of Trimethylamine exhaust

Ailing Ren, Peng Han, Bin Guo, Jing Han & Bei Li

The Institute of Environment Science and Technology, Hebei University of Science and Technology, Shijiazhuang, Hebei Province, China

ABSTRACT: The adsorption characteristics of activated carbon modified by H_2O_2 were investigated. The oxygen-containing functional groups were determined by Boehm's method and the specific surface area was studied by the BET method with N_2 adsorption. And the adsorption properties of the activated carbon modified by H_2O_2 were investigated with trimethylamine adsorption. The experiment studied the static adsorption capacity of trimethylamine, dynamic penetration curve, and the desorption activation energy of activated carbon before and after H_2O_2-modified.

1 INTRODUCTION

Activated carbon is widely used as the adsorbent of toxic and harmful gas in environment. But the specific surface area of activated carbon is not large enough and the adsorption selectivity is not good enough, thus the modification of activated carbon is a hot topic in the world. Wu Wen-yan modified the activated carbon by microwave method, and studied the adsorption of carbon disulfide. Through static and dynamic experiments in room temperature, she obtained that the modified active carbon had better adsorption capacity to carbon disulfide. Tang Hong used ferrous salt and copper salt formulations to modify activated carbon for ammonia and trimethylamine, the result showed that it had better adsorption capacity. Tamon found that the activated carbon modified by different concentrations of nitric acid had a better adsorption property. Soo-Jin Park treated the activated carbon with 30 wt% HCl and 30 wt% NaOH, the modified activated carbon was investigated with CO_2 and NH_3 adsorption, and the result showed that adsorption efficiency was greatly improved than before.

Generally the activated carbon for non-polar compounds has good adsorption effect, while the polar substances (e.g. trimethylamine) adsorption capacity is poor. However, in appropriate conditions after surface treatment with oxidation agent, the surface acidic groups content can be improved, then the surface polarity is enhanced, and it can improve the adsorption capacity of polar substances. There has been some research about the trimethylamine processing at home and abroad, but for the oxidation modified active carbon on trimethylamine adsorbing is reported rarely. This study used hydrogen peroxide on oxidation modification method, through its specific surface area, oxygen-containing functional groups and the adsorption quantity of trimethylamine, studied the properties of hydrogen peroxide modified activated carbon by the adsorptio n of trimethylamine.

2 EXPERIMENTAL CONTENT AND METHOD

2.1 *Modified activated carbon preparation*

Per-treatment: The active carbon was washed 3 times with distilled water, and boiled for 30 minutes, and then the activated carbon was dried in 105°C until the quality did not change.

Hydrogen peroxide modified: 5 g of per-treatment active carbon were mixed with 25 ml different concentrations of H_2O_2 (5%, 10%, 15%, 20%, 25%) respectively, oscillated at room temperature for 1.5 h, suction filtration, the filter cake was dried in the temperature of 105°C for 12 h, then obtained different concentration of modified activated carbon. In the best modified concentration conditions, changed activated carbon H_2O_2-impregnated time (0.5 h, 1.0 h, 1.5 h, 2.0 h, 2.5 h), on oscillator at room temperature, air pump filtration, the filter cake was dried for 12 h in 105°C, then obtained different impregnation time of H_2O_2 modified activated carbon.

2.2 Boehm

The Boehm titration works on the principle that oxygen groups on carbon surfaces have different acidities and can be neutralized by bases of different strengths. A known mass of 1.5 g per-treatment activated carbon was added to 50 ml of one of three reaction bases of 0.05 M concentration: $NaOH$, Na_2CO_3, $NaHCO_3$. Then the samples were agitated by shaking for 24 h and then filtered, and 10 ml filtrate was taken by pipette from the samples. The filtrate of the reaction base $NaHCO_3$ and $NaOH$ was acidified by the addition of 20 ml of 0.05 M HCl, the filtrate of the reaction base Na_2CO_3 was acidified by the addition of 30 ml of 0.05 M HCl. Then acidified solutions were back-titrated with 0.05 M $NaOH$, the titrated base. Then calculate the amount of carbon surface oxygen functional groups.

2.3 BET measurement

BET was manufactured by Quantachrome Instruments, Inc. of U.S., it's the NOVA series of automatic specific surface area and pore size distribution analyzer (NOVA2000e).The measurement principle is based on the BET multi-molecular layer of sorption quantity. Quantitative degassed activated carbon sample is placed under stream of nitrogen, when the activated carbon adsorption nitrogen balance, determination of adsorption quantity, and calculated the specific surface area of activated carbon. BET multi-molecular layer adsorption model can be a more reasonable interpretation of the adsorption isotherm process.

2.4 Determination of static adsorption amount

Take a dryer, the aqueous solution of trimethylamine was put in the bottom layer, and the activated carbon samples (weighing 0.1 g) were put in the upper layer, then capped. The system was placed in the indoor temperature condition, the trimethylamine was naturally evaporated, then formed of a certain concentration of organic gas, and through the activated carbon surface contact adsorb trimethylamine. The quality of activated carbon samples were measured at intervals of a certain time, when activated carbon sample mass (M_1) no longer changed, indicating that the adsorption reached equilibrium.

The adsorption capacity $= (M_1 - M_0)/M_0$.

2.5 Dynamic adsorption breakthrough curves

The dynamic adsorption curve and adsorption desorption activation energy (TP-5080) was studied. The desorption activation energy calculation model, it's a theoretical model which is based on the adsorption kinetics, on the basis of adsorption and desorption rate equation established.

3 RESULTS AND DISCUSSION

3.1 The performance of different concentration H_2O_2 modified activated carbon

Modified by difference concentrations of H_2O_2 (5%, 10%, 15%, 20%, 25%), the performance of modified activated carbon was shown in Table 1.

Table 1. The performance of different concentrations of H_2O_2 solution modified activated carbon and the adsorption of TMA.

Sample	Oxygen functional groups (mmol/g)	Specific surface area (m²/g)	Static adsorption capacity (mg/g)
Base-AC	0.6827	670.958	159.4
5%-AC	1.3702	761.834	309.1
10%-AC	1.5298	793.560	317.7
15%-AC	1.8244	995.236	440.9
20%-AC	1.8373	752.165	301.3
25%-AC	1.8490	747.880	298.9

Table 2. The performance of 15% H_2O_2 concentration, different dipping time modified activated carbon and the adsorption of TMA.

Sample	Oxygen functional groups (mmol/g)	Specific surface area (m²/g)	Static adsorption capacity (mg/g)
15%-0.5 h-AC	1.3208	737.417	353.6
15%-1.0 h-AC	1.6392	775.032	383.9
15%-1.5 h-AC	1.8244	995.236	440.9
15%-2.0 h-AC	1.8298	769.343	368.6
15%-2.5 h-AC	1.8393	727.754	335.4

Table 1 shows that, in the concentration range of 5%–25%, at the concentration of 15%, the static adsorption capacity of carbon to trimethylamine reached 440.9 mg/g which was the highest, increased by 176.6% compared with before. The oxygen-containing functional groups of the activated carbon increased with H_2O_2 concentration increasing. At the concentration of 15%, the oxygen-containing functional groups reached 1.8244 mmol/g, and then changed little with the increasing of the solution concentration. While with the increasing of the concentration, the oxidation of activated carbon was gradually enhanced, the corrosion of the inner surface pore reinforced, thereby the pore collapsed. At the concentration of 15%, the specific surface area reached 995.236 m²/g, and then the surface area reduced with the concentration increasing. In the modified process, the pore surface of activated carbon forms a large amount of oxygen-containing functional groups, and these oxygen-containing functional groups adsorbed on the internal pore surface, resulting in pore stenosis and the specific surface area reduction.

When the concentration of H_2O_2 is 15%, the adsorption efficiency is better.

3.2   The performance of different impregnated time H_2O_2- modified activated carbon

Modified by different impregnated time (0.5 h, 1.0 h, 1.5 h, 2.0 h, 2.5 h) at 15% H_2O_2 solution, the performances of modified activated carbon were shown in Table 2.

Table 2 shows that, the specific surface area of activated carbon is the largest when dipping time is 1.5 h, the adsorption capacity reaches 440.9 mg/g, increased 176.6% compared with unmodified. The oxygen containing functional groups of activated carbon increased with the dipping time increasing, but after 1.5 h, it changed little. The specific surface area reaches the maximum which was 995.236 m²/g at the time of 1.5 h.

In the following section, we studied dynamic adsorption and desorption performance of the modified activated carbon with 15% H_2O_2 solution, dipping time of 1.5 h.

3.3   The dynamic adsorption breakthrough curve

The adsorption breakthrough curve is shown in Figure 1, which shows that the breakthrough time (the breakthrough point is $C/C_0 = 0.25$) of base activated carbon was 10 min, however

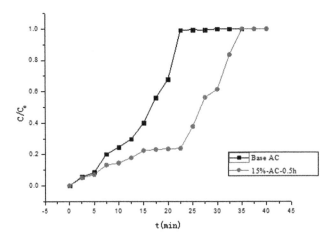

Figure 1.   Adsorption penetration curve for trimethylamine.

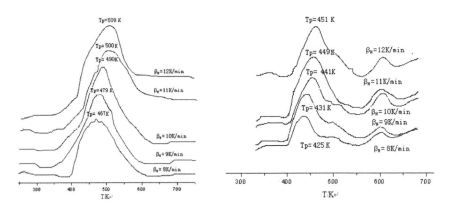

Figure 2.   Trimethylamine adsorption spectrogram on activated carbon at different heating rates.

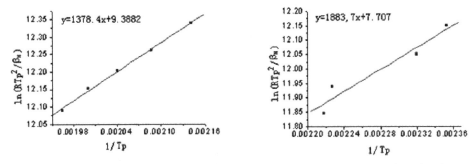

Figure 3.   The linear fit chart of trimethylamine desorption activation energy on activated carbon.

the breakthrough time of modified activated carbon in which the concentration of H_2O_2 solution was 15% impregnated with 1.5 h was 35 min. Observing the graphics area which was surrounded by the curve and the y axis, we can see that the adsorption capacity of the activated carbon modified by hydrogen peroxide is higher than the base activated carbon, and the result matches with the static adsorption capacity determination result.

Table 3. TMA desorption activation energy on modified activated carbons.

Sample	Desorption peak on different heating rates [$T_p(K)$]					Activation energy of desorption (kJ/mol)
	8	9	10	11	12	
Base AC	467	479	490	500	509	11.461
Modified AC	425	431	441	449	451	15.663

3.4 *The desorption activation energy*

The activated carbon desorption activation peaks gradually increased with the increasing heating rate. Figure 2 shows that the desorption spectra of base activated carbon and the modified activated carbon (15%-1.5 h-AC) at different heating rates. Figure 3 shows that the linear fit of desorption peak which obtained at different heating rates. The desorption activation energy which was calculated by the slope of a straight line in Figure 3 was shown in Table 3.

The desorption activation energy of modified activated carbon reflects the interaction force between adsorbent and adsorbate. The desorption activation energy of trimethylamine of the base activated carbon and the modified activated carbon (15%-1.5 h-AC), was 11.461 kJ/mol and 15.663 kJ/mol respectively. The desorption activation energy of modified activated carbon has improved significantly compared with the base activated carbon.

4 CONCLUSIONS

1. The specific surface area of H_2O_2-modified activated carbon can reach 995.236 m^2/g, increased by 28.3% than unmodified. The oxygen-containing functional groups were increased from 0.6827 mmol/g to 1.8490 mmol/g.
2. In the experimental concentration range of 5%–25% H_2O_2 solution, the activated carbon was modified by the 15% H_2O_2 solution and impregnated with 1.5 h has the higher adsorption efficiency for trimethylamine, the adsorption capacity can reach 440.9 mg/g, increased by 176.6%.
3. The dynamic adsorption breakthrough time was longer compared with the unmodified. The desorption activation energy was increased from 11.461 kJ/mol to 15.663 kJ/mol, and the adsorption performance of modified activated carbon was improved.

ACKNOWLEDGEMENT

The research was supported by the commonweal project of national environmental protection (No. 201109004). We also thank the reviewers for helpful comments.

REFERENCES

Goertzen, S.L. Theriault, K.D. & Oickle, A.M. et al. 2010. Standardization of the Boehm titration: Part I. CO$_2$ expulsion and endpoint determination. *Carbon 48*:1252–1261.

Masaki, T. Hiromu, Y. & Nobuo, T. 2007. The effect of treatment of activated carbon by H$_2$O$_2$ or HNO$_3$ on the decomposition of pentachlorobenzene. *Applied Catalysis B: Environmental 74*:179–186.

Park, S.J. & Kim, K.D. 1999. Adsorption behaviors of CO$_2$ and NH$_3$ on chemically surface-treated activated carbons. *Journal of Colloid and Interface Science 212*:186–189.

Tamon, H. & Okazaki, M. 1996. Influence of acidic surface oxides of activated carbon on gas adsorption characteristics. *Carbon 34(6)*:741–746.

Tang, H. Pang, Y.F. & Li, Q.D. 2000. The study of the modified activated carbon adsorption characteristics of ammonia and trimethylamine. *Environmental Chemistry 19(5)*:431–435.

Wu, W.Y. Mei, H. & Yao, H.Q. 2010. Behavior of Adsorbing CS2 on Modified Activated Carbon. *Environmental Science and Technology 33(12)*:156–159.

Hydraulic Engineering – Xie (Ed.)
© 2013 Taylor & Francis Group, London, ISBN 978-1-138-00043-8

Biodiversity of marine life in the Huanghe (Yellow River) estuary and adjacent area

Huanjun Zhang, Zhenbo Lv & Fan Li
Shandong Marine Fisheries Research Institute, Shandong Provincial Key Laboratory of Restoration for Marine Ecology, Yantai, Shandong, China

Liang Zheng & Tiantian Wang
Shandong Marine Fisheries Research Institute, Shandong Provincial Key Laboratory of Restoration for Marine Ecology, Yantai, Shandong, China
Shanghai Ocean University, Institute of Marine science, Shanghai, China

ABSTRACT: To gain a better understanding the inference of the Water and Sediment Discharge Regulation (WSDR) project on the marine biodiversity in the estuary and adjacent area, we investigated surveys at 8 stations in June and July. In each station, the abundance and/or biomass of every species, including Phytoplankton, Zooplankton, Macrobenthos, Ichthyoplankton and Nekton were recorded. The numbers of species (S), abundance (N) and/or biomass (W), Margalef richness index (D), Shannon-Wiener diversity index (H') and Pielou's evenness index (J') were calculated. A t-test was used for testing if there exists significant difference between the two surveys. The result showed that even the values of S, N or W, D', H' and J' changed after WSDR, but no significant difference were found in each index of Zooplankton, Ichthyoplankton and Macrobenthos. Some indexes, such as S, D and J' of Phytoplankton, W of Nekton, has significant difference between the two surveys.

1 INTRODUCTION

The Huanghe (Yellow) River is well known in the world for carrying the highest concentration of silt. It delivered about 560×10^8 m^3 water and 10.8×10^8 tons of sediment annually to the Bohai Sea (Milliman & Meade 1983, Li & Zhang 2001). It is the main nutrient and freshwater input for Bohai Sea. But since the 1950s the Huanghe's water and sediment discharge to the sea has steadily decreased (Yang et al. 1998, Wang et al. 2006). In the 10 years from 1990 through 1999, the average water discharge from the Huanghe to the sea decreased to 13.2×10^9 m^3/year, only 28.7% of its discharge in the 1950s. The average sediment discharge during the period was 389.9×10^6 tons/year, only 29.5% of that in the 1950s (Yang et al. 1998). The decline of the sediment load, as well as synchronous decreases in water discharge has had great physical, ecological effects. The flow broke frequently, the channel shrank, the estuarine environment deteriorated and the fishery productivity decreased (Yang et al. 1998, Jiao et al. 1998). The Huanghe River Conservancy Commission decided to carry out the most significant Water and Sediment Discharge Regulation Project (WSDR) in Chinese river engineering history, to solve the severe problems in the Lower Huanghe River channel. The core objective of the WSDR project was to form an artificial flood peak with limited water to scour the lower channel and protect the riparian and estuarine ecological environment (Ma et al. 2010).

Since the WSDR project was carried out in 2002, water and sediment discharge to Bohai Sea mainly during the WSDR session. From June 19, 2010 to July 7, the tenth WSDR project was carried out alongside scientific adjustment of the discharge time, flow velocity, and water levels of the Wanjiazhai, Sanmenxia and Xiaolangdi reservoirs. Within 19 days, massive freshwater and sediment were put into Huanghe estuary. Would marine life community diversity affect by massive water and sediment discharge in such short period of time? The present work reports on a study on the marine life community diversity in the Huanghe estuary and adjacent area, situated in the Bohai Sea (China), and its main aims were as follows: (a) to identify marine life community diversity in the Huanghe estuary; (b) to distingue community diversity responded to WSDR: threaten or chance?

2 MATERIALS AND METHODS

2.1 *Study area*

The modern Yellow River Delta, located in Shandong Province, China, has developed since 1855 (Chuet al. 2006). Since 1996 August, water and sediment transported into Bohai Sea is via Qingbacha channel other than Qingshuigou promontory. Samples were collected at 8 stations between 119°09′–119°30′E, 37°30′–38°00′N in the Huanghe estuary.

2.2 *Sampling techniques*

At these stations, samples were all conducted by the specification 'specification for oceanographic survey-Part 6: Marine biological survey' (GBT 12763.6-2007). Phytoplankton were sampled by shallow-water plankton net (type III) and zooplankton by type II. Ichthyoplankton samples were collected horizontally with large plankton netat 10 min duration at a speed of 2.0 kn on the sea surface. Macrobenthos samples were taken with a 0.1 m² grab samplers. Nekton samples were collected with bottom trawl at 1 hour at a speed of 3.5 kn.

2.3 *Data analyses*

Differences between two surveys were determined by a *t*-test. Species richness was estimated by a Margalef richness index (*D*) (Margalef 1958); species diversity by a Shannon-Wiener diversity index (*H′*) (Krebs 1989); and Pielou's evenness index (*J′*) (Pielou 1966) were used for evenness. Different formula was used for calculating Shannon-Wiener index. Abundance was used for Phytoplankton, Zooplankton, Ichthyoplankton, and the base was

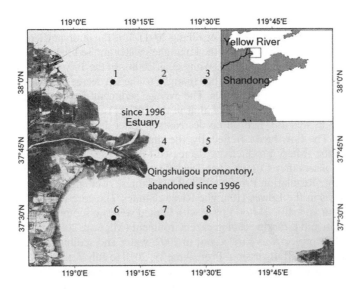

Figure 1.   Sampling stations in the Huanghe estuary and adjacent area.

set to 2. Biomass was used for Macrobenthos and Nekton, and the base was set to natural logarithm (e).

3 RESULTS AND DISCUSSION

3.1 Phytoplankton

The values of numbers of species (S), abundance (N), Margalef index (D) and Shannon-Wiener index (H') declined sharply after WSDR. But the value of Pielou's index (J') increased after WSDR.

3.2 Zooplankton

The mean value of S and N declined after WSDR. But these values did not decline in all station. The value increased in some stations. The value of D, H' and J' increased a little after WSDR.

3.3 Ichthyoplankton

The trend of ichthyoplankton was similar with the trends of Phytoplankton. The value of S, N, D and H' declined, but the value of J' increased.

Table 1.   Diversity of Phytoplankton community in the Huanghe estuaryand adjacent area.

	S		N		D		H'		J'	
Station	BW	AW	BW	AW	BW	AW	BW	AW	BW	AW
1	15	3	30.86	0.81	1.108	0.222	2.235	1.585	0.572	1.000
2	17	2	11.65	0.42	1.372	0.120	2.821	0.918	0.690	0.918
3	13	8	14.56	2.14	1.009	0.702	2.419	2.755	0.654	0.918
4	17	8	56.79	11.40	1.208	0.601	2.970	2.648	0.727	0.883
5	14	11	12.94	13.05	1.104	0.849	3.037	2.459	0.798	0.711
6	20	8	96.12	5.92	1.379	0.637	3.046	2.667	0.705	0.889
7	21	5	23.32	7.97	1.618	0.354	3.384	1.591	0.770	0.685
8	15	20	20.35	93.45	1.145	1.382	1.741	2.822	0.446	0.653
Average	16.5	8.1	33.3	16.9	1.243	0.608	2.707	2.181	0.670	0.832

S, Numbers of species; N, Abundance; D, Margalef index; H', Shannon-Wiener index; J', Pielou's index; BW, Before WSDR; AW, After WSDR.

Table 2.   Diversity of Zooplankton community in the Huanghe estuaryand adjacent area.

	S		N		D		H'		J'	
Station	BW	AW	BW	AW	BW	AW	BW	AW	BW	AW
1	11	11	216.5	50.2	1.860	2.554	2.642	2.419	0.764	0.699
2	11	9	246.9	22.0	1.815	2.588	2.267	2.617	0.655	0.826
3	15	17	124.0	471.2	2.904	2.599	3.008	2.835	0.770	0.694
4	18	12	201.5	998.2	3.204	1.593	2.952	2.322	0.708	0.648
5	12	11	391.6	12.1	1.842	4.011	2.490	3.110	0.695	0.899
6	22	14	598.9	207.7	3.284	2.436	2.244	2.661	0.503	0.699
7	23	19	1291.0	144.9	3.071	3.617	1.989	2.670	0.440	0.629
8	16	16	337.6	85.9	2.576	3.368	2.311	2.979	0.578	0.745
Average	16.0	13.6	426.0	249.0	2.570	2.846	2.488	2.702	0.639	0.730

S, Numbers of species; N, Abundance; D, Margalef index; H', Shannon-Wiener index; J', Pielou's index; BW, Before WSDR; AW, After WSDR.

Table 3. Diversity of Ichthyoplankton community in the Huanghe estuaryand adjacent area.

Station	S BW	S AW	N BW	N AW	D BW	D AW	H' BW	H' AW	J' BW	J' AW
1	7	4	41	6	1.616	1.674	1.743	1.918	0.621	0.959
2	0	1	0	19	0	0	0	0	0	0
3	5	2	59	2	0.981	1.443	1.113	1.000	0.479	1.000
4	2	3	82	141	0.227	0.404	0.421	1.290	0.421	0.814
5	3	0	14	0	0.758	0	1.414	0	0.892	0
6	6	1	89	211	1.114	0	1.970	0	0.762	0
7	6	5	972	20	0.727	1.335	0.750	1.882	0.290	0.811
8	5	2	77	4	0.921	0.721	1.215	1.000	0.523	1.000
Average	4.3	2.3	166.8	50.4	0.793	0.697	1.078	0.886	0.499	0.573

S, Numbers of species; N, Abundance; D, Margalef index; H', Shannon-Wiener index; J', Pielou's index; BW, Before WSDR; AW, After WSDR.

Table 4. Diversity of Macrobenthos community in the Huanghe estuaryand adjacent area.

Station	S BW	S AW	W BW	W AW	D BW	D AW	H' BW	H' AW	J' BW	J' AW
1	7	21	1.4	3.2	1.086	3.367	0.893	1.312	0.459	0.431
2	23	26	53.5	13.4	3.255	2.300	0.710	0.637	0.226	0.195
3	14	18	105.1	1.1	2.486	2.169	1.829	1.824	0.693	0.631
4	17	24	17.5	5.4	1.753	3.428	1.276	2.078	0.451	0.654
5	6	21	0.7	28.9	1.041	3.347	0.758	1.105	0.423	0.363
6	16	22	11.7	4.1	2.031	3.119	1.190	1.428	0.429	0.462
7	30	6	14.2	10.0	3.973	1.191	1.827	0.883	0.537	0.493
8	10	25	2.9	32.4	1.826	3.023	0.675	0.330	0.293	0.103
Average	15.4	20.4	25.9	12.3	2.181	2.743	1.145	1.200	0.439	0.417

S, Numbers of species; W, Biomass; D, Margalef index; H', Shannon-Wiener index; J', Pielou's index; BW, Before WSDR; AW, After WSDR.

3.4 Macrobenthos

The trend of Macrobenthos was different with plankton. Its value of numbers of species declined, but its abundance increased. The large individuals were replaced by small ones. The values of D and H' increased, and the values of J' declined.

3.5 Nekton

The values of S, D, H' and J' were similar between the nekton communities of before-WSDR and after-WSDR. The value of abundance increased sharply after WSDR.

3.6 T-test

T-test shows that no significant difference was found in each index of Zooplankton, Ichthyoplankton and Macrobenthos. Significant difference was only found in phytoplankton's S, D, J' and Nekton's W.

Table 5. Diversity of Nekton community in the Huanghe estuary and adjacent area.

Station	S		W		D		H'		J'	
	BW	AW	BW	AW	BW	AW	BW	AW	BW	AW
1	18	18	9.3	71.7	1.767	1.910	1.353	1.513	0.468	0.523
2	11	20	2.2	69.1	1.570	2.032	1.844	1.638	0.769	0.547
3	14	23	6.2	56.6	1.901	2.275	1.394	1.856	0.528	0.592
4	25	25	12.5	166.6	3.168	2.376	2.435	2.196	0.757	0.682
5	20	26	15.6	239.3	2.208	2.311	1.533	2.210	0.512	0.678
6	19	22	17.8	154.7	2.528	2.136	1.115	1.819	0.379	0.588
7	25	26	24.2	161.3	3.179	2.557	1.976	2.134	0.614	0.655
8	21	25	10.9	289.3	2.636	2.298	2.098	1.640	0.689	0.510
Average	19.1	23.1	12.3	151.1	2.370	2.237	1.719	1.876	0.590	0.597

S, Numbers of species; W, Relative biomass index; D, Margalef index; H', Shannon-Wiener index; J', Pielou's index; BW, Before WSDR; AW, After WSDR.

Table 6. T-test for significant difference between 2 surveys.

Taxa	S	N or W	D	H'	J'
Phytoplankton	3.753*	1.084	4.017*	1.667	−2.653*
Zooplankton	1.148	0.987	−0.778	−1.355	−1.693
Ichthyoplankton	1.948	0.978	0.313	0.519	−0.380
Macrobenthos	−1.370	1.014	−1.231	−0.206	0.264
Nekton	−1.983	−4.665*	0.581	−0.853	−0.133

*Difference is significant ($P < 0.05$).

4 CONCLUSION

Salinity, temperature, oxygen concentration and productivity were the most factors influence the distributions of marine life (Prista et al. 2003). WSDR carried massive freshwater and nutrient into the estuary and adjacent area, which would result in the decline of salinity and the increase of productivity (we has monitored it). This will inevitably affect the distribution of the marine life living in the estuary and adjacent area. And then alter the values and distributions of community diversity of different marine life. Our result showed that even the values of S, N or W, D', H' and J' changed after WSDR, but no significant difference were found in each index of Zooplankton, Ichthyoplankton and Macrobenthos. That is to say, the diversity of Zooplankton, Ichthyoplankton and Macrobenthos is much similar in the two surveys. But some indexes, such as S, D and J' of Phytoplankton, W of Nekton, has significant difference between the two surveys. Phytoplankton would vulnerable to the impact of nutrient changes. The dramatic changes of nutrients lead to changes in the species composition and abundance distribution of phytoplankton. And the diversity index of phytoplankton fluctuations during the period of WSDR. Significant difference of Nekton's W maybe caused by change of environmental factor and/or food organisms. But the reasons leading to this result may also include time difference of migratory fish enter the investigated sea, juveniles added into the fishing net.

This study was only based on small-scale surveys, both in time and space. The conclusion may be biased. Furth study should be held to know more about the effect of WSDR project.

ACKNOWLEDGEMENTS

The authors would like to thank everyone involved insampling surveys and analysis, and the reviewers for their helpful comments and suggestions. This research was supported by the National Marine Public Welfare Research Project (200905019) and fund of 'Taishan scholar of aquatic nutrition and feed'.

REFERENCES

Chu, Z.X., Sun X.G., Zhai, S.K. & Xu, K.H. 2006. Changing pattern of accretion/erosion of the modern Yellow River (Huanghe) subaerial delta, China: based on remote sensing images. *Marine Geology* 227: 13–30.

Jiao, Y.M., Zhang, X.H. & Li, H.X. 1998. Influence on fish diversity in the sea area off the Huanghe river estuary by the cutoff of water supply. *Transactions of Oceanology and Limnology* 4: 48–53.

Krebs, C.J. 1989. *Ecological Methodology*. New York: Harper Collins Publishers.

Li, F. & Zhang, X.R. 2001. Impact of variation of water and sediment fluxes on sustainable use of marine environment and resources in the Huanghe river estuary and adjacent sea II. Variation of marine environment due to water flow cut-off and decrease greatly the volume of water imputed the sea from the Huanghe River. *Studia Marina Sinica* 43: 60–67.

Milliman, J.D. & Meade, R.H. 1983. World-wide delivery of river sediment to the oceans. *The Journal of Geology* 91(1): 1–21.

Margalef, R. 1958. Information theory in ecology. *General Systems* 3: 36–71.

Ma, Y.Y., Li, G.X. & Ye, S.Y. 2010. Response of the distributary channel of the Huanghe River estuary to water and sediment discharge regulation in 2007. *Chinese Journal of Oceanology and Limnology* 28(6): 1362–1370.

Pielou, E.C. 1966. The use of information theory in the study of ecological succession. *Journal of Theoretical Biology* 10: 370–383.

Prista, N., Vasconcelos, R.P., Costa, M.J. & Cabral, H. 2003. The demersal fish assemblage of the coastal area adjacent to the Tagusestuary (Portugal): relationships with environmental conditions. *Oceanologica Acta* 26: 525–536.

Wang, H.J., Yang, Z.S., Saito, Y., Liu, J.P. & Sun, X.X. 2006. Interannual and seasonal variation of the Huanghe (Yellow River) water discharge over the past 50 years: Connections to impacts from ENSO events and dams. *Global and Planetary Change* 50: 212–225.

Yang, Z.S., Milliman, J.D., Galler, J., Liu, J.P. & Sun, X.X. 1998. Yellow River's water and sediment discharge decreasing steadily. *EOS* 79(48): 589–592.

Hydraulic Engineering – Xie (Ed.)
© 2013 Taylor & Francis Group, London, ISBN 978-1-138-00043-8

Influence of summer stratification on the vertical distribution of picoplankton in the Southern Yellow Sea

Pei Qu
College of Environmental Science and Engineering, Ocean University of China, Qingdao, P.R. China
Research Center for Marine Ecology, The First Institute of Oceanography, SOA, Qingdao, P.R. China

Zongling Wang, Min Pang, Mingzhu Fu, Zongjun Xu, Xinming Pu & Ping Sun
Research Center for Marine Ecology, the First Institute of Oceanography, SOA, Qingdao, P.R. China

ABSTRACT: The relative abundance of different picoplankton components and their relationships with environmental factors in the southern Yellow Sea were investigated in August 2011. The average abundance of picoplankton (picoeukaryotes, *Synechococcus* and heterotrophic bacteria) were of the order of 10^3, 10^4 and 10^6 cells/mL, respectively. Two transects (124°E and 35°N) were conducted to reveal the vertical distribution of picoplankton with changing environmental factors, including temperature, salinity, nutrients and chlorophyll. The maximum abundance of picoplankton was in the thermocline. Autotrophic picoplankton made up the majority of total biomass above the thermocline, while heterotrophic bacteria had the highest abundance in the deeper water. The vertical distribution of chlorophyll was similar to that of picophytoplankton, and there was a significant positive correlation ($P < 0.01$) between them. Surprisingly, there was a negative correlation between picoplankton and nutrients. The cytochrome content also varied according to the changes in the environment.

1 INTRODUCTION

Picoplankton (0.2–2 μm) are ubiquitous throughout the world's oceans. They are a prime contributor to productivity, especially in the oligotrophic ocean, since they can use nutrients efficiently. They are primarily composed of *Synechococcus*, *Prochlorococcus* and picoeukaryotes. *Prochlorococcus* is not as ubiquitous as the other two groups because of its low temperature tolerance (Moore et al., 1995). Heterotrophic bacteria are the main heterotrophic components of picoplankton. They are part of the microbial loop, in which they degrade dissolved and particulate organic matter, and recycle inorganic nutrients needed by primary producers (Mukhanov et al., 2007). Both picophytoplankton and heterotrophic bacteria form the base of the food chain, supplying material to higher trophic levels, and their importance has become much more appreciated in recent years (Blanchot, et al., 1992; Blanchot and Rodier 1996; Zubkov et al., 1998; Boyd et al., 1999; Jiao et al., 2005; Zhang et al., 2008).

The aim of the present study was to investigate various environmental effects on picoplankton, especially on their vertical distribution. This research was part of the cooperation between China and Republic of Korea involving the investigation of biodiversity in the ecosystem of the southern Yellow Sea. The southern Yellow Sea is a marginal sea in the northwestern Pacific Ocean. Warming of the surface seawater results in thermal stability so that the deeper water exchange with the upper layer is restricted when there is strong stratification in summer (Ren et al., 2005; Yu et al., 2006). This work mainly focused on the vertical distribution of the picoplankton community and the primary factors impacting it in August 2011. This study will help to reveal the interaction between picoplankton and environmental parameters and provide basic data in support of further research on the large marine ecosystem in the Yellow Sea.

2 METHOD

2.1 Description of the research area

Since solar radiation is strongest in summer, the stratification is also strongest in that season. We occupied 33 stations along four transects in the southern Yellow Sea in August, 2011. Thirteen stations on Transect A were distributed along 124°E, which separates the southern Yellow Sea from the middle, and three other transects were distributed along 34°N, 35°N and 36°N (Fig. 1). Due to their location, Transects A and C were the main focus of this study.

2.2 Sample collection

Seawater samples were collected with a CTD rosette sampler with 12 bottles at 0, 10, 20, 30, 50 and 75 m depth and fixed with 1% paraformaldehyde (Marie et al., 1997) in 5 mL tubes. The samples were quickly frozen in liquid nitrogen on board and stored at −80 °C until analysis in the laboratory Nutrients were analyzed by an automatic analyzer. Hydrographic data including salinity, temperature and density were collected using a CTD (Seabird25, USA) equipped with a Seapoint chlorophyll-*a* detector.

The samples were analyzed with a FACSCalibur™ flow cytometer (Becton-Dickinson, USA) equipped with a 488-nm laser. Autotrophs were detected by their characteristic cytochrome phycobilins (mainly in *Synechococcus*) and chlorophyll (mainly in picoeukaryotes). Plots of SSC (side scatter) vs. FL3 (red fluorescence) and of FL2 (orange fluorescence) vs. FL3 were displayed together. As heterotrophs do not have photosynthetic pigments, we stained their DNA with SYBR Green I, and plots of SSC vs. FL1 (green fluorescence) and FL1 vs. FL3 plots were displayed together (Gasol, 1999; Marie et al., 2004). According to the method described by Jochem (2001), four clusters of bacteria were detected. Clusters B-Ia and B-Ib corresponded to B-I bacteria with low DNA content, while cluster B-II was distinguished by higher DNA content, but lower SSC than that of B-I (Li et al., 1995; Marie et al., 1997).

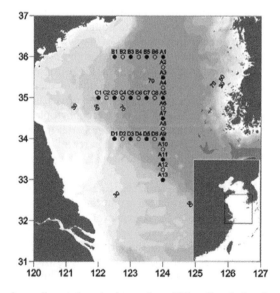

Figure 1.   Location of sampling stations in the southern Yellow Sea during the cruise in August 2011; black dots (●) show the stations where biological, chemical and physical data were collected; white dots (○) show the stations where only chemical and physical data were collected.

A cluster termed B-III by Marie et al. (1997) was not detected in our study area. Instead, B-IV bacteria with higher DNA content, higher SSC were detected.

Carbon biomass was estimated from cell abundance using coefficients of 250, 2100 and 20 fgC/cell for *Synechococcus*, picoeukaryotes and heterotrophic bacteria, respectively (Simek et al., 1999; Moreira-Turcq et al., 2001; Grob et al., 2007; Zhang et al., 2008).

3 RESULTS

3.1 *Physical and chemical conditions*

Temperature is high in the surface layer and therefore there is strong stratification which restricts exchange with the deep water. Salinity and nutrient distribution also show strong stratification. Transect A separates the southern Yellow Sea at 124°E (Fig. 1), and it represents the typical vertical feature (Fig. 2). The thermocline was at 10 to 40 m depth, with the temperature ranging from 23°C (surface) to 6°C and the salinity ranging from 30.7 to 33.6. The maximum concentration of total Dissolved Inorganic Nitrogen (DIN) was 12.7 µM (34.5°N); phosphate was 1.85 µM (34.5°N), and the maximum silicate concentration was 12.4 µM (33.5°N) (Table 1). The nutrient mixed layer was deeper on the northern transect than on the southern transect. The chlorophyll maximum layer was approximately parallel with the seabed, usually at about 20 to 30 m depth, but occasionally at surface.

Transect C was at 35°N and crossed Transect A at Station A5 (Figure 1). As on Transect A, strong stratification occurred on this transect (Fig. 3). The thermocline was at 10 to 30 m depth, and the temperature ranged from 23°C to 7°C. The salinity ranged from 30.4 to 33.6

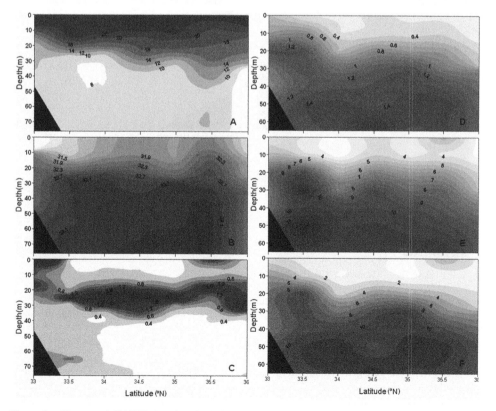

Figure 2.   Transect A (124°E) showing the latitudinal and vertical distribution of (A) temperature (°C), (B) salinity, (C) chlorophyll (µg/L), (D) phosphate (µM), (E) silicate (µM), and (F) total dissolved inorganic nitrogen (DIN, µM).

Table 1.   Cell abundance and environmental variables along Transects A and C and in the total study area. (17 biological stations, 33 physical and chemical stations). Syn. = *Synechococcus*, Euk. = eukaryotic picoplankton, Bac. = heterotrophic bacteria.

	Transect A	Transect C	Total study area
Temperature (°C)	6.2–23.1	7.1–23.4	6.2–23.4
Salinity (psu)	30.7–33.6	30.4–33.6	30.4–33.6
Chlorophyll (µg/L)	0.2–6.0	0.2–2.5	0.2–6.0
DIN (µM)	0.5–12.7	0.5–22.3	0.5–22.3
Phosphates (µM)	0.2–1.9	0.2–1.5	0.2–1.9
Silicate (µM)	3.1–12.4	3.3–19.9	3.1–19.9
Syn. (10^4 cells/mL)	0.39–34.81	0.42–34.36	0.39–34.81
Euk. (10^4 cells/mL)	0.07–5.02	0.08–3.58	0.05–5.02
Bac. (10^4 cells/mL)	55.64–163.39	57.53–154.20	51.93–224.99

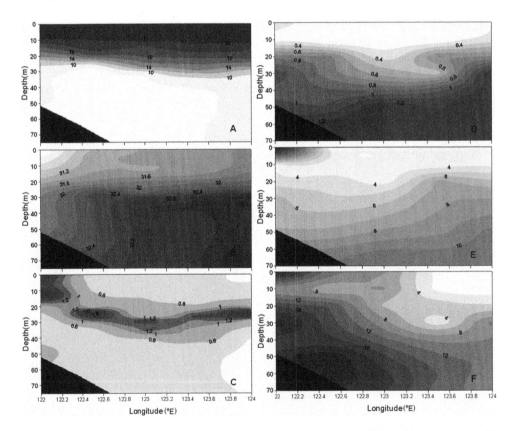

Figure 3.   Transect C (35°N) showing the latitudinal and vertical distribution of (A) temperature (°C), (B) salinity, (C) chlorophyll (µg /L), (D) phosphate (µM), (E) silicate (µM), and (F) total dissolved inorganic nitrogen (DIN, µM).

(Table 1), and there were gradients from the coast to the offshore deeper water. The salinity data indicated that in the western part of Transect C, environmental factors were strongly affected by the coastal current, and the concentrations of nutrients were abnormally high, in some cases reaching eutrophic conditions. Excepting in the western area, nutrients increased significantly with depth ($P < 0.01$). The chlorophyll maximum layer was around 30 m depth, and it was also influenced by abnormally high nutrient concentrations in the western part of the transect.

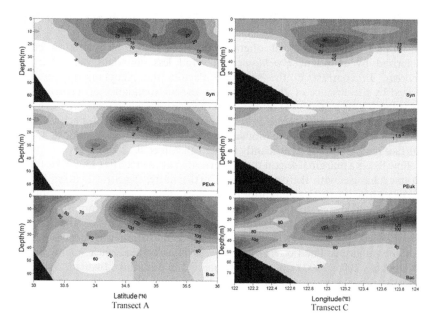

Figure 4. Vertical distribution of picoplankton (10⁴ cells/mL) along Transect A (124°E) and C (35°N).

3.2 Vertical distribution of picoplankton

Although *Prochlorococcus* is ubiquitous in tropical and subtropical oceans, it was not detected in the southern part of our study area, in contract to previous research (Jiao et al, 2005). Although temperature can temporarily reach their growth requirements, the higher temperatures do not last long enough to maintain long-term survival in this environment. The other picoplankton groups, *Synechococcus,* picoeukaryotes and heterotrophic bacteria, were widely distributed (Fig. 4). Picophytoplankton were distributed mainly around the thermocline (above 40 m). *Synechococcus* dominated at the thermocline, and their numbers were close to one order of magnitude greater than those of picoeukaryotes (Table 1). The distribution of *Synechococcus* did not correlate with nutrient concentrations. Paradoxically, its highest abundance (35×10^4 cells/mL) was observed in the relatively lower nutrient layer at Stn. A7 (DIN = 1.6 µM, P = 0.3 µM, Si = 4.5 µM at 10 m depth; 124°N, 34.5°N) (Figure 4, Table 1), while the minimum (0.4×10^4 cells/mL) was observed at 65 m depth, where nutrient concentrations were high (DIN = 10.8 µM, P = 1.4 µM, Si = 11.2 µM). The distribution of picoeukaryotes was similar to that of *Synechococcus*. Their highest abundance (5×10^4 cells/mL) was also observed at Stn. A7 (124°E, 34.5°N) (Figure 4, Table 1).

The distribution of heterotrophic bacteria (Figure 4) was different from that of picophytoplankton. The highest abundance was at Stn. D5 (225×10^4 cells/mL; 10 m; 123.5°E, 34°N) on Transect D. Their abundance was one order of magnitude higher than that of total picophytoplankton. Although they had a high abundance ($>100 \times 10^4$ cells/mL) in the thermocline, there were still appreciable quantities ($>50 \times 10^4$ cells/mL) in the cold deep water mass.

4 DISCUSSION

4.1 Picoplankton community composition

The picoplankton were grouped into autotrophic and heterotrophic components by flow cytometry (Gasol, 1999). The vast majority of the primary productivity, as inferred from size-fractionated chlorophyll concentrations, was attributed to picoplankton (73%, Fu et al., 2009).

In the tropical and subtropical oligotrophic Pacific, the picoplankton community is characterized by abundant *Prochlorococcus*, and less abundant picoeukaryotes and *Synechococcus*. In contrast in the subarctic Pacific, the picoplankton community is characterized by the absence of *Prochlorococcus* and an abundance of *Synechococcus* and picoeukaryotes (Zhang et al., 2008). Previous studies have shown that the northern boundary of the occurrence of *Prochlorococcus* is at 45°N in the subarctic North Pacific (Boyd et al., 1999; Obayashi et al., 2001), and the biomass is very low even in the Atlantic at 45°N, where the sea surface temperature is significantly higher (Tarran et al., 2006).

The Yellow Sea is a marginal sea on the western edge of the Pacific (33–37°N, 121–124°E) and seldom exchanges with the open ocean. Furthermore, the Stratification forms every summer, and prevents vertical mixing. These distinct characteristics contribute to the special biological features that determine the picoplankton community composition. Although our study area in the Yellow Sea is not part of any of the Pacific areas previously studied, the picoplankton community composition was surprisingly similar to that of the subarctic Pacific. *Prochlorococcus* is associated with warm oligotrophic waters (Jiao et al., 2005, 2007) and was not detected in our study. Instead, *Synechococcus* dominated the picoplankton and was one order of magnitude more abundant than the picoeukaryotes. Temperature is the main limiting factor for *Prochlorococcus* growth. When the temperature is >28 or <12.5°C (Moore et al., 1995), *Prochlorococcus* will not grow. In the southern Yellow Sea, the annual temperature changes significantly with temperatures < 12°C for several months (Ren et al., 2005; Yu et al., 2006), and therefore it is not suitable for *Prochlorococcus* growth. In contrast, *Synechococcus* can tolerate a large range of temperatures and can even exist at 2°C or as high as 40°C (Shapiro et al., 1988; Jiao et al., 2002). There were some differences between the picoplankton community in our study and that of the subarctic Pacific.

The vertical distribution of the heterotrophic plankton was different than the autotrophic plankton since their abundance changed little with depth (Table 2). However, some of these sub-groups of bacteria did change with depth and there were differences in LBac (Low DNA Bacteria) clusters between upper and deep waters, as revealed by flow cytometry. In the upper water samples, LBac was divided into two clusters, B-Ia and B-Ib (Frank et al., 2001), but there was only one cluster (Fig. 5) detected in the deep water (>20 m).

Although the heterotrophic plankton were higher in numerical abundance than the autotrophs, their carbon biomass was significantly less (<30% of the total) than that of the autotrophs in the upper 30 m layer (Figure 5). The vertical distribution indicates that primary productivity dominated and secondary productivity was weaker in the upper layer, while in the deeper layer (>30 m) carbon recycling was more dominant. Therefore, primary productivity dominated the euphotic layer and secondary productivity dominated at depth.

Table 2. Average of picoplankton cell abundance at different depths in total study area. Syn. = *Synechococcus*, PEuk. = picoeukaryotes, LBac = Low DNA Bacteria, HBac = High DNA Bacteria.

| Depth (m) | Average (10^4 cells/mL) | | | | |
	Syn.	PEuk.	LBac.	HBac.	B-IV
0	15.72	1.16	59.27	41.93	11.96
10	18.31	1.68	60.51	52.80	15.33
20	11.56	1.55	49.13	52.88	9.26
30	4.79	0.79	39.22	42.34	6.18
50	0.87	0.16	31.44	33.32	3.40

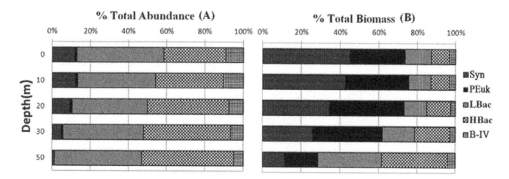

Figure 5. Proportion of total picoplankton abundance (A) and biomass (B) in each layer. (From 17 stations of the total study area) Syn. = Synechococcus, PEuk. = picoeukaryotes, LBac = Low DNA Bacteria, HBac = High DNA Bacteria.

4.2 *Influence of environmental factors on abundance*

Our correlation analysis showed a correlation between *Synechococcus* abundance and chlorophyll that is significant (r = 0.21, $P < 0.05$, n = 90, Table 3), but the correlation for the picoeukaryotes was more significant (r = 0.39, $P < 0.01$, n = 90, Table 3) because picoeukaryotic photosynthesis primarily produces chlorophyll. Although picoeukaryotes have high biodiversity, their main cytochrome is chlorophyll.

From the temperature data (Figure 2), it can be seen that the layer above the thermocline was characterized by low nutrients and high abundance, while the layer below the thermocline was characterized by high nutrients and low abundance. These characteristics were similar to other oceanic areas (Dandonneau et al., 2006; Silovi et al., 2011) and lead to a poor correlation between abundance and nutrients. The results of our study indicated that the pico-autotrophs could endure low nutrient concentrations, and therefore nutrients were not likely the factor limiting their growth.

In contrast, there were significant correlations between the abundance of autotrophs and temperature (*Synechococcus*, r = 0.84, $P < 0.01$, n = 90; picoeukaryotes, r = 0.71, $P < 0.01$, n = 90). Similar results were found in the southern Mid-Atlantic Bight (Moisan et al., 2010), USA. According to the data collected during our previous research cruise, the euphotic layer was above 37 m in the southern Yellow Sea (Li et al., 2008). Below the euphotic zone, biomass was relatively low since the autotrophs were limited by light.

4.3 *Picophytoplankton cytochromes and photo-adaptation*

The Mean Fluorescence Intensity (MFI) is a parameter obtained from flow cytometery. It reflects the cytochrome fluorescence under 488 nm excitation light. This parameter is an estimate of the photo-adaptation characteristics of autotrophs. The two phytoplankton groups have a similar photo-adaptation feature that increases with depth in the euphotic layer (Blanchot et al., 1996). Within the 37 m euphotic zone, light decreases with depth, while the fluorescence associated with photo-adaptation increases. Below the euphotic layer, the fluorescence intensity does not increase, and the abundance begins to decrease. For example, at Stn A1, photo-adaptation begins at the surfaceand the fluorescence intensity at 10 m is twice that at the surface (Fig. 7). Between 10 and 30 m, the fluorescence intensity increased 5-fold. Below 30 m, the increase in fluorescence intensity was slower, but cell abundance decreased significantly. There was a similar distribution of picoeukaryotes, and the correlation between their fluorescence and depth was significant (*Synechococcus*, r = 0.75, $P < 0.01$, n = 90; picoeukaryotes, r = 0.42, $P < 0.01$, n = 90). The change in picoeukaryotes was relatively slow conidering their higher biodiversity. Autotrophs increased their cytochrome concentration to adapt the low light level.

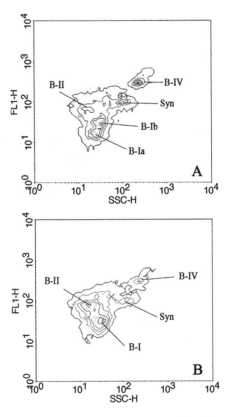

Figure 6. Cytometric 2-parameter plots (green fluorescence = FL1-H vs. side scatter = SSC) reveal 4 clusters of bacteria characterized by their DNA content. The typical flow cytometric cytograms of heterotrophic bacteria in upper (A, 10 m) and deeper (B, 30 m) water samples at Stn. A7. Cell types are different in these two layers. Abbreviations: Syn = *Synechococcus*, B-Ia and B-1b = Low DNA Bacteria, B-II = High DNA Bacteria.

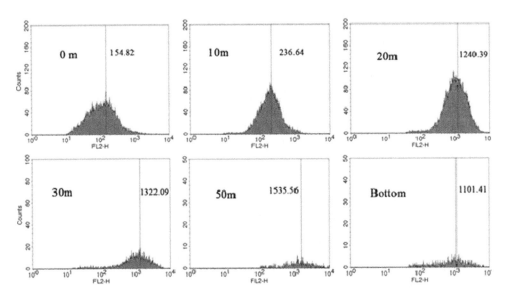

Figure 7. Frequency distribution of typical phycobilin fluorescence for *Synechococcus* in six different layers at Stn. A1. The mean fluorescence intensities of phycobilin are shown.

ACKNOWLEDGEMENTS

We are grateful to the Laboratory of Physical Oceanography and chemical group for measuring the environmental parameters. This study was supported by the National Key Basic Research Program of China (973 Program, Grant No. 2010CB428703 & 2010CB428701), Basic Scientific Fund for National Public Research Institutes of China (2011T07, 2011T01) and China-Korea Cooperation Program on the Yellow Sea Cold Water Mass.

REFERENCES

Blanchot, J., Rodier, M. and Bouteiller, A.L., (1992) Effect of El Niño Southern Oscillation events on the distribution and abundance of phytoplankton in the Western Pacific Tropical Ocean along 165°E. *J. Plankton Res.*, 14,137–156.

Blanchot, J. and Rodier, M., (1996) Picophytoplankton abundance and biomass in the western tropical Pacific Ocean during the 1992 El Niño year: results from flow cytometry. *Deep-sea Res.* Pt. I, 43,877–895.

Boyd, P.W., Harrison, P.J., (1999) Phytoplankton dynamics in the NE subarctic Pacific. *Deep-Sea Res. Pt. II*, 46, 2405–2432.

Dandonneau, Y., Montel, Y., Blanchot J., Giraudeau, J., Neveux, J., (2006) Temporal variability in phytoplankton pigments, picoplankton and coccolithophores along a transect through the North Atlantic and tropical southwestern Pacific. *Deep-sea Res. Pt. I.* 53, 689–712.

Frank, J.J., (2001) Morphology and DNA content of bacterioplankton in the northern Gulf of Mexico analysis by epifluorescence microscopy and flow cytometry. *Aquat. Microbial. Ecol.*, 25,179–194.

Fu, M.Z., Wang Z.L., Yan L., Li, R.X., Sun P., Wei X.H., Lin X.Z., Guo, J.S., (2009) Phytoplankton biomass size structure and its regulation in the southern Yellow Sea Seasonal variability. *Cont. Shelf Res.*, 29, 2178–2194.

Gasol, J.M., (1999) How to count picoalgae and bacteria with the FACS calibur flow cytometer. Http://www.marbef.org/training/FlowCytometry/Lectures/Gasol2.pdf.

Glazer, A.N., (1999) Cyanobacterial photosynthetic apparatus: an overview. IN: Harpy, L. (Eds) Marine Cyanobacteria. *Institute Oceangraphique and ORSTOM,* Paris. pp. 419–430.

Grob, C., Ulloa, O., Li, W., Alarcón, G., Fukasawa, M., Watanabe, S. (2007) Picoplankton abundance and biomass across the eastern South Pacific Ocean along latitude 32.5°S. *Mar. Ecol. Prog. Ser.*, 332, 53–62.

Jiao, N.Z., Yang, Y.H., Koshikawa, H. and Watanabe, M (2002) Influence of hydrographic conditions on picoplankton distribution in the East China Sea. *Aquat. Microbial. Ecol.*, 30, 31–78.

Jiao, N.Z., Yang, Y.H., Hong, N., Ma, Y., Harada, S., Koshikawa, H., Watanabe, M., (2005) Dynamics of autotrophic picoplankton and heterotrophic bacteria in the East China Sea. *Cont. Shelf Res.*, 25, 1265–1279.

Jiao, N.Z., Zhang, Y., Zeng, Y.H., Gardner, W.D., Mishonov, A.V., Richardson, M.J., Hong, N., Pan, D.L., Yan, X.H., Jo, Y.H., Chen, C.T.A., Wang, P.X., Chen, Y.Y., Hong, H.S., Bai, Y., Chen, X.H., Huang, B.Q., Deng, H., Shi, Y., Yang, D.C., (2007). Early impacts of the three gorges dam on the East China Sea. *Water Res.*, 41, 1287–1293.

Jochem, F.J., (2001) Morphology and DNA content of bacterioplankton in the northern Gulf of Mexico: analysis by epifluorescence microscopy and flow cytometry. *Aquat. Microbial. Ecol.*, 25, 179–194.

Li, W.K.W., Jellett, J.F., Dickie, P.M., (1995) DNA distributions in planktonic bacteria stained with TOTO or TO-PRO. *Limnol. Oceanogr.*, 40:1485–1495.

Li, Z., Zhang, X.L., Wang, Z.L., Sun, P. and Xu Z.J. (2011) Preliminary study on vertical distribution pattern of chlorophyll-*a* in south Yellow Sea in summer 2008. *Adv. Mar. Sci.*, 29, 82–89.

Marie, D., Partensky, F., Jacquet, S., and Vaulot, D. (1997) Enumeration and cell cycle analysis of natural populations of marine picoplankton by flow cytometry using the nucleic acid stain SYBR Green-I. *Appl. Environ. Microb.*, 93, 186–193.

Marie, D., Simon, N., Vaulot, D., (2004) Phytoplankton Cell Counting by Flow Cytometry. In: Andersen RA (Eds) Algal Culturing Techniques, Academic Press, London, pp. 1–15.

Moisan, T.A., Blattner, K., and Makinen, C., (2010) Influences of temperature and nutrients on *Synechococcus* abundance and biomass in the southern Mid-Atlantic Bight. *Cont. Shelf Res.*, 30, 1275–1282.

Moore, L.R., Goericke, R. and Chisholm, S.W., (1995) Comparative physiology of *Synechococcus* and *Prochlorococcus*: influence of light and temperature on growth, pigment, fluorescence and absorptive properties. *Mar. Ecol. Prog Ser.*, 116, 259–275.

Moreira-Turcq, P.F., Cauwet, G. and Martin, J.M. (2001) Contribution of flow cytometry to estimate pico-plankton biomass in estuarine systems. *Hydrobiologia*, 462,157–168.

Mukhanov, V.S., Naidanova, O.G., Lopukhina, O.A., Kemp, R.B., (2007) Cell-, biovolume- and biosurface-specific energy fluxes through marine picoplankton as a function of the assemblage size structure. *Thermochim. Acta* 458,23–33.

Obayashi, Y., Tanoue, E., Suzuki, K., Handa, N., Nojiri, Y., Wong, C.S., (2001) Spatial and temporal variabilities of phytoplankton community structure in the northern North Pacific as determined by phytoplankton pigments. *Deep-sea Res. Pt. I,* 48,439–469.

Ren, H. and Zhan, J., (2005) A numerical study on the seasonal variability of the Yellow Sea cold water mass and the related dynamics. *J. Hydrodyn,* 20,887–896.

Shapiro, L.P. and Haugen, E.M., (1988) Seasonal distribution of *Synechococcus* in Boothbay Harbor. Marine. *Estuar. Coast. Shelf Sci.,* 26,517–525.

Silovic, T., Ljubesic Z., Mihanovic, H., Olujic, G., Terzic, S., Jaksic, Z. and Vilicic, D. (2011) Picoplankton composition related to thermohaline circulation: The Albanian boundary zone (southern Adriatic) in late spring. *Estuar. Coast. Shelf Sci.,* 91,519–525.

Simek, K., Kojecka, P., Nedoma, J., Hartman, P., Vrba, J., Dolan, J.R., (1999) Shifts in bacterial community composition associated with different microzooplankton size fractions in a eutrophic reservoir. *Limnol. Oceanogr.,* 44, 1634–1644.

Tarran, G.A., Heywood, J.L., Zubkov, M.V., (2006) Latitudinal changes in the standing stocks of nano- and picoeukaryotic phytoplankton in the Atlantic Ocean. *Deep-Sea Res. Pt. II,* 53, 1516–1529.

Thomsen, H.A., (1986) A survey of the smallest eukaryotic organisms of the marine phytoplankton. *Can. Bull. Fish. Aquat. Sci.,* 214,121–158.

Zhang Y., Jiao N.Z. and Hong N., (2008) Comparative study of picoplankton biomass and community structure in different provinces from subarctic to subtropical ocean. *Deep-Sea Res. Pt. II,* 55, 1605–1614.

Zubkov, M.V., Sleigh, M.A., Tarran, Gl. A., Burkill, P.H., Leakey, R.J.G. (1998) Picoplanktonic community structure on an Atlantic transect from 50°N to 50°S. *Deep-Sea Res. Pt. I,* 45, 1339–1355.

Hydraulic Engineering – Xie (Ed.)
© 2013 Taylor & Francis Group, London, ISBN 978-1-138-00043-8

Preliminary studies of aeromonas hydrophila diseases in Rana spinosa David and whole bacteria inactivated vaccine preparation method

Jun-Qiang Qiu
Shanghai Ocean University, National Pathogen Collection Center for Aquatic Animals, Shanghai, China

Jian-Ping Wang
Ningbo Ocean & Fishery Institute, Ningbo, Zhejiang, China

Yun Le
Ningbo Ocean & Fishery Enforcement Detachment, Ningbo, Zhejiang, China

Qi Sun & Xian-Le Yang
Shanghai Ocean University, National Pathogen Collection Center for Aquatic Animals, Shanghai, China

Gui-Fang Lin
Ningbo Ocean & Fishery Institute, Ningbo, Zhejiang, China

Wei Fang
Aquatic Animal Epidemic Disease Prevention and Control Center in Guangdong Province, Guangzhou, China

Wei Tang
Anjiyuda Animal Husbandry Development Co., Ltd., Anj, China

ABSTRACT: [Objective]In order to carry on effective medication and immune prevention and control, pathogen and etiology of a *Rana spinosa David* disease in a *Rana spinosa David* breeding company in Zhejiang was researched from 2009 to 2010. [Methods] Conventional bacteria separation and purification technology was used to purify pathogenic bacteria, regression infection and toxicity test was used to determine its virulence, physiological and biochemical and molecular biology identification was carried out to the the pathogen, k-b paper method was adopted for the susceptibility test, normal paraffin wax flaking technology was used for pathology observations, finally, the inactivation conditions of pathogenic bacteria was explored. [Results] The pathogenic bacteria separated from sick *Rana spinosa David* body had Median Lethal Concentration (LC_{50}) of 5.62×10^5 cfu/ml; the strain was identified as *Aeromonas hydrophila* by ATB bacteria identification instrument and 16s rRNA sequence analysis (GenBank login number: HQ322682); histopathology observation results showed that the disease had main symptoms of the liver and kidney damage, inflammatory cells increase. Susceptibility test showed that the strain was highly sensitive to gentamicin, norfloxacin, florfenicol, enrofloxacin etc. [Conclusions] 1% formaldehyde could well inactivate the pathogenic bacteria at 60°C for 24 h, which would provide technical reference for whole bacteria inactivated vaccine preparation.

1 INTRODUCTION

Rana spinosa David, also known as chukars chukar chicken, pit, belongs to *Rana, Ranidae, Anura, Amphibia, Vertebrata, Chordata* in zootaxy, is mainly distributed in the ravine stream of deep mountains and forests in southern provinces in our country, and is in a wild state for a long time. Food value and medical value are both high due to its tender flesh, rich nutrition, delicious taste, and efficacy of clearing away heat and toxic material, nourishing and strengthening the body, it is one of the rare aquatic products in the southern hills and mountains. In recent years, a certain size of attempted breeding has been carried out in some areas, and has achieved some success. Artificial domestication of stone frogs is a new breeding industry, has very high economic benefit. The breeding of stone frogs has advantages such as less investment, saving feed, short breeding cycle, saving labor, without pollution etc., it is a special economic animal species with high breeding potential (Deng 2009).

Due to the lack of systematic basic research and related experience in the prevention and control of diseases, and the insufficient researches on stone frog physiology and artificial breeding technology, stone frogs frequently get sick in recent years, which does more and more serious harm to the breeding industry. Among them, the "rotten mouth disease" (also called lousy skin disease) is the most common disease, sick frogs have serious symptoms of revealing nasal bone, getting slow in action, lassitude, stopping feeding, presenting decay on body surface, and even death, which seriously influences stone frogs health and raise quality (Liu 2000, Wang 2008). In 2010, the disease broke out in a stone frog breeding company in Anji, Zhejiang. In order to woke out the pathogeny and pathogenesis of sick stone frogs, and look for effective drugs and immune control method, conventional bacteria separation and purification technology was used to purify pathogenic bacteria, regression infection and toxicity test was used to determine its virulence, physiological and biochemical and molecular biology identification was carried out to the the pathogen, k-b paper method was adopted for the susceptibility test, and finally formaldehyde was used to inactivate pure culture of the pathogen, to test the inactivation effects under different inactivating conditions, so as to lay foundation for researches on stone frog "rotten mouth disease" and other frogs diseases.

2   MATERIALS AND METHODS

2.1   *Stone frog source*

Naturally ill and healthy stone frogs, weighing 5 to 10 g for each, were all offered by a stone frog breeding company in Anji county, Zhejiang province.

2.2   *Pathogen separation and purification*

Stone frogs with typical natural symptoms were chosen, and repeatedly wiped the body surface with 75% alcohol cotton, sick frog lesions tissues such as liver and kidney and a small amount of ascites were taken under aseptic technique, respectively scribing on common nutrient AGAR medium plate for separation, cultivated at 28°C for 24 h, and advantage single colony was picked out for scribing purification, and inoculated on common nutrient AGAR medium slant, saved at 4°C refrigerator.

Aeromonas hydrophila standard strain ATCC 7966 was presented by teacher Qian Dong from Zhejiang freshwater aquatic products research institute.

2.3   *Artificial regression infection test of pathogenic bacteria*

Lawn dragged down from each cant preserved separated strain was respectively diluted with stroke-physiological saline solution to suspension concentration of 2.2×10^7 cfu/ml, then healthy stone frogs were infected with celiac lateral injection, each had an injected dose of 0.1 ml. Healthy stone frogs injected with the same amount of stroke-physiological saline solution were for the control group. There were 8 frogs for injection test in each group, the room temperature was from 19 to 22°C. Symptoms and death numbers of test frogs were observed and recorded for continuous 7 d, at the same time, the dying test frogs infected with artificial regression infection disease were conducted with pathogen separation again to observe whether the separated strain and the original separated strain were consistent in shape and physical and chemical characteristics.

2.4   Toxicity test of pathogenic strain

Suspension liquid of the separated pathogenic strains was diluted into 2.2×10^4 cfu/ml, 2.2×10^5 cfu/ml, 2.2×10^6 cfu/ml and 2.2×10^7 cfu/ml, then infected healthy stone frogs with celiac lateral injection, each had an injection dose of 0.1 ml. Healthy stone frogs injected with the same amount of stroke-physiological saline solution were for the control group. There were 8 frogs for injection test in each group, the room temperature was from 19 to 22°C. Death numbers of test frogs were observed and recorded for continuous 7 d, and the Median Lethal Concentration (LC_{50}) was calculated with probability unit graphical method (Xu et al. 2002).

2.5   Morphology observation and biological characteristics detection

The separated pathogenic strain was inoculated on rabbit blood AGAR plate, cultivated at 28°C for 24 h to observe whether there were hemolysis rings; at the same time, fresh cultured pathogenic strains were conducted with gram staining. In addition, the ATB bacteria identification instrument was used for physiological and biochemical identification of separated pathogenic strains.

2.6   PCR detection

The primer was designed according to the whole sequence of Genbank reported *Aeromonas hydrophila* (ATCC7966): P1: 5′-GAAAGGTTGATGCCTAATACGTA-3′; P2: 5′-CGTGCTGGCAACAAAGGACAG-3″.

The pathogenic bacteria strain and standard strain were prepared into bacteria liquid, the genomic DNA was extracted for PCR amplification, the PCR products were done with tapping recovery, then through 5′ end phosphorylation and treated with carrier, connected with the carrier pSIMPLE-18 ECoR/BAP and converted into a competent cell *E.coli* JM109 (the above operations were all according to related products specification of Baoshengwu (Dalian) co., LTD. Blue-white spot screening was done through the IPTG and X-gal, blue spot cloning was picked out and added into LB medium with AMP, cultured for the night, Miniprep plasmid extraction kit of Baoshengwu (Dalian) co., LTD. was used for extracting plasmid, double enzyme digestion was conducted to determine a successful connection, the inserted DNA fragments were finally sequenced by Shanghai Yingweijieji trade co., LTD.

2.7   Pathological section making and testing

Tissues like liver, kidney, muscle and lung of sick frogs were chosen and fixed with Bouin'S fluid, paraffin embedded after the series alcohol dehydration, the slice was 5 um thick, tissue lesion process was observed after HE staining.

2.8   Susceptibility test

Common nutrient AGAR was used, pathogenic bacteria strain was used as the tested strain, k-b paper method was adopted for the susceptibility test. Sets of susceptibility paper were produced and provided by Hangzhou microbiological reagents co., LTD.

2.9   Pathogenic bacteria whole bacteria inactivated vaccine preparation

Bacteria liquid cultivated under the same condition was inactivated under different inactivation conditions, processing temperature were respectively 30°C and 60°C. Formaldehyde concentrations were respectively 0, 0.25%, 0.5% and 1%; inactivation time was respectively 2 h, 4 h, 12 h, 24 h, 36 h and 48 h. Then the inactivation effect was tested to get the inactivation method of optimal effect (Xiao 2004, Xue 2004, Zhang et al. 2009). Inactivated under the best conditions, inspected under aseptic conditions, put at 4°C for use.

3 RESULTS

3.1 *Bacteria separation, purification and artificial infection results*

Morphology of advantage strains separated from each viscera naturally pathogenetic stone frogs was slightly the same, strains isolated from liver, kidney, lung and ascites were respectively temporarily named BYK2010RS-1, BYK2010RS-2, BYK2010RS-3 and BYK2010RS-4. The artificial regression infection tests showed that only strain BYK2010RS-1 had pathogenicity to stone frogs, 7 d after artificial injection infected stone frogs, stone frogs mortality rate was as high as 100% (Table 1), testing sick frogs also showed red and swollen, fester in oronasal place and leg, abdomen was full of blood, intestinal hyperemia, or hemorrhage etc. symptoms, and strains had the same morphological characteristics and physical and chemical properties with strain BYK2010RS-1 was separated from artificial regression infection dying frog bodies. According to the established test frog mortality (%) and bacteria liquid concentration logarithm (log (cfu/ml)) curve: $y = 36.25x-158.58$ (Fig. 1), Median Lethal Concentration (LC_{50}) of strain BYK2010RS-1 to stone frogs was 5.62×10^5 cfu/ml, according to Devesa division standard[6] for fish pathogenic bacteria virulence, it could be determined that strain BYK2010RS-1 had strong virulence to stone frogs.

3.2 *Pathogenic bacteria BYK2010RS-1 morphology and physiological and biochemical identification results*

Gram negative bacillus, somatic size for $(1.0 \sim 1.2)$ μm $\times$ $(2.1 \sim 2.4)$ μm, with lateral flagella growing around, obvious beta-hemolytic ring could be produced in rabbit blood AGAR plate. The physiological and biochemical identification results of ATB bacteria identification instrument to the strain showed that (Table 2), this strain was *Aeromonas hydrophila*, the credibility of the Identification result was 99%.

Table 1. Results of artificial infection test of strains.

Strain	Concentration cfu/ml	Number	Death number 1 d	2 d	3 d	4 d	5 d	6 d	7 d	Mortality %
BYK2010RS-1	2.2×10^7	8	2	4	6	6	7	8	8	100
BYK2010RS-2	2.2×10^7	8	0	0	0	0	0	0	0	0
BYK2010RS-3	2.2×10^7	8	0	0	0	0	0	0	0	0
BYK2010RS-4	2.2×10^7	8	0	0	0	0	0	0	0	0

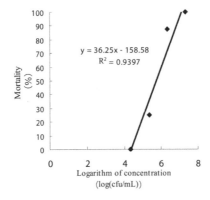

Figure 1. The relation curve of *Rana Spinosa* on mortality with logarithm of bacterial concentration.

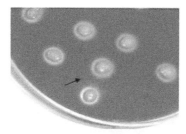

Figure 2.   The beta hemolysis zone of strain BYK2010RS-1 on rabbit-blood agar plate.

Table 2.   The identification of strain BYK2010RS-1 by ATB bacteria identification instrument.

Test item	Result	Test item	Result	Test item	Result
OX	+	CIT	−	SAC	+
GLU	+	HIS	+	MAL	+
SAL	+	2KG	−	ITA	−
MEL	−	30BU	−	SUB	−
FUC	−	5- keto—gluconate	−	MNT	−
SOR	−	30BU	−	ACE	−
ARA	+	RHA	+	LAT	+
PROP	−	NAG	+	ALA	+
CAP	−	PRO	+	GLYG	+
VALT	+	INO	−		

Note: "+" denotes positivity, "−"denotes negativity.

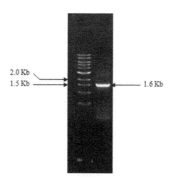

Figure 3.   The result of PCR for gene 16s rRNA of strain BYK2010RS-1.

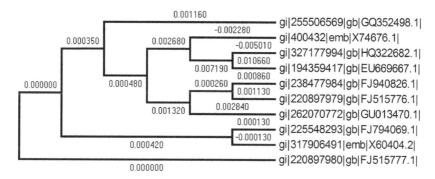

Figure 4.   Phylogenetic tree of gene 16s rRNA of BYK2010RS-1 and other published bacteria's genes 16s rRNA.

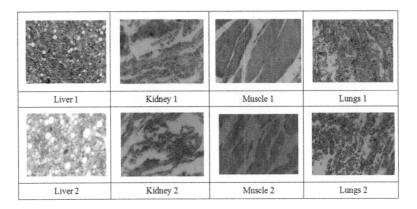

Liver 1	Kidney 1	Muscle 1	Lungs 1
Liver 2	Kidney 2	Muscle 2	Lungs 2

Figure 5.   Ology test of tissue.

Table 3.   The results of the sensitivity test.

Antibiotics	Sensitivity	Antibiotics	Sensitivity
Penicillin G	R	Norfloxacin	S
Ampicillin	I	Polymyxin B	I
Cefradine (VI)	S	Bactrim	S
Cefotaxime	I	Nitrofurantoin	S
Gentamicin	S	Foroxone	I
Kanamycin	S	Amoxicillin	R
Streptomycin	S	Rifampin	R
Neomycin	S	Chloromycetin	S
Doxycycline	I	Enrofloxacin	S
Erythromycin	S	Florfenicol	S

"S" denotes high sensitivity (d > 15 mm); "I" denotes moderate sensitivity (10 mm ≤ d ≤ 15 mm); "R" denotes low or no sensitivity (0 mm ≤ d < 10 mm).

Table 4.   The number of survival pathogen in all inactivation methods.

	2 h		4 h		12 h		24 h		36 h		48 h	
	30°C	60°C	30°C	60°C	30°C	60°C	30°C	60°C	30°C	60°C	30°C	60°C
0	>100	>100	>100	>100	>100	>100	>100	>100	>100	>100	>100	>100
0.25%	>100	>100	>100	>100	68	64	53	56	25	20	10	0
0.5%	>100	>100	>100	>100	34	39	18	12	10	8	0	0
1%	>100	>100	>100	>100	22	18	0	0	0	0	0	0

3.3   Pathology detection results

Viewing from the pathology examination result, stone frogs mainly displayed symptoms of liver and kidney damage, and inflammatory cells increased. Many fat granules and cavitations appeared in liver. There was no obvious damage in lung and muscle. Just as shown in the chart below:

3.4   Results of susceptibility test

Viewing from the results of susceptibility test, the strain separated was sensitive to fluoroquinolones like norfloxacin and enrofloxacin, and was also sensitive to the nitrofurans and chloramphenicol. Therefore, fluoroquinolones were suitable for prevention and treatment in clinical use. The specific results were shown in Table 3.

It was found that basically no viable organism would exist when inactivated in 1% formaldehyde at 60°C for above 24 h. Therefore, in this experiment, being inactivated with 1% formaldehyde at 60°C for 36 h was used as the preparation conditions of tissue pulp inactivated vaccine and whole bacteria inactivated vaccine.

4 DISCUSSIONS

Aeromonas hydrophila is *a* pathogen for person, beast and fish, is widespread in the fresh water, sewage, silt, soil and human fece (Wu et al. 2010), its pathogenic strains can infect fish, amphibians, reptiles, birds and mammals and other animals. The main feature in clinical is acute hemorrhagic septicaemia. Chronic infection mainly shows skin ulcer or enteritis (Song 2001, Song et al. 2009).

Frogs diseases involve various sorts of pathogens, including *Aeromonas hydrophila*, pitting air cell bacteria, reduction bacteria, fluorescent pseudomonad and proteus, etc., in which *Aeromonas hydrophila* is the main pathogen. Skin ulcerative disease of frogs has features like onset suddenly, spreading fast, high incidence and mortality, if it can be found early, timely scientific prevention and control measures are taken, economic loss can be reduced. In recent years, frogs breeding has achieved rapid development in our country, has become an important economic industry, frog disease prevention has become more and more important, and has become one of the most important research subjects of scientific and technical workers (Xi 2004, Xue 2010).

In this study, a large number of pathogenic bacteria were separated from Anji pathogenetic stone frog bodies. The bacteria morphology and molecular biology identification showed that one of the strains was the pathogenic *Aeromonas hydrophila*. The bacterium was sensitive to fluoroquinolones and nitrofurans drugs. Pathological examination results showed that the disease mainly displayed symptoms of the liver and kidney damage. The results further enriched the *Aeromonas hydrophila* research results, providing a certain technical reference to prevention and control of the stone frog bacterial "bad mouth" disease.

It is a widely used method in the production to apply antibiotics to bacterial disease prevention and control. However, due to the lack of some necessary stone frog pharmacology research results, the use of antibiotic to control stone frog bacterial "rotten mouth disease" is easy to cause problems of unreasonable use of antibiotic drugs, such as the unconspicuous disease control effect, the adverse impact on the environment and influencing the stone frogs quality, etc. Therefore, this study also tried to prepare whole bacteria vaccine, in order to provide technical references for the immune prevention and control.

At present, as dozens of complex antigenic of *Aeromonas hydrophila* was found, each genotype and phenotype also includes many different O antigens; the diversity of pathogen antigen undoubtedly caused difficulty to vaccine development. In addition, due to the influence of environmental factors, frequently variations happened to pathogens, further hindered the progress of vaccine development. A large amount of manpower and material resources were invested in this kind of pathogenic bacteria vaccine research in our country, but still no prominent progress (Yang 2006). Whole bacteria inactivated vaccine of the pathogen was basically used for immune prevention and control in production, commonly known as "indigenous vaccine".

In this study, the method inactivated with 1% formaldehyde at 60°C for 24 h, can well inactivate pathogens, can provide technical references for whole bacteria inactivated vaccine. Whole bacteria inactivated vaccine is simple in operation, low in cost, and convenient to use, therefore it has a certain application prospect. When batches of stone frogs received soak immune with "indigenous vaccine" achieved by the method in this study from fall of 2010 to 2011 summer, morbidity and mortality were obviously improved compared with the control group (data is being systemized, to be published).

Compared with attenuated live vaccine, although inactivated vaccine is secure, convenient to save, relatively simple to produce etc., there are also shortcomings like the short immune reaction duration, low induced immune response level, and the need for additional adjuvant (Qiu et al. 2009) therefore, the future fundamental and applied researches relating inactivated vaccine and other "rotten mouth disease" immune prevention still need to be continued.

At present, there are few researches on stone frogs related disease in our country, when faced with similar problems, the only way was to turn to and refer to research materials of wood frogs and bullfrogs, thus researches related to basic physiology and disease remain to be further strengthened.

ACKNOWLEDGMENTS

Special funds (NYCYTX-49-17) of modern agriculture industry technology system construction; open issues (BM2007-07) of aquatic animal genetics, breeding and breeding biology key open laboratory, the ministry of agriculture; the Major Science and Technology Project of Ningbo City, China (2008C10022); the Project of Agricultural Key Programs for Science and Technology Development of Ningbo (2011C11006).

REFERENCES

Deng, D.F., Liu, Z.Q. 2009. Prevention and control countermeasures of rana spinosa rotten skin disease[J]. Scientific Fish Farming (5): 60. (in Chinese).

Liu, T., Xu D.S. 2000. Pathogen isolation and identification of Red leg sickness[J]. Chinese Journal of Veterinary Drug. 34(4): 39–41. (in Chinese).

Qiu, J.Q., Yang, X.L., Cheng X.J. 2009. Advances in the characteristics and pathogenic mechanisms of virulence factors in Aeromonas hydrophila[J]. Journal of Pathogen Biology (8): 616–619. (in Chinese).

Song, C.T., Zhu, H.Y. 2001. Frogs disease diagnosis and prevention[J]. China Animal Health (3): 21–22. (in Chinese).

Song, Z.F., Wu, J.W., Xiao M., et al. 2009. Isolation, Identification and Selection of Preventable Drugs of One Strain of Pathogen from Rotten-mouth-diseases in Rana[J]. Journal of Anhui Agricultural Sciences 37 (12): 16884–16885,16889. (in Chinese).

Wang, X.W. 2008. Disease control and prevention of Chinese forest frog (below)[J]. New Agriculture (8):45. (in Chinese).

Wu, Q., Yan, F., Liu, F.B. 2010. Progress in studies on Aeromonas hydrophila[J]. Livestock and poultry industry (2): 28–31. (in Chinese).

Xiao, L.L., Zhang, Q.H. 2004. Test on the immune enhancement of propolis-adjuvant inactivated vaccine against Aeromonas hydrophila of Carassius auratus gibelio[J]. Marine Tieheries 26(4): 295–299. (in Chinese).]

Xi, Z.J. 2004. The common frog disease and prevention countermeasures[J]. Inland Fisheries, 29(1): 35–36. (in Chinese).

Xu, S.Y., Bian, R.L., Chen, X. 2002. Pharmacological experimental methodology[M]. Beijing; People Hygiene Press 1651–1654. (in Chinese).

Xue, H. 2010. Prevention and control of bullfrog diseases (below)[J]. Farmhouse becomes rich (2): 46–47. (in Chinese).

Yang, X.L., Cao, H.P. 2006. The development of aquaculture vaccines in China[J]. Journal of Fisheries of China 30(2):264–271. (in Chinese).

Zhang, Y.F., Zhang, X.J., He, S.M. 2009. A Study on Cultural Condition and Inactivating Method for Whole Cell Bacteria Vaccine from Aeromonas hydrophila[J]. Acta Agriculturae Universitatis Jiangxiensis 31(1): 31–34. (in Chinese).

Hydraulic Engineering – Xie (Ed.)
© 2013 Taylor & Francis Group, London, ISBN 978-1-138-00043-8

Characterization of Cr(VI) removal and Cr equilibrium adsorption by sulfate reducing granular sludge in stimulant wastewater

Zhang Jia-Lin
College of Environmental Science and Engineering, SYSU, Guangzhou, China

Luo Jun, Pang Zhi-Hua, Lin Fang-Min & Wang Zhen-Xing
South China Institute of Environmental Sciences, Ministry of Environmental Protection of the People's Republic of China, Guangzhou, P.R. China

ABSTRACT: Sulfate reducing granular sludge cultivated in small scale EGSB reactor was used to Cr(VI) removing. The characterization of Cr(VI) removal and Cr equilibrium adsorption was studied, and the adsorption isotherm was fitted. The Cr(VI) removal rate increased with the dosage of granular sludge; the increasing of oscillation speed and temperature could enhance Cr(VI) removal and Cr adsorption, but while the oscillation speed reached 150 r·min^{-1} or the temperature came to 40°C, the physical structure of granular sludge would be affected and discrete, and Cr equilibrium adsorption decreased. The maximum adsorption of Cr by granular sludge was 6.84 mg·g^{-1}, and the Cr adsorbing process fitted in with Langmuir adsorption isotherm.

1 INTRODUCTION

Cr(VI) is recognized as a powerful carcinogenesis, teratogenesis and mutagenesis material (Garbisu, et al., 1998, Murti et al., 1991). It is also one of the 129 kinds of priority-control pollutants recognized by U.S. EPA. Researchers have been searching for the economical and effective methods to remove Cr(VI), and biological methods are currently the focus and hot spots of this field. Within them, yeasts, molds, algae are the most widely studied (He, et al., 2007, Luo, et al., 2007, Peng, et al., 2008 and Park, et al., 2005a). The microbial Cr(VI) removal process is generally considered including the reduction process and the adsorption process. The respective proportion of these two processes depends on the nature of the different microorganisms (Park, et al., 2005b).

Sulfate-reducing bacteria are morphological and nutritional diversification, and they are also a kind of strict anaerobes which use sulfate as an electron acceptor for organic matter catabolism. The use of Sulfate-Reducing Bacteria (SRB) to remove Cr(VI) has been reported. For example, Qu Jianguo, etc. isolated the SRB from the soil and used them to remove Cr(VI). They could completely reduce the Cr(VI) of 800 mg·L^{-1} in solution within 36 hours (Qu, et al., 2005). Wu Shuhang found that the SRB had a good repair effect on the soil contaminated by Cr(VI), and that their conversion rate of Cr(VI) could reach 75.3% after 10 days (Wu, et al., 2007). Ma Xiaozhen etc. studied on acid desulfovibrio desulfuricans to remove Cr(VI) in solution. She pointed out that the electron transport pathway could not dominate the reduction process of Cr(VI), and the removal efficiency of Cr(VI) was 51.42% in 24 h. In contrast, the H$_2$S pathway dominated in the reduction process of Cr(VI), and the removal efficiency of Cr(VI) was 78.02% (Ma, et al., 2009). Recently, Somasundaram studied on Cr reducing bacteria, sulfate-reducing bacteria and iron-reducing bacteria to remove Cr(VI), and their mathematical model of these processes (Somasundaram, et al., 2009). However, in these existing studies, Cr(VI) was only removed by reduction with SRB. Cr(VI) is converted to the low toxicity of Cr(III) and remained in the environment. Thus, there is always the risk of environmental re-oxidized to Cr(VI). Research in SRB's adsorption of Cr is relatively rare in domestic, and the adsorption of sulfate-reducing granular sludge of Cr has not been reported basically.

We succeeded in using SRB and culturing sulfate-reducing granular sludge (hereinafter referred to this as "granular sludge") through the EGSB reactor which designed by ourselves. Based on the early study of *Aspergillus fumisynnenatus* mycelial pellets removing Cr(VI), the granular sludge's removal of Cr(VI) and the analysis of the characteristics of its adsorption would be focal point in this study. Also, the initial concentration of Cr(VI) in solution, the dosage of granular sludge, the speed of oscillation, temperature and pH value would be studied. At the same time, the maximum adsorption capacity of the granular sludge would be confirmed through the equilibrium adsorption experiments of Cr and the fitting of the adsorption isotherm. Therefore, the granular sludge removal of Cr(VI) could be determined, and the mechanism of its action could be described.

2 MATERIALS AND METHODS

2.1 Materials in the experiment

The granular sludge was cultured by the EGSB reactor which developed by the laboratory (as shown in Table 1 and Fig. 1). $K_2Cr_2O_7$ (AR. Sinopharm Chemical Reagent Co., Ltd.) was dissolved in deionized water to prepare stock solution with Cr(VI) concentration of 1000 $mg \cdot L^{-1}$. The solution of other concentration was diluted by the stock solution. H_2SO_4 and NaOH were used to adjust the pH of the solution in the experiments.

2.2 Experimental methods

Erlenmeyer flask of 250 mL and rubber stopper were used in the experiment. Each Erlenmeyer flask had 100 mL of Cr(VI) solution and different amount of granular sludge. Then they were sealed with rubber stopper and were placed into thermostatic oscillation incubator. The speed and temperature in the experiment were regulated according to the need of the experiments. The samples had to be analyzed the remaining Cr(VI) concentration and total concentration of Cr at regular intervals.

Table 1. Character of sulfate reducing granular sludge.

Color	Shape	VSS/TSS/ %	Rate of water conten/%	Particle size distribution/mm	Density/ $g \cdot mL^{-1}$	Sulphide content/$mg \cdot g$
Black	Elliptical particle	80.07%	90.40%	0.5–2.0	1.061	98.68

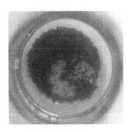

Figure 1. The picture of sulfate reducing granule sludge.

268

Instruments	Instrument models	Manufacturers
Thermostatic oscillation incubator	THZ-03M2R	Changzhou NoKi Co, Ltd
Scanning electron microscope	H-3000N	Hitachi of Japan

2.4   *Analytical methods*

The concentration of Cr(VI) was determined by diphenyl carbon acyl hydrazine spectrophotometry, and the concentration of total Cr was determined by flame atomic absorption. The surface morphology of the granular sludge was characterized by Scanning Electron Microscopy (SEM).

3   RESULTS AND DISCUSSIONS

3.1   *The morphology characterization of the granular sludge*

The surface morphology of the granular sludge was observed by Scanning Electron Microscopy (SEM). Observed from the morphology, a large number of micro-organisms were distributed on the surface of the granular sludge. Bacilli accounted for the most part, still there was a small amount of filamentous bacteria and vibrio. The particle size of the micro-organisms was about 1 μm, and the length of the micro-organisms did not exceed 2 μm. Secretions which produced by the microbial cells was relatively few, only scattered on the surface of the granular sludge.

3.2   *The influence of the dosage of granular sludge*

When Cr(VI) concentrations were 50, 100, 200 and 400 mg·L^{-1}, the dosage of the granular sludge taken in the experiment was respectively 1.00, 2.00, 4.00 and 8.00 g. The results were shown in Figures 3 to 10.

Figure 2.   Scan electronic microscope picture of sulfate reducing granular sludge.

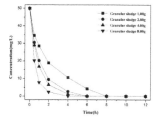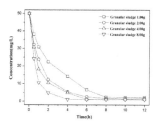

Figure 3.   Removal of Cr(VI) and adsorption of total Cr with granular sludge, Cr(VI) 50 mg/L (a, b).

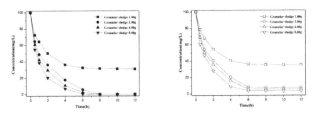

Figure 4. Removal of Cr(VI) and adsorption of total Cr with granular sludge, Cr(VI) 100 mg/L (a, b).

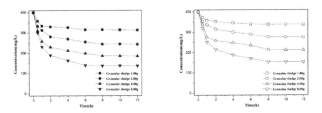

Figure 5. Removal of Cr(VI) and adsorption of total Cr with granular sludge, Cr(VI) 200 mg/L (a, b).

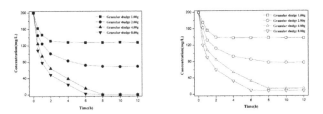

Figure 6. Removal of Cr(VI) and adsorption of total Cr with granular sludge, Cr(VI) 400 mg/L (a, b).

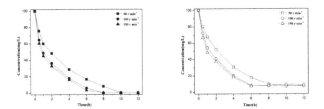

Figure 7. Removal of Cr(VI) and adsorption of Cr with granular sludge in various shaking velocity (a, b).

The pH value of this process was approximately 6.0. In this situation, the reduction was weak and the adsorption accounted for a major role (Parsons, et al., 2002). When the dosage of granular sludge was 1.00 g, the concentration of 50 mg · L^{-1} of Cr could be basically adsorbed in 8 hours. The removal efficiency of Cr(VI) increased with the dosage of granular sludge. For the same dosage of granular sludge, the higher concentration of solution of Cr(VI), the removal rate of Cr(VI) would be faster. However, a certain amount of granular sludge could only correspond to the removal of a certain amount of Cr(VI), which might be related to the saturated adsorption capacity.

The two to four hours after the start of the experiment was the rapid removal process of Cr(VI), then the removal rate decreased significantly. The removal role of the granular sludge was onset of the first 8 hours after the start of the experiment. After that, the concentration of Cr(VI) and total Cr in the solution was basically stabilized. It could be considered that the removal and adsorption of Cr reached a dynamic equilibrium after 8 hours.

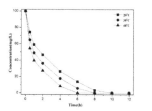 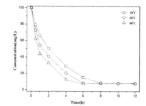

Figure 8. Removal of Cr(VI) and adsorption of Cr with granular sludge in various temperature (a, b).

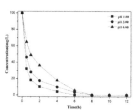 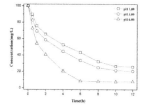

Figure 9. Removal of Cr(VI) and adsorption of Cr with granular sludge in acidic pH (a, b).

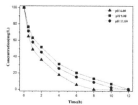 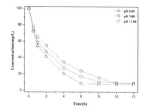

Figure 10. Removal of Cr(VI) and adsorption of Cr with granular sludge in alkalic pH (a, b).

3.3 The influence of the speed of oscillation

When the initial concentration of Cr(VI) was 100 mg·L^{-1} and the dosage of granular sludge was 2.00 g, the speed of oscillation in the experiment was respectively 50, 100 and 150 r·min^{-1}. The results were shown in Figure 7.

When the oscillation speeded, the rate of granular sludge adsorption of Cr(VI) and the removal efficiency of Cr increased. But the oscillation speed did not influence the equilibrium adsorption amount of the granular sludge on Cr. The oscillation speed impacted on the solid-liquid mass transfer rate and thus affected the rate of adsorption of Cr and the removal efficiency of Cr(VI).

The granular sludge was in high moisture content and had no obvious polymerization. As the oscillation speed increased, the granular sludge was more discrete broken. But the integrity of the structure of granular sludge had a great influence on the adsorption of Cr. With the oscillation speed reached 150 r·min^{-1}, the broken degree of granular sludge significantly increased. At the same time, the adsorption of Cr appeared fluctuations at the six hours after the start of the experiment. The equilibrium adsorption capacity at this situation was even slightly lower than the oscillation speed of 100 r·min^{-1}. Since the structure of granular sludge always maintained a relatively complete state, the adsorption effect of the total Cr were relatively stable when the oscillation speed were 50 r·min^{-1} and 100 r·min^{-1}.

3.4 The influence of the temperature

When the initial concentration of Cr(VI) was 100 mg·L^{-1} and the dosage of granular sludge was 2.00 g, the temperature in the experiment was respectively 20°C, 30°C and 40°C.

The results were shown in Figure 8. The temperature of the system had greater impact on the removal efficiency of Cr. When the system temperature were 20°C, 30°C and 40°C, the complete removal time of Cr in solution were respectively 10 h, 8 h and 6 h.

The adsorption of granular sludge was an endothermic process (Arıca, et al., 2005), the rises of temperature could speed the granular sludge to its balance, but temperature had little effect on the equilibrium adsorption amount of Cr. Granular sludge was more sensitive to the change of temperature. When the system temperature reached 40°C, the granular sludge began to appear a certain degree of discrete phenomena, thereby affecting the adsorption of Cr. The sort of equilibrium adsorption amount of different temperature was about: 20°C ≈ 30°C > 40°C.

3.5 The influence of the pH value

When the initial concentration of Cr(VI) was 100 mg·L⁻¹ and the dosage of granular sludge was 2.00 g, the pH value in the experiment was respectively 1.00, 3.00, 9.00 and 11.00, and compared with the natural pH value which was 6.00.

When the pH value was 1.00 or 3.00, the solution released large amounts of H_2S gases within 1 minute after dosing the granular sludge. At the same time, the solution became opacity and pale green emulsion. The removal efficiency of Cr(VI) was shown in Figure 9. When the pH value was 1.00 and 3.00, the removal of Cr(VI) was higher than the natural pH value, but the adsorption of Cr was lower than it. This result indicated that reduction had a large proportion in the removal of Cr. Because Cr(VI) had strong oxidation at low pH and it had redox with the granular sludge, Cr(VI) was deoxidized to Cr(III) and the solution became light green.

When the solution was alkaline, there was no H_2S gases released during the experiment. The removal of Cr(VI) and the adsorption effect of total Cr were shown in Figure 10. The removal efficiency of Cr(VI) and the adsorption effect of total Cr in the pH value of 9.00 and 10.00 were lower than those in the natural pH value. When the solution was alkaline, the amount of Cr(III) was less. This indicated that reduction in alkaline was weaker than that in the natural pH and acid. Granular sludge could adapt to the alkaline environment, and the sulfide were not affected by alkaline solution, so the effect of the adsorption of total Cr was better than that in the acidic solution. However, the removal efficiency of Cr(VI) was lower than that in the natural pH value, resulting in the removal efficiency of total Cr was lower than that in the natural pH value.

3.6 The adsorption equilibrium experiment

When the dosage of granular sludge was 2.00 g and the adsorption time was 12 h, the initial concentrations of Cr(VI) solution in the adsorption equilibrium experiment were respectively 50, 100, 200 and 400 mg·L⁻¹.

The equilibrium adsorption results of different initial Cr(VI) concentration were shown in Figure 11. The results respectively used the Langmuir model and Freundlich model to fitting, and the fitting curve were shown in Figure 12. From Figure 11, it showed that the adsorption capacity of granular sludge increased with the steady increase of concentration basically,

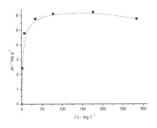

Figure 11.   Cr equilibrium adsorption by granule sludge.

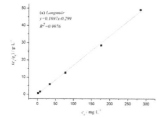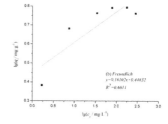

Figure 12. Isotherm model of Cr biosorption by granule sludge.

and up until the saturated adsorption capacity of granular sludge. The maximum adsorption measured in the experiment was about to be 6.34 mg/g. The obtained fitting equation using Langmuir model was: $Ce/qe = 0.1697Ce - 0.299$, $R^2 = 0.9976$, $K_L = 0.358$. The obtained fitting equation using Freundlich model was: $\lg qe = 0.16102\lg Ce + 0.44832$, $R^2 = 0.6811$. The results showed that the fitting equation using Langmuir model had a good correlation. Whether it was in the low concentration or the high concentration of the experiments, the fitting equation using Langmuir model could always fit the measured results.

4 DISCUSSION

4.1 The influence of the granular sludge dosage and the influence of the initial concentration of Cr(VI)

When the granular sludge dosage increased, the adsorption sites on the surface increased, thus the removal efficiency of Cr and Cr(VI) increased. But the high concentration of adsorbent might lead the adsorption sites of biosorbent to be unsaturated, and this caused the aggregation of adsorbent and the reduction of the adsorption effective area (Wang, et al., 2008). Therefore, the adsorption of Cr(VI) of the solution increased, but the unit particle adsorption amount of Cr(VI) decreased. When the initial concentration of Cr(VI) increased, the unit particle adsorption amount of Cr(VI) and Cr increased. But the total removal efficiency Cr(VI) and Cr would reduce.

4.2 The influence of the speed of oscillation and the temperature

The electron micrographs of the granular sludge showed that there were a lot of micro pores on the surface. These pores could supply the interception places for Cr, and this was the scope of physical adsorption. When the oscillation speed and the temperature were too high, the structure of granular sludge would disperse. At the same time, the micro pores would be damaged, and some of the Cr which was adsorbed by the granular sludge would be released back into the solution. Thus, the granular sludge adsorption efficiency decreased and this was accorded with the results that studied on *Aspergillus fumisynnenatus* mycelial pellets removing Cr(VI) (Luo, et al., 2007).

4.3 The adsorption equilibrium

Langmuir model and Freundlich model were respectively used to simulate the process of the granular sludge adsorption of Cr. The R^2 values fitting by Langmuir model and Freundlich model were 0.9976 and 0.6811 respectively, and this indicated that Langmuir model described the granular sludge adsorption equilibrium of Cr more accurately. q_n was the maximum adsorption capacity of the granular sludge and an important indicator of the adsorption properties of Cr. The maximum adsorption measured in the experiment was about to be 6.34 mg/g and this showed its good adsorption properties of total Cr. $K_L = 0.358$ and it showed the better affinity of the granular sludge and Cr. The granular sludge was the intermittent product of the EGSB reactor and it was easy to obtain. If it was used in the adsorption of Cr in wastewater, it would have a good effect, so it had good application prospects.

5 CONCLUSION

1. Granular sludge could remove Cr(VI) effectively. When the concentration of Cr(VI) was 50 mg/L and the pH value was 6.0 (natural pH value), the removal efficiency of Cr(VI) and the adsorption efficiency of Cr were nearly 100%.
2. The speed of oscillation and the temperature could affect the mass transfer process between granular sludge and Cr(VI), thus affecting the removal of Cr(VI). And they also impacted on the physical structure of granular sludge to affect the adsorption of Cr.
3. The maximum adsorption of granular sludge of total Cr was about 6.34 mg/g. The obtained fitting equation using Langmuir model was: $Ce/qe = 0.1697Ce - 0.299$, $R^2 = 0.9976$, $K_L = 0.358$. This showed its good adsorption properties of total Cr.

ACKNOWLEDGEMENT

The paper was supported by Environmental nonprofit industry special research project: No. 201209048-2, National Natural Science Foundation of China: No. 51204074.

REFERENCES

Arıca M.Y., Tüzün İ., Yalçın E., et al. Utilisation of native, heat and acid-treated microalgae *Chlamydomonas reinhardtii* preparations for biosorption of Cr(VI) ions. Precess Biochemistry, 2005, 40: 2351–2358.

Garbisu C, Alkorta I, Llama M J, et al. Aerobic chromate reduction by *Bacillus subtilis* [J]. Biodegradation, 1998, 9 (2):1332141.

He BY, Yin H, Peng H, et al. Physiology metabolism and cell morphology analysis of chromium biosorption by Yeast [J]. Environmental Science, 2007, 28(1):194–198.

Luo J, Hu YY, Zhong HT. Removal of Cr(VI) in solution by *A spergillus fum isynnem a tus* mycelia: reduction and biosorption [J]. Ac ta Scientiae Circumstantiae, 2007, 27(10):1585–1592.

Ma XZ, Fei BJ, Jin N, et al. Reduction Characteristics of the Desulfovibrio SRB7 of Cr(VI) [J]. Microbiology, 2009, 36(9):1324–1328.

Murti C.R.K., Viswanathan P.(Ed.). Toxic Metal in the Indian environment [M]. Tata McGraw-Hill, New Delhi, 1991, pp. 7.

Park D., Yun Y-S., Park JM. Studies on hexavalent chromium biosorption by chemically treated biomass of *Ecklonia* sp. [J]. Chemosphere, 2005a, 60:1356–1364.

Park D., Yun Y-S, Jo J.H., et al. Mechanism of hexavalent chromium removal by dead fungal biomass of *Aspergillus niger* [J]. Water Research, 2005b, 39:533–540.

Parsons J.G., Hejazi M., Tiemann K.J., et al. An XPS study of the binding of copper(II), zinc(II), chromium(III) and chromium(VI) to hops biomass [J]. Microchem J, 2002, 71:211–219.

Peng K, Hu YY, Wang BE. Mechanism of chromium removal by pretreated dead *A spergillus fum isynnem a tus* biomass [J]. Ac ta Scientiae Circumstantiae, 2008. 28 (9): 1751–1757.

Qu JG, Shen RX, Xu BX, et al. Preliminary study of sulfate-reducing bacteria to restore Cr(VI) [J]. Journal of East China Normal University (natural sciences), 2005, 1:105–110.

Somasundaram V., Philip L, Bhallamudi S.M. Experimental and mathematical modeling studies on Cr(VI) reduction by CRB, SRB and IRB, individually and in combination [J]. Journal of Hazardous Materials, 2009, 172:606–617.

Wang BE, Hu YY, Xie L, et al. Biosorption of reactive brilliant blue KN-R by inactive *A spergillus fum iga tus* immobilized on CMC beads: Batch studies and thermodynamics [J]. Ac ta Scientice Circumstantiae, 2008, 28(1):83–88.

Wu SH, Zhou DP, Lv WG, et al. Study of sulfate-reducing bacteria repairing the contaminated soil by Cr(VI) [J]. Journal of Agro-Environment Science, 2007, 26(2):467–471.

Hydraulic Engineering – Xie (Ed.)
© 2013 Taylor & Francis Group, London, ISBN 978-1-138-00043-8

The effect of arsenic removal of high arsenic spring water

Liu Yanan, Liu Zhihui, Wang Juan & Zhang Shu
Yunnan University, Institute of Engineering Technology, Kunming, China

ABSTRACT: High arsenic hot spring water samples are processed by precipitation and flocculation in this article, arming to reduce the arsenic content. Flocculant was 1.50% $FeCl_3$ solution in the experiment, arsenic removal rate is up to 98.19%. Through experimental analysis, while the ratio of hot spring water samples and flocculant is 100:1, hot spring water can reach II Grade; experiments show that the effect of arsenic removal in water samples is not obvious in the range of 30°C~60°C. Water arsenic concentration can be controllable and adjustable by this method.

1 INTRODUCTION

Arsenic-containing waste water, waste gas and waste residue are produced in mining and smelting of arsenic and arsenic metal and the manufacture of glass, pigments, pharmaceuticals and paper production with arsenic or arsenic compounds as raw materials as well as the combustion of coal [1–3]. As arsenic compounds has a strong stability and is difficult to decompose under natural conditions as well as is easy to produce residual accumulation, thus will destroy the ecology, cause environmental pollution [4–9]. In order to reduce the hazards of arsenic on human health and harm, looking arsenic pollution situation, the arsenic pollution control of water bodies has been urgent, therefore, development of efficient, user-friendly and cost-effective arsenic removal treatment technology in high-arsenic water is of great practical significance. Currently, the main methods are precipitation, adsorption, biological, a membrane separation, etc [10–16], for the removal of arsenic in water at home and abroad. Although various methods have better treatment effect, there are different limitations and deficiencies in a certain extent.

This article is mainly to remove arsenic in water by precipitation and flocculation. by mixing flocculant (flocculant was 1.50% $FeCl_3$ solution in the experiment) samples to high arsenic hot spring water samples, the influence of flocculant ratio and temperature on the rate of arsenic removal is studied, in order to obtain optimized parameters and arsenic removal effect of high arsenic spring water treatment by precipitation and flocculation and achieve the purpose of efficient, streamlined arsenic removal.

2 EXPERIMENT METHOD

High arsenic water is taken directly from a hot spring water in Yunnan in the experiments, the arsenic concentration of 1.27 mg/L, flocculant was 1.50% $FeCl_3$ solution in the experiment. First hot spring water samples are heated to 30°C, 60°C and 100°C, then mix flocculant to hot spring water samples in accordance with the proportion of hot spring water samples and flocculant of 5000:1, 4000:1, 3000:1, 2500:1, 2000:1, 1000:1, 500:1, 100:1. We analysis the influence of the flocculant proportion and the temperature on the rate of removal of arsenic by comparativing experimental data, experimental samples were detected with AFS-930 atomic fluorescence spectrometer.

3 EXPERIMENTAL DATA AND ANALYSIS

3.1 *The influence of the flocculant proportion on the arsenic removal rate*

The hot spring water samples are heated to 30°C, then mix flocculant to hot spring water samples in accordance with the proportion of hot spring water samples and flocculant of 5000:1, 4000:1, 3000:1, 2500:1, 2000:1, 1000:1, 500:1, 100:1, determine changes in the arsenic content and discuss the removal of arsenic in the case of different proportion of flocculants in 30°C, see Table 1.

Mapping according to table1 as follows, see Figure 1.

As can be clearly seen from Figure 1, As the ratio of hot spring water samples and flocculant is decreasing, arsenic content presents a gradual downward trend, in the ratio of 2000:1, The curve appears suddenly dropped. in the ratio of 500:1, arsenic content is reduced to 0.200 mg/L from 1.270 mg/L, arsenic removal rate is up to 84.25%. in the ratio of 100:1, arsenic content is reduced to 0.023 mg/L from 1.270 mg/L.

Table 1.   30°C arsenic content in the case of the proportion of different flocculants.

Proportion	Temperature (°C)	Arsenic content (mg/L)	Riginal arsenic content (mg/L)	Arsenic removal rate (%)
5000:1	30	0.768	1.270	39.53%
4000:1	30	0.762	1.270	40.00%
3000:1	30	0.751	1.270	70.81%
2500:1	30	0.705	1.270	44.49%
2000:1	30	0.704	1.270	44.57%
1000:1	30	0.372	1.270	70.71%
500:1	30	0.200	1.270	84.25%
100:1	30	0.023	1.270	98.19%

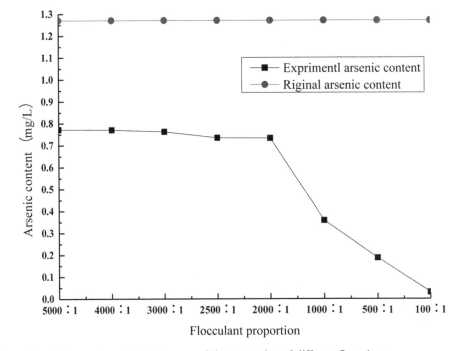

Figure 1.   30°C arsenic content in the case of the proportion of different flocculants.

3.2 *The influence of the temperature on the arsenic removal rate*

In order to study effect of temperature on removal of high arsenic hot spring water samples, the hot spring water samples are heated to different temperature for experimental study. the hot spring water samples are heated to 60°C, then mix flocculant to hot spring water samples in accordance with the proportion of hot spring water samples and flocculant of 5000:1, 4000:1, 3000:1, 2500:1, 2000:1, 1000:1, 500:1, 100:1, and determine changes in the arsenic content and discuss the removal of arsenic in the case of different proportion of flocculants in 60°C, see Table 2.

Mapping according to table1 as follows, see Figure 2.

As can be seen from Figure 2, the overall change of arsenic content shows a gradually declining trend with the reduction of the proportion of the hot spring water samples and flocculant in this group of experiments. Compared data of Figure 1, in the ratio of 500:1, arsenic content is reduced to 0.187 mg/L from 1.270 mg/L, arsenic removal rate is up to 85.28%. in the ratio of 100:1, Arsenic removal rate decreased by 0.63% compared to Figure 1.

Table 2.   60°C arsenic content in the case of the proportion of different flocculants.

Proportion	Temperature (mg/L)	Arsenic content (mg/L)	Riginal arsenic content (%)	Arsenic removal rate (°C)
5000:1	60	0.772	1.270	39.21%
4000:1	60	0.770	1.270	39.37%
3000:1	60	0.762	1.270	40.00%
2500:1	60	0.734	1.270	42.20%
2000:1	60	0.733	1.270	42.28%
1000:1	60	0.358	1.270	71.28%
500:1	60	0.187	1.270	85.28%
100:1	60	0.031	1.270	97.56%

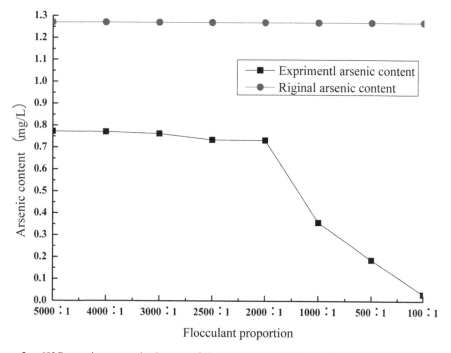

Figure 2.   60°C arsenic content in the case of the proportion of different flocculants.

The hot spring water samples are heated to boiling (t = 92°C), then mix flocculant to hot spring water samples in accordance with the proportion of hot spring water samples and flocculant of 5000:1, 4000:1, 3000:1, 2500:1, 2000:1, 1000:1, 500:1, 100:1, and determine changes in the arsenic content and discuss the removal of arsenic in the case of different proportion of flocculants in 92°C, see Table 3.

Mapping according to table1 as follows, see Figure 3.

As can be seen from Figure 3, arsenic levels downward trend becomes flatten, in the ratio of 500:1, arsenic content is reduced to 0.359 mg/L from 1.270 mg/L, arsenic removal rate is up to 71.73%, in the ratio of 100:1, Arsenic removal rate is less than in Figure 1, only 0.16%.

Through the above experimental analysis, in the ratio of 100:1, arsenic removal rate is up to 98.19%. Arsenic removal effect is very significant, we list of arsenic content in the case of the proportion of different flocculants and different temperatures, see Table 4.

Table 3.   92°C arsenic content in the case of the proportion of different flocculants.

Proportion	Temperature (°C)	Arsenic content (mg/L)	Riginal arsenic content (mg/L)	Arsenic removal rate (%)
5000:1	92	0.629	1.270	50.47%
4000:1	92	0.622	1.270	51.02%
3000:1	92	0.618	1.270	51.34%
2500:1	92	0.609	1.270	52.05%
2000:1	92	0.512	1.270	59.69%
1000:1	92	0.465	1.270	63.39%
500:1	92	0.359	1.270	71.73%
100:1	92	0.025	1.270	98.03%

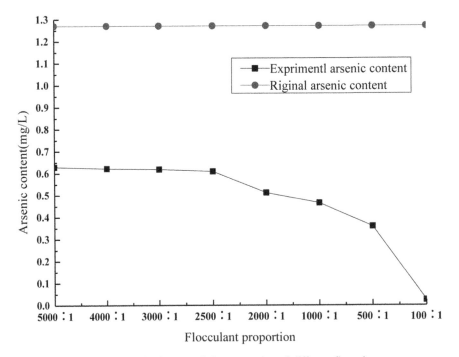

Figure 3.   92°C arsenic content in the case of the proportion of different flocculants.

Table 4. Arsenic content in the case of different flocculants proportion and different temperatures.

Temperature (°C)	Arsenic removal rate (%)							
	(100:1)	(500:1)	(1000:1)	(2000:1)	(2500:1)	(3000:1)	(4000:1)	(5000:1)
30°C	98.19	84.25	70.71	44.41	44.49	40.87	40.00	39.53
60°C	97.56	85.28	71.81	42.20	42.28	40.00	39.37	39.21
92°C	98.03	71.73	63.39	59.69	52.05	51.34	51.02	50.47

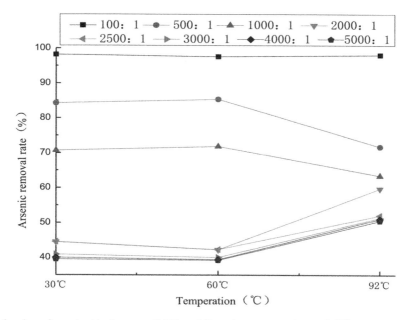

Figure 4. Arsenic content in the case of different flocculants proportion and different temperatures.

Mapping according to Table 1 as follows, see Figure 3.

The impact of temperature on arsenic removal rate is not great as can be seen from Figure 4. In the temperature range of 30°C~60°C, the temperature does not basically affect the water arsenic removal rate; in the temperature range of 60°C~92°C (boiling), proportion of hot spring water samples and flocculant is smaller, the temperature does not basically affect the water arsenic removal rate; in the ratio of (500~1000):1, removal rate decreases with increasing temperature; in the ratio of (500~1000):1, Removal rate increases with increasing temperature.

4 CONCLUSION

Studies have shown that effect of precipitation and flocculation is very significantly on the high-arsenic water arsenic removal, the experimental analysis shows that:

1. Arsenic content gradually decreases with the decreasing of the proportion of the hot spring water and flocculation, in the ratio of 2000:1, curve shows an inflection point, arsenic content has a sharp drop in water.
2. The proportion of the hot spring water and flocculation is 100:1, arsenic content is reduced to 0.023 mg/L from 1.270 mg/L, arsenic removal rate is up to 98.19%, precipitation and flocculation method has high efficiency and good effect for arsenic removal.

3. In the temperature range of 30°C~60°C, the temperature does not basically affect arsenic removal rate in the water; in the temperature range of 60°C~92°C (boiling), the effect of temperature on the arsenic removal rate has fluctuation.

In summary, precipitation and flocculation method for arsenic removal has high efficiency, good effect, low cost and controllable and adjustable on request for the arsenic content in the water.

ACKNOWLEDGEMENTS

This paper has been supported by Provincial Department of Education Fund (Grant No. 2011Y113). Thank you for communication author Zhang Shu researcher and teacher Wang Juan's care and guidance, as well as being quoted articles author support, thank you.

REFERENCES

Juan Wang, Zhiling Xu, Yanan Liu, Shu Zhang. 2011. The effect of different electric field strengths on waste water with high arsenic [J]. *Conference on Environmental Pollution and Public Health*. (2): 1068–1071.

Jun Gun, Juan Wang. 2011. Research on arsenic removal with heterosexual electrode heterogeneous materials in water [J]. *International Conference on Electric Information and Control Engineering*, (6): 5235–5236.

Li Xiao-bo, Hu Bao-an, Gu Ping. 2007. Application of pressure-driven membrane technologies for the removal of arsenicfrom drinking water [J]. *Journal of Hygiene Research*. 36(3): 395–398.

Li Ya-lin, Huang Yu, Du Dong-yun. 2008. Recovering arsenicfrom waste acid water with ferrous sulfide in producing vit-riol [J]. *Journal of Chemical Industry and Engineering*. 59(5):1294–1298.

Liu Jian-tong, Xiao Bang-ding, Chen Zhu-jin, et al. 1997. Study on As removal from basic wastewater using combinedaeration and flocculation method [J]. *China Environmental Science*. 17(2):184–187.

Mandal BK, Suzuki KT. 2002. Arsenic round the world: a review [J]. *Talanta*. 58(1): 202–209.

Song S, Lopez-Valdivieso A, Hernandez-Campos DJ, etal. 2006. Arsenic removal from high-arsenic water by enhanced coagulation with ferric ions and coarse calcite [J]. *Water Research*. 40(2):364–372.

Wang Ying, Lv Sidan, Li Xin, Wu Yingjie. 2010. Research progress and Prospect of arsenic Removal in water [J], *Environmental Science and Technology*. (33): 103–107.

Wang Juan, Zhang Shu. 2010. Different material electrode remove arsenic function in electric field [J]. *Conference on Environmental Pollution and Public Health*. (2): 853–856.

Wei Dacheng. 2003. Sources of arsenic in the environment [J]. *Foreign Medical Sciences Section*. 12 (24): 4.

Wu Zhao-qing, Chen Liao-yuan, Xu Guo-qiang, et al. 2003. Studyon treating high-arsenic wastewater from sulfuric acid plantwith lime-ferrate process [J]. *Mining and Metallurgy*. 12(1):79–81.

Xu Gen-fu. 2002. Chemical precipitation methods for treatment ofhigh-arsenic concentration industrial effluents [J]. *Hydromet-allurgy of China*. 8(1):13–17.

Xu Zhiling, Liu Yanan, Wang Juan, Zhang Shu. 2012. The effect of cathode using different material on phosphorus removal rate from phosphorus-rich [J]. *International Conference on Environmental Pollution and Public Health*. (1): 98–101.

Yuan Bao-ling, Li Kun-lin, Deng Lin-li, etal.,2006. Removal ofarsenic (III) by ferrate oxidation-coagulation from drinking water [J]. *Environmental Science*. 27(2):281–284.

Zhao Weimei. 2010. Sources and effect of arsenic on environment [J]. *Pollution prevention and controll*. (3): 146.

Zhiling Xu, Juan Wang, Yanan Liu, Shu Zhang. 2011. Research on the effect electric field to phosphorus removal in industrial phosphorus-rich wastewater [J]. *Conference on Environmental Pollution and Public Health*. (2): 1072–1075.

Hydraulic Engineering – Xie (Ed.)
© 2013 Taylor & Francis Group, London, ISBN 978-1-138-00043-8

Comparison experimental study on electric field method for arsenic removal

Wang Juan, Liu Yanan, Liu Zhihui & Zhang Shu
Yunnan University, Institute of Engineering Technology, Kunming, China

ABSTRACT: A series of experiment on electric field method for arsenic removal efficiency by regulating the different action time parameters were developed, based on DC electric field method and intermittent electric field method for arsenic pollution water treatment. The experiments results showed that the arsenic removal efficiency of DC electric field and intermittent electric field could resach detection limits low, nearly 100%. Intermittent electric field method for arsenic removal effect was better than that of DC electric field method, so that it could be provided a new train of thought for physical method for arsenic removal.

1 INTRODUCTION

Arsenic is a kind of endocrine disruptors, its compounds are toxic, and carcinogenic, teratogenic, mutagenic effect to people and livestock [1–5]. The toxicity of arsenic and arsenic compounds has bearing on arsenic presence morphology, generally speaking, inorganic compound toxicity is greater than that of organic compounds. Although inorganic arsenic exists in the form of pentavalent arsenic and trivalent arsenic, trivalent arsenic toxicity is 60 times higher than that of pentavalent arsenic [6–9]. In recent years, with the rapid development of economy, human production activity makes the problem of arsenic pollution in water body more severely. Whereas the toxicity and carcinogenicity of arsenic and accumulation in body, more and more people pay much attention to the protection of water resources.

At present, the methods disposed arsenic pollution in water are chemical method, biological method, physical method [10] and interdisciplinary arsenic removing method. Although the above arsenic removal methods have good processing effects, there exists some deficiencies and defects. In this paper, the author had a in-depth discussion and research, based on DC electric field method and intermittent electric field method for arsenic pollution water treatment, developed a series of experiment on electric field method for arsenic removal efficiency by regulating the different action time parameters and hoped to find an efficient, fast, safe, low carbon, low energy consumption method for removing arsenic.

2 EXPERIMENT METHOD

The equipment in this test for removing arsenic was home-made electric field device. The study of this task group in this year showed that iron material, which was selected as polar material in this experiment, was a kind of high arsenic removing rate, economic, environment friendly polar material. The water samples were taken from yangzonghai arsenic pollution water body, in which arsenic concentration was 66 µg/L, PH = 8.5~9.0. The experimental conditions were that each experiment water content was the same, the voltage was 50 V, and the distance between two electrode plates was 9 cm. In the DC electric field effect and intermittent electric field effect, the author mostly discussed the influence of the electric field method for arsenic removal rate by the time parameter.

3 EXPERIMENT RESULTS AND ANALYSIS

3.1 The results and ananlysis of the DC electric field method removing arsenic

Under the action of DC electric field, the author carried this study for removing arsenic through different reaction time, 10 s, 20 s, 30 s, 40 s, 50 s, 60 s, 70 s, 80 s, 90 s, 100 s, 110 s, 120 s, 150 s, respectively. see Table 1.

Which can be seen in Table 1, DC electric field on the water arsenic removal effect was good, the effect of 10 s can make the arsenic removal rate reach 53%, (see Fig. 1).

Figure 1 showed that the effect of DC electric field, whose polar plates were iron polar plate, was significant, the variation of arsenic content was larger in the action time of 10 s–30 s, 10 s arsenic removal rate was 53%, 30 s arsenic removal rate was 86.4%, respectively, a difference of 33.4%. While the variation of arsenic content was in fluctuations among 30 s–70 s to a certain extent, it illustrated that the variation of arsenic in water body became more complex. After 70 s, the arsenic content in water tended to slow down, and removing arsenic rate tended to reach 100% after120 s.

The DC electric field, whose polar plate was iron plate, for removing arsenic confirmed that the electric field method for removing arsenic was a kind of high efficiency method.

3.2 The results and analysis of intermittent electric field method removing arsenic

Under the action of intermittent electric field, the author carried this study for removing arsenic through different reaction time, 30 s, 60 s, 90 s, 120 s, 150 s, respectively, and the intermittent time was 10 s, see Table 2.

Table 1. Arsenic removing effect of DC electric field.

Time (s)	10	20	30	40	50	60	70	80	90	100	120	150
Sample (µg/L)	31	22	9	11	5	6	3	2	2	1	0.5 L	0.5 L
Original sample (µg/L)	66	66	66	66	66	66	66	66	66	66	66	66

Note: The "L" means more than the detection limit in detection value of arsenic content.

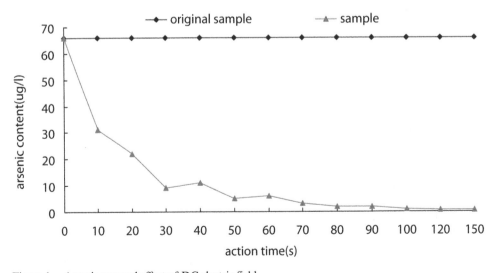

Figure 1. Arsenic removal effect of DC electric field.

Table 2. Arsenic removing effect of intermittent electric field.

Reaction time (s)	30	60	90	120	150
Sample (µg/L)	9	1	0.5 L	0.5 L	0.5 L
Original sample (µg/L)	66	66	66	66	66

Note: The "L" means more than the detection limit in detection value of arsenic content.

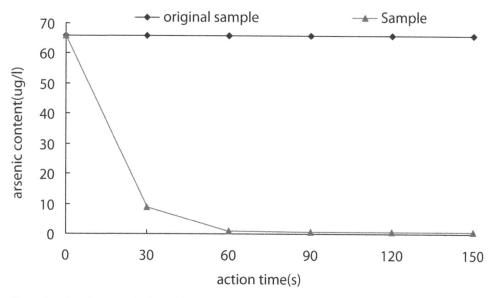

Figure 2. Arsenic removal effect of intermittent electric field.

From the date of Table 2, we could see that the removing efficiency in the intermittent electric field was high, specially, reach 100% in 90 s. According to the data in the table to make a curve for better analysis, (see Fig. 2).

From Figure 2, we could get it that the intermittent electric field removing arsenic rate could reach 86.4% in the action time of 30 s, and the figure could reach 98.5% in 60 s, 100% in 90 s, respectively. These dates showed that the removing arsenic effect of the intermittent electric field was very good. Take the water sample whose arsenic concentration was 66 $\mu g/L$ for example, it could reduce the arsenic content to reach the national drinking water standard in the reaction time of 30 s.

The removing arsenic effect of the intermittent electric field proved sufficiently the feasibility of electric field method removing arsenic and the validity of removing arsenic mechanism.

3.3 Comparison analysis

Make a comparison chart by arsenic content variation of DC electric field and intermittent electric field, (see Fig. 3).

From Figure 3, in the same point, intermittent electric field have a good effect than DC electric field in the 60 s, Arsenic content is almost zero when the time is 90 s in intermittent electric field but DC electric field is 120 s. So the intermittent electric field of arsenic removal takes less time, shows more fast, high efficiency, low carbon concept.

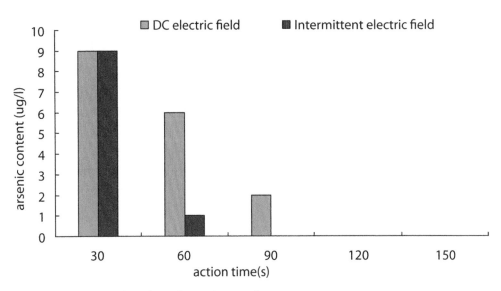

Figure 3. The comparison chart of removing arsenic.

4 CONCLUSION

The research showed that the removing arsenic effect of electric field method was significant, action fast, high removing arsenic rate. We could see the arsenic variation from Figures 1–3:

1. The removing arsenic efficiency of electric field method was high, nearly 100%, and the arsenic content value was under the detection limit.
2. Although the removing arsenic rate of DC electric field was higher than that of intermittent electric field in the beginning of action time, the latter's rate was much better than the former along with time prolonging.
3. In intermittent electric field, ionized iron ions, ferric hydroxide and the role of arsenic had been fully reactive, improved the efficiency of removal arsenic, and reflected the fast, efficient, low carbon concept.
4. Due to the adoption of an intermittent electric field, and the role of time was significantly shortened, thus effectively reduced energy consumption.
5. This research called physical method removing arsenic provided a new train of thought, and it had a very good application prospect.

The research results would lay an important experiment foundation for a more in-dept removing arsenic study, and have a very real practice role.

ACKNOWLEDGEMENTS

This paper has been supported by Provincial Department of Education Fund (Grant No. 2011Y113). Thank you for communication author Zhang Shu researcher and teacher Wang Juan's care and guidance, as well as being quoted articles author support, thank you.

REFERENCES

Juan Wang, Zhiling Xu, Yanan Liu, Shu Zhang. 2011. The effect of different electric field strengths on waste water with high arsenic [J]. *Conference on Environmental Pollution and Public Health.* 2: 1068–1071.

Li Jie, Li Jincheng, Li Wei et al. 2010. Adsorption of As (III) from water by ferric hydroxide [J]. *Water technogy.* 10:17–20.

Liu Qiaoling. 2009. The Study of Arsenic Remove in Drinking Water [J]. *Building Materials and Decorations.* 10(58):132–133.

Teng Botao. 2010. He Jiabing the investigation report on arsenic contamination in groundwater [J]. *Science & Technology Information.* 31:773&817.

Thomas MG, Gregory KD, M ccleskey RB, et al. 2001. Rapid arsenite oxidation by the musquaticus and the mus the rmophilus field and laboratory investigation [J]. *Environ Sci Technol.* 35:3857–3862.

Wang Juan, Zhang Shu. 2010. Different material electrode remove arsenic function in electric field [J]. *Conference on Environmental Pollution and Public Health.* 2: 853–856.

Yang Jinhui, Wang Jinsong, Chen Siguang et al. 2010. Study on the As(III) Removal by Magnesium Hydroxide in Wastewater Treatm ent [J]. *Metal Mine.* 10:169–171.

Yang Jie, Gu Haihong, Zhao Hao, Xu Yanhua. 2003. Review of arsenic-contaminated wastewater treatment [J]. *Industrial Water Treatment.* 23(6):14–18.

Yu Qingyuan, Wang Ling, Zhang Baowei. 2007. On the elimination of arsenic in water [J]. *Shanxi Arch IT Ecture.* 33(1):182–184.

Zhang Xuehong, Zhu Yijun, Liu Huili. 2009. Environmental Chemistry of Arsenic Process [M]. *Beijing: Science Press.*

Hydraulic Engineering – Xie (Ed.)
© *2013 Taylor & Francis Group, London, ISBN 978-1-138-00043-8*

Comparative study on the two kinds of *Litopenaeus vannamei* culture modes

Wang Yangcai, Jin Zhongwen & Zheng Wenbing
Ningbo Academy of Ocean and Fishery, Ningbo, Zhejiang, China

ABSTRACT: Bottom aeration mulch pond culture mode is a new one practiced in *Litopenaeus vannamei* culture. It can improve yield and prevent the shrimp disease effectively comparing the earthen pond mode. In this study, the dissolved oxygen is evenly distributed in the whole mulch pond water, the mean value of NO_3^-, NO_2^-, NH_4^+, PO_4^{3-} and water temperature in mulch ponds is higher than that in earthen ponds, but SiO_4^{4-} and S^{2-} is lower. A total of 62 phytoplankton taxa were identified, and the phytoplankton community structure is more complicated in mulch ponds. Principal Response Curves (PRC) show significant difference in two culture modes, and present four key species of Chlorophyta in mulch ponds. The species with high weight of PRC in chlorophyta and the more stable phytoplankton community structure might play an important role in *Litopenaeus vannamei* disease resistant in the process of the mulch pond culture.

1 INTRODUCTION

A variety of cultivation modes are implemented in *Litopenaeus vannamei* culture, such as earthen pond culture, deep pool high dam culture, indoor industrial culture and various forms of polyculture modes. A mulch culture mode was practiced successively in Thailand and in Guangdong province China in the 1990s (Wu QS 1995). In this mode, plastic mulch is spread in the bottom of the pond and isolated the pond soil, the waste concentrated in the bottom of the pool can be suck out, and management is the same as the intensive culture, moreover, the plastic mulch can be used again after harvest. Bottom aeration technology, with aeration pipes installed in the pond bottom and pumped by air blower, has been used in plastic mulch pond in recent years in Ningbo, Zhejiang province. This culture mode improves yield, prevents the cultured species disease effectively, and has done well enough to make it rapid promotion in this area as well (Jin ZW et al. 2009, 2010). This present paper compares the bottom aeration plastic mulch culture mode with the traditional earthen pond culture mode, analyzes the different changes of physico-chemical parameter and phytoplankton community structure in two culture modes, contributes to accumulating the basic data for high-yielding, disease prevention of *Litopenaeus vannamei* culture.

2 MATERIALS AND METHODS

2.1 *Study site and culture description*

Seven ponds in this study are located in Xian Xiang Town, Yinzhou District, Ningbo City, Zhejiang Province, China, 1#–5# for mulch pond, and 6–7# for earthen pond.

Aeration pipes 10 mm in diameter are installed in the bottom of the mulch pond 4 m in spacing. Pipes are perforated with 0.6 mm in diameter and 2 m in interval. An air blower is equipped with power of 2.5 kw in each mulch pond, and a 1 kw waterwheel aerator in each earthen pond.

Shrimps were stocked form July 10 to 13, 2011, 70,000 individuals in each mulch pond, 35,000 individuals in each earthen pond, and harvested successively from September 7, to October 15.

Compound feed are used in breeding procession, and microorganism preparation and fertilizer are used according to the water quality. No disease happened in mulch ponds, and the average yield was 1.00 ± 0.11 kg/m^2 with the size of 102 individual/kg, but disease happened in earthen ponds about 50 days after stocking, and the average yield was 0.30 ± 0.04 kg/m^2 with the size of 221 individual/kg. The characteristics of sampled ponds see Table 1.

2.2 Collections and analyses

Phytoplankton samples were collected in 7 ponds at an interval of 10 days from July 15, to September 26. 8 times were done. Mixed samples were collected with 2 L organic glass hydrophore from four in sides and one in middle of the pond. For quantitative estimation of phytoplankton, 1 L duplicate water samples were preserved with Lugol's solution. After 24 h of sedimentation, 50 ml concentrated samples were collected. Plankton were identified and enumerated in a 0.1 ml counting chamber. Water samples for physico-chemical parameter were collected simultaneously to the phytoplankton sampling. Dissolved oxygen concentration and water temperature were measured in the field with Thermo Orion portable dissolved oxygen instrument. All other physico-chemical parameters, dissolved inorganic phosphate (PO_4^{3-}), dissolved inorganic nitrogen (NH_4^+, NO_2^-, NO_3^-), dissolved silicate (SiO_4^{4-}) and sulfide (S^{2-}), were followed in accordance with the relevant national standards (SOC 1991).

2.3 Numerical and statistical procedures

The dominance (Y) of dominant species was calculated as $Y = (ni/N) \cdot fi$, Where ni is the abundance of taxon i, fi is the occurrence frequency of taxon i in every sample at the same sampling time, N is the total phytoplankton abundance. The dominant species is only $Y \geq 0.02$ (Xu R 2009).

Principal response curves was conducted using three data files: (1) species variables (there are physico-chemical variables and phytoplankton abundance variables in this paper), species data need to be filtered as: the frequently occurring species > 12.5%, the relative abundance of the species at least at one sample $\geq$ 1% (Lopes MRM et al 2005, Muylaert K et al 2000). The physico-chemical variables and phytoplankton abundance variables were transformed using $\ln(x+1)$ (Flores LN et al 1998); (2)explanatory variables, this experiment consists of two treatment: mulch pond (t group), earthen pond (c group). The treatment t is five replicated, whereas c is replicated. (3)time covariables, 8 data sets. explanatory variables and covariables are dummy variables, not quantitative, and they are assume a value of either 0 or 1, depending on whether the variable is present or absent in a given sample (Van den Brink PJ et al 1999). To assess the statistical significance of t group and c group differences, a Monte Carlo permutation test (Van den Brink PJ et al 1998) was conducted at $p \leq 0.05$. All analyses were conducted using CANOCO.version4.5.

Table 1. The characteristics of sampled ponds.

Pond number	1#	2#	3#	4#	5#	6#	7#
Pond area (m^2)	612	589.3	619.8	723.8	620.8	450	450
Product (kg)	641.5	622.5	690.5	600.0	597.5	123.5	144.0
Stocking quantity (individual)	61018	60430	70250	55990	72720	29640	29430
Survival percentage (%)	87.2	86.3	100.0	80.0	100.0	84.7	84.1
Per unit yield (kg/m^2)	1.05	1.06	1.11	0.83	0.96	0.27	0.32

3 RESULTS

3.1 *Physico-chemical parameter*

Table 2 and Figure 1 present the results of physico-chemical parameter. The water temperature is between 23.50°C–34.03°C in the culture period, and the change trend of temperature was basically uniform in all the ponds. The water temperature in mulch ponds is always higher than that of earthen ponds, and the mean temperature was 30.53°C in mulch ponds, 0.44°C higher than that in earthen ponds. The surface dissolved oxygen in mulch pond and earthen pond is 5.76 mg/L, 7.55 mg/L, and 5.08 mg/L, 5.62 mg/L in bottom water respectively.

The mean value of dissolved inorganic nitrogen (NO_3^-, NO_2^-, NH_4^+) and dissolved inorganic phosphate (PO_4^{3-}) is 1.19 mg/L, 0.15 mg/L, 0.54 mg/L, 1.51 mg/L in mulch ponds, and 0.64 mg/L, 0.13 mg/L, 0.21 mg/L, 0.76 mg/L more than that in earthen ponds separately. The mean different of NO_3^-, NO_2^-, NH_4^+ and PO_4^{3-} between the mulch ponds and the earthen ponds is 0.08 mg/L, 0.01 mg/L, 0.08 mg/L and 0.11 mg/L before August 25, and 1.58 mg/L, 0.34 mg/L, 0.55 mg/L and 1.85 mg/L respectively after September 6.

Dissolved silicate (SiO_4^{4-}) and sulfide (S^{2-}) in mulch ponds is always lower than that in earthen ponds, the mean concentration is 1.27 mg/L, 0.04 mg/L in mulch ponds, and 2.60 mg/L, 0.13 mg/L in earthen ponds. The concentration of dissolved silicate decreased from 1.55 mg/L, 3.73 mg/L on July 15 to 0.67 mg/L, 1.38 mg/L on September 26 in mulch ponds and earthen ponds, fell by 57.32%, 62.92% separately. The concentration of sulfide increased from 0.01 mg/L, 0.03 mg/L to 0.03 mg/L, 0.12 mg/L, with respective growth rates of 236.17%, 347.06%.

The response of the physico-chemical parameter based on the PRC analysis is presented in Figure 2. In total, the PRC explained 50.1% of the total variation, with the first and second axes accounting for 31.9 and 11.9%, respectively. Time (difference between sampling dates) and treatment accounted for 63.1 and 23.6%, respectively, of the 99.1% variance explained. Monte Carlo permutation tests were statistically significant ($P = 0.002$) for the first PRC.

The diagram shows the deviations in time of the mulch ponds compared to the earthen ponds. The larger the deviation, the more effect of the different culture mode on the physico-chemical parameter. The deviations were small from July 15 to August 25, and larger from September 5 on. The species weight shown on the right side of the diagram can be interpreted as the affinity of each physical chemical parameter with the response given in the diagram.

Species with a high positive weight conform to the principal response pattern, a negative one are inferred to show the opposite pattern, whereas species with near zero weight either show no response or a response that is unrelated to the pattern shown by the PRC. The higher the weight, the more pronounced the actual response pattern of the species is likely to fol-

Table 2. Physico-chemical parameters in each pond.

Pond number	1#	2#	3#	4#	5#	6#	7#
T (°C)	31.18 ± 2.21	31.23 ± 2.12	31.16 ± 2.14	31.20 ± 2.14	31.07 ± 2.29	30.83 ± 2.16	30.71 ± 2.15
DO	5.69 ± 1.76	5.63 ± 1.34	5.50 ± 1.61	5.68 ± 3.37	5.67 ± 1.53	7.17 ± 2.62	7.94 ± 3.70
NO_3^-	1.29 ± 1.32	1.48 ± 1.64	1.04 ± 0.87	1.06 ± 0.91	1.05 ± 1.44	0.54 ± 0.52	0.56 ± 0.53
NO_2^-	0.15 ± 0.24	0.12 ± 0.21	0.17 ± 0.23	0.17 ± 0.26	0.17 ± 0.25	0.02 ± 0.04	0.02 ± 0.03
NH_4^+	0.56 ± 0.61	0.48 ± 0.50	0.54 ± 0.73	0.46 ± 0.40	0.77 ± 1.15	0.33 ± 0.42	0.34 ± 0.68
PO_4^{3-}	1.55 ± 1.43	1.43 ± 0.87	1.31 ± 0.88	1.15 ± 0.68	1.98 ± 1.75	0.64 ± 0.59	0.86 ± 1.08
SiO_4^{4-}	1.41 ± 0.99	1.00 ± 0.82	0.93 ± 0.70	1.65 ± 1.19	1.43 ± 1.18	2.47 ± 1.54	2.73 ± 1.58
S^{2-}	0.03 ± 0.02	0.04 ± 0.03	0.03 ± 0.02	0.04 ± 0.02	0.04 ± 0.04	0.09 ± 0.04	0.16 ± 0.16

Values are expressed as means and standard deviations over the experimental period (mg/L)

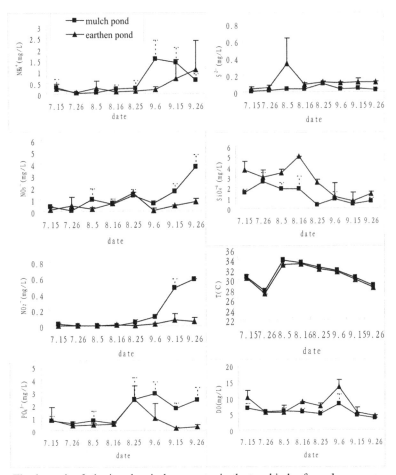

Figure 1.   The dynamic of physico-chemical parameter in the two kinds of ponds.

low the pattern in the PRC (Van den Brink PJ et al 1998, 1999). The species with the highest scores is sulfide, and PO_4^{3-}, NO_2^-, and NO_3^- are species with the highest negative scores showing a significant in crease in mulch ponds compared to earthen ponds.

3.2   *Phytoplankton community structure*

A total of 62 phytoplankton taxa were identified during this study belonging to seven phyla: Cyanophyta, Chlorophyta, Bacillariophyta, Pyrrophyta, Cryptophyta, Euglenophyta and Chrysophyta. Chlorophyta represented the most diverse group with 26 species. Bacillario-phyta was the second most diverse group with 15 species. The most abundant phytoplankton group was Cyanophyta, with a mean abundance of 3.52×10^8 cell/L, accounting for 71.43% of the total abundance, and the second was Chlorophyceae, with 5.67×10^7 cell/L, account-ing for 16.12% of the total abundance, and the third was Diatom, accounting for 9.43%. Chrysophyta was detected just on the first sample on July 15, Cryptophyta and Euglenophyta appeared on the samples after September 5, Pyrrophyta was rare all the samples. Of all the species with occurrence frequency above 60%, 5 species belong to Cyanophyta, including *Dactylococcopsis rhaphidioides*, *Chroococcus minor*, *Oscillatoria tenuis*, *Tetrapedia gothica*, *Merismopedia tenuissima*, with occurrence frequency 80.00%, 69.09%, 69.09%, 67.27% and 60.00% respectively, 2 species each belong to Chlorophyta and bacillariophyta, including *Oocystis elliptica* 74.55%, *Chlamydomonas ovalis*, 69.09%, and *Cyclotella sp.* 65.45%, *Navicula simoles*, 61.82%. The dominant species (Y ≥ 0. 02)changed constantly in the culture process.

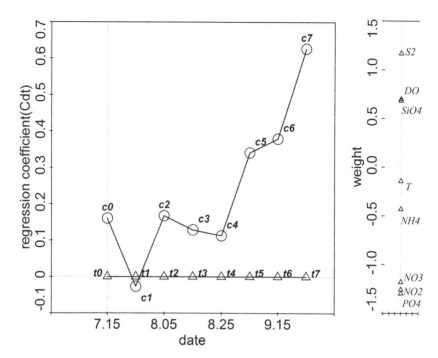

Figure 2.　PRCs resulting from the analysis of the physico-chemical data set.

5 dominant species belong to Cyanophyta, Chlorophyta, Pyrrophyta and Chrysophta on July 15, with wide distribution, but dominant species just belong to Cyanophyto and Chlorophyta on August 16. 3 dominant species left on September 6, on which phytoplankton composition was simple, especially eutrophic species, and *Oscillatoria tenuis* was the first dominant species (Y = 0.82), with a mean abundance of 1.20×10^9 cell/L, accounting for 83.05% of the total abundance. Although the dominance of *Oscillatoria tenuis* dropped to 0.251 on September 26, the dominant species were still from Cyanophyta and Chlorophyta.

There is a relatively uniform composition of phytoplankton species between the mulch ponds and the earthenponds, but the phytoplankton community structure is different. The proportion of Cyanophyta fluctuated wildly from 5.36% on July 15 to 92.37% on September 15 in the mulch ponds, but that was over 65% in the earthen pond all the time except August 5. In the respect of phytoplankton abundance, the changes were the same in two culture modes before August 16, and after then, phytoplankton abundance increased in mulch ponds, with the peak abundance of 3.65×10^8 cell/L on September 15, but decreased in earthen ponds to the lowest abundance 3.88×10^7 cell/L on September 6, 46.25% lower than on August 25, 16.60% compared with the same sampling time in mulch ponds, and then increased to 4.37×10^8 cell/L at the end of sampling time.

The response of the phytoplankton parameter based on the PRC analysis is presented in Figure 5. In total, the PRC explained 23.6% of the total variation, with the first and second axes accounting for 10.6 and 5.5%, respectively. Time and treatment accounted for 38.2 and 18.6%, respectively, of the 84.6% variance explained. Monte Carlo permutation tests were statistically significant (P = 0.002) for the first PRC.

Diagram shows the differences of the PRC in species composition between the mulch and earthen ponds. The first large deviation was seen before shrimp disease on August 16, but shrank on September 6, and a larger one from then on. After filter, 23 phytoplankton species were analyzed as species variables with PRC, among them, 6 in Cyanophyta, 7 in Chlorophyta and Bacillariophyta each, 2 in Pyrrophyta and 1 in Cryptophyta. 60.9% of species with positive weights conform to the earthen response pattern indicated by the PRCs, whereas 39.1% with negative weights exhibit the opposite pattern. The species with weights

Table 3. Dominant species and dominance (Y) of phytoplankton.

Date	7.15	7.26	8.05	8.16	8.25	9.06	9.15	9.26
Merismopedia tenuissima	0.105	0.049	0.054				0.031	
Dactylococcopsis rhaphidioides				0.043	0.072	0.035		0.021
Chroococcus minor			0.080				0.026	
Chrooloccus sp.	0.064		0.106					
Tetrapedia gothica		0.082	0.080	0.333	0.023			
Oscillatoria tenuis		0.263		0.127	0.297	0.820	0.793	0.251
Microcystis incerta								0.040
Oocystis elliptica	0.063	0.066	0.038	0.166	0.029			0.026
Chlamydomonas ovalis		0.030		0.041	0.023			
Scenedesmus quadricanda								0.213
Stephanodiscus astraea						0.034		
Cyclotella sp.					0.074			
Navicula simoles			0.085					
Cryptomonas ovata						0.055		0.029
Gymnodinium coeruleum	0.040							
Chromulina ovalis	0.208							

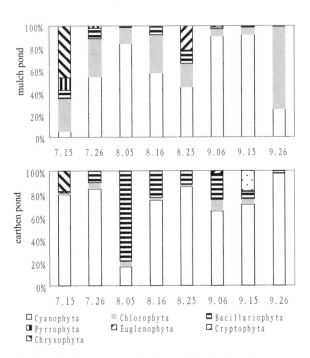

Figure 3. Changes of phytoplankton composition in two kinds of ponds.

between −1 and +1 are not shown, because these are likely to show either a weak response or a response that is unrelated to that shown in Figure 5. There are only 9 species left on the diagram, which is indicative of an impact due to culture mode. Five species with a positive weight in the diagram, *Dactylococcopsis rhaphidioides, Merismopedia tenuissima, Chrooloccus sp.* are belonged to Cyanophyta, and *Navicula* si*moles, Nitzschia pungens* to Bacillari-ophyta, while the four species with a highest negative scores are Chlorophyta. Some advance species in abundance and dominance, such as *Oscillatoria tenuis* and *Tetrapedia gothica*, are

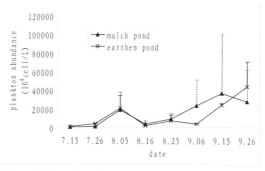

Figure 4. Changes of phytoplankton abundance in two kinds of ponds.

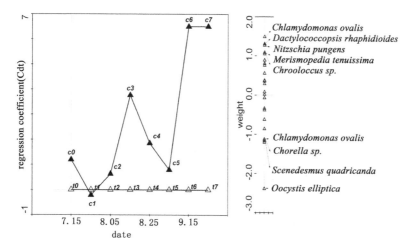

Figure 5. PRCs resulting from the analysis of the phytoplankton data set.

not present on the diagram, but *Navicula simoles, Nitzschia pungens* present, with disadvantage in abundance and dominance.

4 DISCUSSION

Bottom aeration technology is effective in enriching the oxygen of the culture water. The air was pumped into the culture water through aeration pipes installed in the bottom of the pond by the air blower, and the dissolved oxygen improved while the water was stirred up-and-down. Despite of high shrimp density and great oxygen consumption, the difference of dissolved oxygen between surface and bottom water is small in the mulch ponds, and the dissolved oxygen is evenly distributed in the whole culture water. Dissolved inorganic nitrogen is from the decomposition of the feed undigested, biologic metabolism and other inputs (Funge SJ et al 1999). In the period of July 15 to August 25, shrimp individual was small, artificial feed was less, and dissolved inorganic nitrogen maintained low. In the mid to late culture, along with the shrimp growing and the feed addition, the dissolved inorganic nitrogen increased in mulch ponds, but kept low lever in earthen ponds, because of less feed and low shrimp density on account of shrimp disease and harvest in succession. Samocha report that the Lower culture density can make total nitrogen concentration keep low level in culture pond (Samocha TM et al 2004). Dissolved inorganic phosphate mainly comes from artificial feed and external nutrient in culture pond (Cremen MCM et al 2007). In this study, dissolved inorganic phosphate kept a proper lever in the early days because of external nutrient before shrimps stocking, increased on August 25 with the increase of feed and metabolic product, and kept higher lever to the end of experiment in mulch ponds, but decreased in earthen ponds as less feed and more consumption of growing phytoplankton.

Playing an important role in the biological community of shrimp pond, phytoplankton community is associated intimately with water quality and shrimp survival (Feuga AM 2000). In the beginning of the culture, phytoplankton community structure was complicated, dominant species was belong to 5 phyla, *Chromulina ovalis* was the most dominant taxa, and more algae were eatable for shrimps. In the mid to end of culture, with the feed addition and accumulation of shrimp excretion, water was over-nutrition, and that lead to some algae grow rapidly, such as *Cryptomonas ovata*, *Oscillatoria tenuis*, dominant species changed from complexity to simplicity, Cyanophta became the most dominant, environment pressure increased on shrimps. Benjamas reported that the quality of culture environment was determined by dominant algae, phytoplankton density and the stability of phytoplankton community (Benjamas C et al 2003) Comparing the phytoplankton changes of mulch and earthen ponds before shrimp disease shows that: (1) phytoplankton abundance descend, with the decline in mulch and earthen ponds 52.27% and 64.57% respectively. High abundance is the feature of low salty pond, and one of the conditions to keep ecosystem balance in high culture density pond. Algae play a dominant role in intaking nutrients in pond water, restraining the growth and reproduction of deleterious microbe (Benjamas C et al 2003), the decrease of phytoplankton abundance might lead to the decline in the capacity of restraining deleterious microbe; (2) phytoplankton community structure different, phytoplankton composition changed from 4 phyla to 5 phyla in mulch ponds. The abundance proportion of Cranophyta dropped from 83.83% to 44.81%, and that of Chlorophyta rose from 14.83% to 21.35%, other phyla also increased accordingly. Phytoplankton community structures were more stability. On the contrary, the abundance proportion of Cranophyta rose from 16.43% to 85.94% in earthen ponds, with no changes in phytoplankton phyla, phytoplankton community structure were simpler. For the mulch ponds, the increase of phytoplankton community stability might counterbalance the disadvantage of the drops of phytoplankton abundance, and low phytoplankton abundance might latently increase the risk of shrimp disease (Peng CC 2011). For the earthen ponds, the simpler phytoplankton community structure might further reduce disease resistance.

The principal response curve method is based on the redundancy analysis ordination technique, which is the constrained form of principal component analysis (Kristie N et al 2008). The method distills the complexity of time-dependent, community-level effects of culture modes to a graphic form that is easy to comprehend. Moreover, the responses of individual species to the culture mode can be inferred from the PRC curves.

S^{2-}, PO_4^{3-}, NO_2^- and NO_3^- are the indicative of an impact due to culture mode. S^{2-} with high positive weight from degradation of sediment organism is response to earthen ponds, PO_4^{3-}, NO_2^- and NO_3^- with high negative species weight are dramatically following the increase of feed and biologic metabolism in mulch ponds, but the deviation of PRCs in physico-chemical parameter are large in later period of culture, and small before the shrimp disease. In addition to the later period of culture, the deviation of PRCs in phytoplankton species is large before the shrimp disease as well. *Dactylococcopsis rhaphidioides*, *Merismopedia tenuissima*, *Chroolocus sp.* *Navicula simoles*, *Nitzschia pungens*, are response to earthen ponds, and *Oocystis elliptica*, *Scenedesmus quadricanda*, *Chlorella sp.*, *Chlamydomonas ovalis*, with high negative speices weight, are the key species of mulch ponds. Huang reported that water quality got better after importing *Oocystis borgei* and *Nannochloris oculata*, and index in disease resistant improved as well (Huang XH et al 2005). In this study, the species with high weight in chlorophyta and the more stable phytoplankton community structures might play an important role in *Litopenaeus vannamei* disease resistant in the process of the mulch pond culture.

5 CONCLUSIONS

Bottom aeration technology is effective in enriching the oxygen of the culture water, and the dissolved oxygen is evenly distributed in the whole culture water.

Phytoplankton community structures were more stability in mulch ponds comparing with earthen ponds.

The key species of chlorophyta and the more stable phytoplankton community structures might play an important role in *Litopenaeus vannamei* disease resistant in the process of the mulch pond culture.

ACKNOWLEDGEMENTS

This research was funded by Marine Fishery Resources Sustainable Development & Integrated Utilization Project (2008C10021) and Agriculture & Social Development Major Project (2012C10023) of Ningbo Science and Technology Bureau.

REFERENCES

Benjamas, C Sorawit, P & Piamsak, M. 2003. Water quality control using Spirulina platensis in shrimp culture tanks. *Aquaculture,* 220: 355–366.

Cremen M.C.M, Martinez-Goss M.R & Corre Jr V.L et al. 2007. Phytoplankton bloom in commercial shrimp ponds using green-water technology. *J Appl Phycol,* 19:615–624.

Feuga A.M, 2000. The role of microalgae in aquaculture: situation and trends. *J Appl Phycol.* 12:527–534.

Flores L.N & Barone R. 1998, Phytoplankton dynamics in two reservoirs with different trophic state (Lake Rosamarina and Lake Arancio, Sicily, Italy). *Hydrobiologia,* 369(370): 163–178.

Funge-smith S.J. & Briggs M.R.P. 1999, Nutrient budgets in intensive shrimp ponds: implications for sustainability. *Aquaculture.* 164:117–133.

Huang X.G. Li C.L. & Zheng L. et al. 2005. Studies on immobilized microalgae improving culture water quality and strengthening anti-disease ability of prawn. *Marine Science Bulletin.* 24 (2): 57–62.

Jin Z.W, Hua J.Q. & Dai H.P. et al 2009. Configuration and application of pond-bottom rechargeable aerobic facilities. *Fishery Modernization.* 36(5):27–31.

Jin Z.W, Z Heng Z.M. & Wu S.J. et al 2010. Preliminary study on improvement of pond water quality by bottom aeration. *South china fisheries science,* 6(12):20–25.

Kristie N. Sven U. & Robert H. 2008. Is light the limiting factor for the distribution of benthic symbiont bearing foraminifera on the GreatBarrierReef. Journal of Experimental Marine *Biology and Ecology,* 363: 48–57.

Lopes M.R.M, Bicudo C.E.M.& Ferragut M.C. 2005. Short term spatial and temporal variation of phytoplankton in a shallow tropical oligotrophic reservoir, southeast Brazil. *Hydrobiologia,* 542: 235–247.

Muylaert K, Sabbe K, & Vyverman W. 2000, Spatial and temporal dynamics of phytoplankton communities in a freshwater tidal estuary. *Estuarine, Coastal and Shelf Science,* 50: 673–687.

Peng C.C, Li Z.J. & Cao Y.H, et al. 2011, Change of dominant species of planktonic microalgae in Litopenaeus vannamei's semi-intensive culture ponds and its impact on the culture environment. *Marine environmental science.* 30(2):193–198.

Samocha T.M., Lopez I.M. & Jones E.R., et al. 2004, Characterization of intake and effluent waters from intensive and semi-intensive shrimp farms in Texas. *Aquacultrue Research.* 35:321–339.

SOC (State Ocean China),1991, The criterion of marine monitoring. Beijing: *Oceanographical Press,* 205–282.

Van den Brink PJ. & Ter Braak CJF. 1999. Principal response curves: Analysis of time-dependent multivariate responses of biological community to stress. *Environ Toxicol Chem* 18:138–148.

Van den Brink PJ. & Ter Braak CJF. 1998. Multivariate analysis of stress in experimental ecosystems by principal response curvesand similarity analysis. *Aquat Ecol* 32:163–178.

Wu Q.S. 1995. Several modes of industrial shrimp culture in China and abroad. *Fishery machinery and instrument,* 1:7–9.

Xu R. Li Y.H. & Li Z.E. 2009. Quantitative comparison of zooplankton in different habitats of the Changjiang Estuary. *Acta Ecologica Sinica,* 29(4):1688–1696.

Hydraulic Engineering – Xie (Ed.)
© 2013 Taylor & Francis Group, London, ISBN 978-1-138-00043-8

Fire gases detection system based on photoacoustic spectroscopy principle

Ya-Long Jiang & Jie Zhang
Civil and Environmental Engineering School, Anhui Xinhua University, Hefei, Anhui, China

Jin-Jun Wang
State Key Laboratory of Fire Science, University of Science and Technology of China, Hefei, Anhui, China

ABSTRACT: A fire gas detection system is designed based on photoacoustic principle using the self-made photoacoustic cell combined with the wavelength modulation technique, while the CO concentrations generated from the combustions of polyurethane foam and n-heptane are measured. The experiments show that the detection system responds well to CO and has a good linear relationship. The system is characterized with stability and reliability over a long period of time, which is able to meet the need of fire detection.

1   INTRODUCTION

The Fire is one of the frequently occurring disasters in the world today, and also spans a wide range both temporally and geographically. Not only does it bring to the humans and societies huge casualties and property losses, but also has great impact on the ecological environment. It is understood that during the "Eleventh Five-Year" period, 793,000 fires were reported across China, causing 7,202 deaths and 4,501 injuries, resulting in direct property losses worth 7.2 billion yuan (Liu, Y-y et al. 2011). Fires give rise to a variety of disasters, and therefore people are always committed to the study of fire detection technologies. The current fire prevention researches are shifting from passive firefighting to proactive detection and prevention.

In most cases of fire, gas is the first to appear (Hagen B et al. 2000), and therefore the use of a gas for fire detection seems to be the most appropriate method. Unlike the smoke particles in the fire, gases produced in the fire can be driven upward very fast by much less heat. Certain gases, if lighter than the air, can be easily dispersed even in the absence of heat. This is very important for the deployment of fire detectors and early capture of fire information. In fact, fire generates an array of gases (e.g., CO, CO_2, HCN, H_2S, HCL), but CO is usually chosen as target gas. Some gases do not exist in the normal state (i.e., non-fire state). In other words, the concentration of these gases largely reflects the occurrence or non-occurrence of a fire.

There are several ways to detect gas. However, conventional gas detectors are disadvantageous because of their low sensitivity, less selectivity and slow response, and also because they are prone to false positives. For these reasons, they often fail to meet the needs for fire detection. This is where gas detection based photoacoustic spectroscopy comes in, with high sensitivity, good selectivity, fast response and good stability. With this approach, it is easy to implement multi-component gas detection, and it has already caught the attention of many researchers. The method has the following advantages (Liu X-l et al. 2011): (1) direct measurement of the optical absorption of gas with high accuracy and reliability; (2) virtually no zero drift, and acoustic signals generated only when gas is measured; (3) simultaneous detection of multi-component gas; (4) photoacoustic signal detection used to improve the quality of the detector, resulting in good stability while no correction or calibration is required; and (5) high sensitivity and selectivity.

For example (Kerr E.L et al. 1968, Lu M-h et al. 2011, Rooth R.A et al. 1990, Sigrist M.W et al. 1994), Kerr and Atwood first reported the detection of trace gases based on photoacoustic spectroscopy using laser as a light source in 1968. In 1971, Kreuzer analyzed theoretically the limit of detection (*LOD*) of the photoacoustic detection technique using high-sensitivity microphones and dye laser as light source. For trace gas analysis, *LOD* could reach up to 0.1 *ppt* magnitude. In 1995, F. G. C. Bijnen et al. designed the first longitudinal mode photoacoustic cell, which was small in size but highly sensitive, built in the chamber of a waveguide *CO* laser device. Its sensitivity in detecting ethylene was up to 6 *ppt*. In 1997, M. A. Gondal designed a set of photoacoustic system for long-distance and real-time detection of air pollutants, which could be used for remote detection of pollutants in automobile exhaust.

In particular, as laser light sources of high power and good monochromatic properties and high sensitive microphones emerge in the modern times, the detection sensitivity of the photoacoustic spectroscopy technique is further improved, and has become an emerging area of research, widely used in the fields of physics, chemistry, biology and materials, and recognized as an popular topic for international researches (Bicanic D. D. 2011, Kuusela T et al. 2009, Peng Y et al. 2009, Sthel M.S et al. 2011, Teodoro C.G et al. 2010).

2 THEORY OF PHOTOACOUSTICS

In photoacoustic spectroscopy measurements, study samples are usually placed in a closed cell along with a high sensitive microphone, where they are radiated with monochromatic light after intensity modulation. If any incident photon is absorbed by the sample, the sample's internal energy can be excited to upper states. Subsequently, during the de-excitation process of upper states, some or all of the absorbed light energy is transformed into heat energy through non-radioactive de-excitation. Since the incident light is intensity-modulated, the heating process inside the sample is modulated as well. This temperature rise leads to gas expansion. As the cell is sealed, the resulting periodic pressure fluctuations can be detected with a microphone. The signal intensity is related to the gas concentration and therefore can be used for gas detection.

Acoustic disturbance in the gas can be expressed with sound pressure $\bar{P}$ as follows:

$$\bar{p} = P - P_0 \tag{1}$$

where P and P_0 are the total and the average of sound pressures.

As mentioned above, the temperature rises after the sample gas absorbs the modulated light energy in the photoacousitc cell, generating a source of Joule-heat power density H($\bar{r}$,t), which will excite sound waves in the cell. Assuming that the sample gas in the cell is approximated to an ideal gas, the excited acoustic signal can be described in the following wave equation:

$$\nabla^2 \bar{p} - \frac{1}{c^2}\frac{\partial^2 \bar{p}}{\partial t^2} = -\frac{(\gamma-1)}{c^2}\frac{\partial \bar{H}}{\partial t} \tag{2}$$

where ∇^2 is the Laplace operator, c is the speed of sound in gas, $\gamma = C_p/C_V$ is the heat capacity ratio of the gas, and C_p and C_V are the constant-pressure molar heat capacity and constant volume molar heat capacity respectively.

The formula above is treated with Fourier transform, combined with other conditions. Ultimately, the relationship between the amplitude of the acoustic signal $A_j(\omega)$ and the optical inputs can be obtained:

$$A_j(\omega) = \frac{-i\omega}{\omega_j^2} \frac{\frac{\beta(\gamma-1)}{\Lambda_j v}\iiint_V p_j^*(\mathbf{r})I(\mathbf{r},\omega)dV}{\left[1-\left(\frac{\omega}{\omega_j}\right)^2 - \left(i\frac{\omega}{\omega_j Q_j}\right)\right]} \tag{3}$$

where ω is the angular frequency of the modulated light, ω_j and $p_j^*(r)$ are the normal frequency and the complex conjugate mode of the normal mode p_j, $I(r,\omega)$ is the intensity of the incident light, β is the absorption coefficient of the sample, V is the volume of the photoacoustic cell, and Q_j is the quality factor.

Formula (3) reflects the relationship between the amplitude of the acoustic signal and the intensity of the input light in the photoacoustic cell. When the modulation frequency $\omega/2\pi$ of the resonant photoacoustic cell equals the j-order normal frequency $\omega_j/2\pi$, the amplitude of the photoacoustic signal $A_j(\omega_{jj})$ becomes the maximum value, expressed as follows:

$$A_j(\omega_j) = \frac{\beta Q_j(\gamma-1)}{A_j V(\omega_j)} \int\int_V\int p_j^*(\mathbf{r}) I(\mathbf{r},\omega) dV \qquad (4)$$

3 EXPERIMENTAL DEVICE

In the photoacoustic detection system, the acoustic wave in the photoacoustic cell is excited by the modulation of the intensity of the incident light source, which requires modulation techniques. The most commonly used of such techniques is mechanical chopping modulation, which works as a light beam passes through a mechanical chopper. Mechanical chopper modulator is easy to use and delivers efficient modulation, but it affects the detection a great deal when the modulation frequency is high, due to the mechanical vibration and air noise generated by the high-speed rotation. For this reason, the system uses a wavelength modulation technique, where a continuously tunable DFB diode laser is used as light source. Low-frequency sawtooth signals produced by the signal generator are superimposed with high frequency sine waves generated by a lock-in amplifier, and then injected into the laser controller for laser modulation and scanning to obtain a modulated incident light source.

The entire detection system consists of signal generator (*TFG3050*), laser, laser controller (*LDC3724B, ILX Light Wave*), absorption path, photoacoustic cell, microphone (*B & K, 4192 type*), pre-amplifier (*B & K* company 2669L), lock-in amplifier (*7265, Signal Recovery*) and data acquisition system (*INV303B*), as shown in Figure 1.

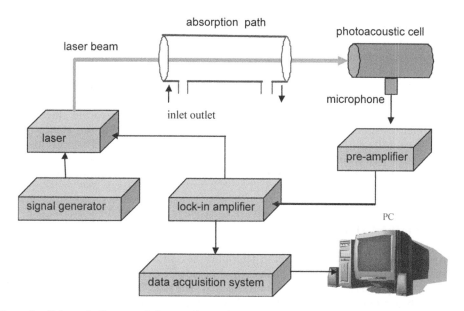

Figure 1.   Schematic diagram of the experimental setup.

In order to prevent combustion products from polluting the wall of the cell, the photoa-cousitc cell is not designed to be an open one. Instead, an absorption path is added between the light source and the photoacoustic cell. The open design of the absorption path allows combustion products in the experiment to enter the absorption path through the inlet, where they are tested. The combustion products in a fire are in fact a mixture of gases and smoke particles, and the latter would cause light attenuation during scattering and absorption. Therefore, the smoke is filtered before entering the absorption path. The photoacoustic cell is filled with high purity CO gas, which generates an acoustic signal in the photoacoustic cell once having absorbed the incident light. The acoustic signal is then picked up by a high-sensi-tivity microphone. After going through filtering amplification and frequency lock amplifica-tion with the preamplifier, conditional amplifier and lock-in amplifier, the signal enters the multi-channel signal collector before eventually making its way into the computer for data recording and processing.

In the photoacoustic detection system, the design of the photoacoustic cell is one of the most critical parts of all. As the signal source, the photoacoustic cell, depending on its per-formance, largely determines the resolution, the limit of detection and other detection capa-bilities of the entire system.

The (pmn)th order normal mode of the resonant photoacousitc cell can be expressed as $P_{pmn}(r)$. The photoacousitc cell herein is designed to be in a first-order radial resonant state (001), therefore $p = m = 0, n = 1$, where the normal mode and the normal frequency are simplified as:

$$P_{001}(r) = J_0\left(\frac{\pi\alpha_{01}}{R}r\right) \tag{5}$$

$$f_{001} = \frac{v}{2}\frac{\alpha_{01}}{R} \tag{6}$$

The value of $\pi\alpha_{01}$ is 3.832 with reference to the Table, and therefore $\alpha_{01} = 1.2197$, which, substituting (6), returns the normal frequency as $2243\,Hz$.

For the 0th order Bessel function, $J_0(0)$ is the maximum value, so $J_0(\pi\alpha_{01}r/R)$ reaches the maximum when $r = 0$. The beam $I(r, \omega_j)$ must be incident along $r = 0$ to have the strongest coupling between the beam and the normal mode p_{001}. Because the location where $r = R$ is also acoustic pressure antinode of radial resonant, it is suitable to place the microphone on the surface of the cell wall where $r = R$.

In addition, as the background signal generated as a result of the transmittance window and the wall of the photoacousitc cell absorbing modulated light energy has the same time-varying characteristics with the photoacoustic signal, it cannot be eliminated by the use of other instruments like the lock-in amplifier, and the sensitivity of photoacoustic detection would be subject to great restrictions. In order to remove the impact this background noise has on the measurement, we have designed a photoacousitc cell with acoustic filtering. The acoustic filter consists of three parts, the intermediate portion with a radius of $1.8\ cm$, and the portions on the left and right side each with a radius of $0.3\ cm$. The corresponding maximum noise reduction amounts to $25\ dB$. On either end of the photoacousitc cell is an infrared transmitting glass window $8\ mm$ in radius and $2\ mm$ in thickness, made of CaF_2 material.

4 EXPERIMENTAL RESULTS

Appropriate temperature values are set to determine the center frequency of the laser accord-ing to the CO characteristic absorption lines (in the case of CO, the characteristic gas in fires, the absorption spectral lines near $1579\ nm$ is selected). The laser frequency is made to scan the entire CO absorption line by the injection of low-frequency sawtooth waves, and thereby the singlet absorption spectrum is obtained. The singlet feature of the absorption spectral lines can avoid other component gases cross absorbing and interfering with the tested gas, so that accurate measurements can be ensured.

Before the experiment, the photoacousitc cell is fixed on the optical bench, and the laser is drawn from the laser device through an optical fiber, with the optical fiber splice fixed on the optical bench with a bracket. Then turn on the vacuum gauge and start the vacuum pump to make the photoacousitc cell into a vacuum state (pressure can be read out on the pressure gauge). Open the *CO* cylinder to pass the 99.99% pure *CO* gas into the photoacousitc cell. When the manometer readings show one atmospheric pressure, close the gas cylinder, and then turn off the intake valve of the photoacousitc cell with the remaining *CO* in the tube pumped out. After inflation is completed, adjust the frequency of the signal generator while tuning the full-scale sensitivity, time constant and other parameters of the lock-in amplifier so that it corresponds to the maximum photoacoustic output. Connect the outputs of the lock-in amplifier and the photoelectric receiver to the first and the second inputs end of the data acquisition device, while running the *DASP2003* system software and setting up the appropriate sampling channels and frequencies. The experiments starts after all parameters of the electronic equipment are set in the best condition and stablized for 30 minutes. Red light is use to calibrate the light path before infrared is applied to ensure that the beam passes through the center of the photoacousitc cell and the absorption path. After the light path is calibrated, the optical fiber is shifted to the infrared output port of the laser controller.

Two sets of simulated experiments are conducted with the photoacoustic fire detection system. The combustion products in the experiment are drawn into the absorption path through a PVC pipe. The first set of experiment uses polyurethane foam, where a *25 cm × 25 cm × 3 cm* flexible polyurethane foam, which is free of flame retardant and has the specific gravity of about 40 *kg/m³*, is placed in the combustion dish. A small dish *5 cm* in diameter is filled with 5 mL of methylated spirits and placed under one corner of the polyurethane foam, and this is the corner that is to be lit. The *CO* concentration is shown in Figure 2. As can be seen from the Figure, the *CO* concentration increases in the course of the experiment until reaching a maximum value of *65 ppm* when *t = 119s*.

The second set of experiments uses n-heptane as raw materials. The n-heptane added with 3% toluene is poured into a round container with the bottom area of *1089 cm²* (*33 cm × 33 cm*) and the height of *5 cm*, made from *2 mm* thick steel plates, and then ignited with a flame. The concentration of *CO* is shown in Figure 3. As can be seen from the Figure, the *CO* concentration reaches *10 ppm* when *t = 19s*, and continues to rise until hitting a maximum value of *59 ppm* when *t = 117s*.

Figure 4 shows a linear fitting relationship between the measured values and the true values of the *CO* gas. The fitting coefficient *R* is *0.99*. It can be seen that the concentration values

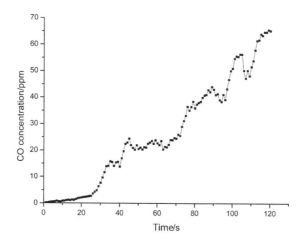

Figure 2. Variation curve of carbon monoxide concentration in polyurethane foam experiment.

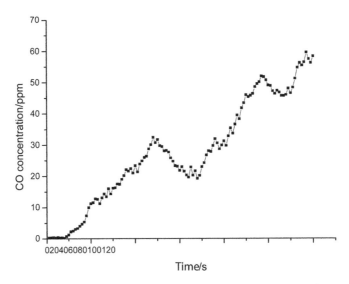

Figure 3.   Variation curve of carbon monoxide concentration in n-heptane experiment.

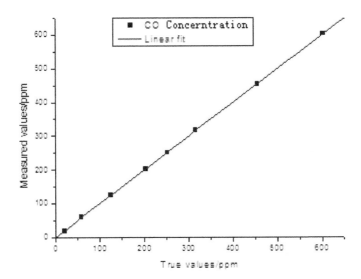

Figure 4.   The linear relation between the measured and the true values of *CO* concentration.

measured by the system has a good linear relationship with the configured concentration values. Its good reproducibility is demonstrated by a number of tests under the same conditions, and the maximum deviation between the signals is controlled in a small range.

5   CONCLUSION

A fire gas detection system is designed based on the theory of photoacoustic spectroscopy. The results show that the system can be used to effectively measure the concentration of *CO* gas generated from the burning of polyurethane foam and n-heptane. Fire alarm can be achieved through appropriately configured thresholds. The concentration values measured by the system have a good linear relationship with the true values, and the system, running with stability on a long-term basis, can meet the needs for fire detection. Based on the high sensitivity, good selectivity and a wide range of dynamic detection, the photoacoustic spectroscopy technology can also be applied in the detection of coal mine methane gas and environmental pollutants, if diode lasers of different wavelength are used.

ACKNOWLEDGEMENTS

This research was funded by open fund of the state key laboratory of fire science (*NO. HZ2010-KF07*) and Natural Science Foundation of Anhui Province (*NO.KJ2008B047*).

REFERENCES

Bicanic D. D. 2011. On the photoacoustic, photothermal and colorimetric quantification of carotenoids and other phytonutrients in some foods: a review. Journal of Molecular Structure, (993):9–14.

Hagen B, Milke J.A. 2000. The use of gaseous fire signatures as a means to detect fires. Fire Safety Journal, (34):55–67.

Kerr E.L. Atwood J.G. 1968. The laser illuminated absorptivity spectrophone:a method for measurement of weak absorptivity in gases at laser wavelengths. Applied Optics: (7):915–921.

Kuusela T. Peura J. Matveev B.A. *et al.* 2009. Photoacoustic gas detection using a cantilever microphone and *III-V* mid-IR *LEDs*. Vibrational Spectroscopy, (51):289–293.

Liu X-l. Luo R-h. ZHANG C. *et al.* 2011. The Research Progress of Microphone based Photoacoustic Spectroscopy Gas Detection. ELECTRONICS & PACKAGING, 11(2):34–38. (In Chinese).

Liu Y-y. Wang F. 2011. China is still in the period of high risk of fire[*EB/OL*] http://www.chinesetoday.com/zh/article/528450.

Lu M-h. Hao R-y. Wang Z-j. *et al.* 2011. Research on the Photoacoustic Spectroscopy for C_2H_4 Gas Detection and Applications. Journal of Changzhi University, 28(5):29–32.

Peng Y. Zhang W. Li L. *et al.* 2009. Tunable fiber laser and fiber amplifier based photoacoustic spectrometer for trace gas detection. Spectrochimica Acta Part A, (74):924–927.

Rooth R.A. Verhage A.J. Wouters L.W. 1990. Photoacoustic Measurement of Ammonia in the Atmosphere:Influence of Water Vapor and Carbon Dioxide. Appl. Opt, (29):3643–3653.

Sigrist M.W. 1994. Air Monitoring by Spectroscopic techniques. New York.

Sthel M.S. Schramm D.U. Lima G.R. *et al.* 2011. CO_2 laser photoacoustic detection of ammonia emitted by ceramic industries. Spectrochimica Acta Part A, (78):458–462.

Teodoro C.G. Schramm D.U. Sthel M.S. *et al.* 2010. CO_2 laser photoacoustic detection of ethylene emitted by diesel engines used in urban public transports. Infrared Physics & Technology, (53):151–155.

Hydraulic Engineering – Xie (Ed.)
© 2013 Taylor & Francis Group, London, ISBN 978-1-138-00043-8

Experimental study on treatment of ammonia nitrogen in landfill leachate flowing from MBR using catalytic wet peroxide oxidation

Li-Hua Teng
Institute of Biology and Environmental Science of Zhejiang Wanli University, Ningbo, Zhejiang, China

Jian-Ping Wang
Ningbo Ocean & Fishery Institute, Ningbo, Zhejiang, China

Qian-Guang Mao
Ningbo Lvzhou Energy Utilization Co., Ltd. Ningbo, Zhejiang, China

Yun Le
Ningbo Ocean and Fishery Enforcement Detachment, Ningbo, Zhejiang, China

ABSTRACT: Active iron catalysts with 5A molecular sieve as the carrier were prepared, and then were used in the treatment of ammonia nitrogen in landfill leachate pretreated by MBR by using CWPO. The results show that the preparation process of catalysts and assistants had great effects on catalytic activity; when cerium was used as an assistant, Fe-Ce/5A catalyst roasted for 3 h at 400°C had a good catalytic effect. As 10 g of Fe-Ce/5A catalyst was added to water sample, and landfill leachate pretreated by MBR reacted with 15 mL of H_2O_2 for 30 min at 60°C, the removal rate of ammonia nitrogen was up to 90.8%, that is, ammonia nitrogen concentration decreased from 253 to 23 mg/L, reaching the national emission standard. Besides, the kinetic analysis of ammonia nitrogen removal reveals that the removal reaction of ammonia nitrogen conformed with pseudo first order kinetic equation. Thus, it is feasible to use this method to deeply treat landfill leachate pretreated by MBR.

Sanitary landfill of rubbish can produce large amounts of landfill leachate, which has been treated mainly by biological methods at present, and Membrane Bio-Reactor (MBR) has been widely applied (Halil et al. 2009). However, after landfill leachate is treated by MBR, COD, ammonia nitrogen, heavy metals and other pollutants still exist in effluent water, so physical and chemical methods also should be adopted to deeply treat landfill leachate to improve the biodegradability of waste water or make effluent water reach national standard (Tang & Wu 2012, Wei 2010, Castrillon et al. 2010, Xu et al. 2010). Catalytic Wet Peroxide Oxidation (CWPO)(Cezar 2003, Niu 2011, Xie 2002, Zhang 2011, Guelou 2003, Masende 2003), an advanced oxidation technique which can effectively treat poisonous, harmful and hardly degradable waste water, is derived from traditional catalytic wet oxidation, that is, liquid oxidizer H_2O_2 replaces oxygen to make oxidation reaction carry out under mild conditions. Presently, there are large quantities of studies on the treatment of landfill leachate using this method, but COD and other hardly degradable pollutants have been paid more attention to (Luo 2009, Wu 2009, Li 2011, Rossignol 1999, Bamwenda 2000). Here, active iron catalysts with 5A molecular sieve as the carrier were prepared firstly, and then the removal effect of little ammonia nitrogen in landfill leachate pretreated by MBR was analyzed by using CWPO, so as to provide theoretical foundations for the deep treatment of landfill leachate treated by MBR.

1 MATERIALS AND METHODS

1.1 *Experimental water sample*

Landfill leachate treated by MBR was collected from a landfill, and its ammonia nitrogen concentration was measured by Nessler reagent spectrophotometry, namely 253 mg/L.

1.2 *Preparation of the catalyst*

Taking 5A-calcium molecular sieve as a carrier and Fe as an active component, we prepared an active iron catalyst by using impregnation method. In detail, certain amounts of pre-treated 5A molecular sieve was immersed in certain concentrations of ferric nitrate solution for 6 h. Afterwards, suction filtration was carried out, and then they were dried at 110°C. The next day, they were roasted in a box-type resistance furnace, and then were cooled down. The catalyst was called Fe/5A molecular sieve.

1.3 *Treatment of waste water*

Firstly, 50 ml of landfill leachate was poured into a conical beaker with total volume of 250 ml, and then certain quantities of H_2O_2 and catalyst were added to the conical beaker. Hereafter, the conical beaker was put in a water bath, and the reaction carried out for 30 min under a constant temperature. Finally, ammonia nitrogen concentration in the filtrate was measured, and removal rate of ammonia nitrogen could be calculated based on ammonia nitrogen concentration in the original water sample.

2 RESULTS AND ANALYSES

2.1 *Effects of preparation process of catalysts and assistants on the removal rate of ammonia nitrogen*

2.1.1 *Influences of steeping fluid concentration on catalyst activity*
Concentration of the steeping fluid could affect the quantity of the active component loaded the catalyst. Six catalysts loading various contents of active component were prepared when ferric nitrate solution concentration was 1.0, 1.5, 2.0, 2.5 and 3.0 mol/L respectively, and it was roasted for 4 h at 400°C. In addition, 10 g of catalysts, 10 ml of H_2O_2 and 50 ml of water sample reacted for 30 min at 60°C (the same below). As shown in Figure 1, as the concentration of the steeping fluid was lower than 2 mol/L, with the increase of steeping fluid concentration, the removal rate of ammonia nitrogen increased to 70.1%. Afterwards, its removal rate went down as the increase of steeping fluid concentration. The results indicate that the amount of the active component loaded by certain quantities of carrier was

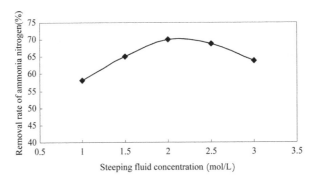

Figure 1. Effects of steeping fluid concentration on the removal rate of ammonia nitrogen.

limited, and low concentrations of active component could be loaded on the carrier evenly through internal and external diffusion, which was beneficial for the improvement of catalyst activity. While its concentration was high, the active component lost and could not be loaded completely, and distributed unevenly on the carrier; its activer centers also distributed unevenly after roasting, so that its catalytic activity reduced, and thereby the removal rate of ammonia nitrogen decreased.

2.1.2 *Impacts of roasting temperature on catalyst activity*

When steeping fluid concentration was 2.0 mol/L, five catalysts were roasted for 4 h at 200, 300, 400, 500 and 600°C respectively. Afterwards, they were used in the treatment experiments of waste water. From Figure 2, we could see that catalyst activity varied greatly with roasting temperature, and the best roasting temperature was 400°C. According to catalysis theory, crystal grains of active component in a catalyst form and grow during catalyst roasting, which is one of key factors affecting catalyst activity. When roasting temperature was low, ferric nitrate on the carrier was not completely oxidized to ferric oxide, and crystal grains did not form; as the increase of roasting temperature, crystal grains grew gradually. However, as roasting temperature was higher than 400°C, crystal grains gathered and were sintered, so that the dispersion of active component on the carrier decreased. Additionally, high temperatures also damaged skeleton structure of the carrier, so that the catalytic effect of the catalyst decreased.

2.1.3 *Effects of roasting time on catalyst activity*

As steeping fluid concentration was 2.0 mol/L, four catalysts were roasted for 2, 4, 5 and 6 h at 400°C respectively, and then they were used in the treatment experiments of waste water. According to Figure 3, roasting time had great effects on catalyst activity, that is, catalyst

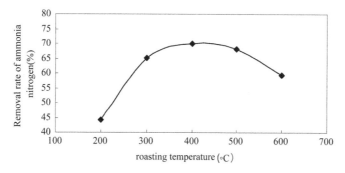

Figure 2.   Influences of roasting temperature on the removal rate of ammonia nitrogen.

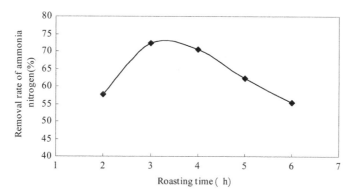

Figure 3.   Impacts of roasting time on the removal rate of ammonia nitrogen.

activity increased firstly and then decreased with the increase of roasting time. While the catalyst was roasted for 3 h, the removal rate of ammonia nitrogen was up to 72.2%. It indicates that besides being affected by roasting temperature, crystal formation and growth of active component were limited by roasting time. That is, ferric nitrate was not oxidized completely in a short term, and there were no active crystal grains; when roasting time was too long, crystal grains gathered, and the dispersion of active component on the carrier became poor, affecting catalyst activity. Therefore, reasonable control of roasting time is the key to preparation of highly active catalysts.

2.1.4 *Influences of assistants on catalyst activity*

The carrier 5A molecular sieve was steeped in two mixtures composed of ferric nitrate and cerium nitrate as well as ferric nitrate and lanthanum nitrate respectively, wherein ferric nitrate concentration was 2.0 mol/L, while both cerium nitrate and lanthanum nitrate concentration were 0.01 mol/L. Hereafter, after they were roasted for 3 h at 400°C, two catalysts containing Cerium (Ce) and Lanthanum (La) respectively were obtained, which were called Fe-Ce/5A molecular sieve and Fe-La/5A molecular sieve separately. Finally, the two catalysts were used in the treatment experiments of waste water. The results reveal that the addition of assistants affected catalyst activity obviously, and a suitable assistant was beneficial to the dispersion of active component on the carrier to improve catalyst activity. When Fe/5A molecular sieve, Fe-Ce/5A molecular sieve and Fe-La/5A molecular sieve were used as catalysts, the removal rates of ammonia nitrogen were 72.2%, 88.9% and 82.9 respectively, so cerium was more suitable than lanthanum for the improvement of catalyst activity. At present, there have been many successful cases of adding CeO_2 to solid catalysts to improve catalyst activity (Rossignol 1999, Bamwenda 2000, Arena 2008), and CeO_2 could be used as a structure, modulation and poison assistant. Functions of the two assistants in our study need to be analyzed through structure characterization.

2.2 *Effects of reaction conditions on catalyst activity*

2.2.1 *Influences of reaction temperature on the removal rate of ammonia nitrogen*

When 10 g of Fe-Ce/5A molecular sieve was used as catalyst, 50 ml of water sample and 10 ml H_2O_2 reacted for 60 min at 30, 40, 50, 60, 70 and 80°C respectively. As shown in Figure 4, the removal rate of ammonia nitrogen went up as the increase of reaction temperature from 30 to 80°C. While reaction temperature was lower than 60°C, the increase was obvious, and the removal rate of ammonia nitrogen reached 88.9% as it rose to 60°C. Hereafter, there was no significant increase in the removal rate of ammonia nitrogen. It is because that the higher the reaction temperature, the more the energy absorbed by H_2O_2, so O–O parted more easily to produce more HO·, which was more favorable to the degradation of ammonia nitrogen, but H_2O_2 was evaporated more quickly under a high temperature (Wu et al. 2004).

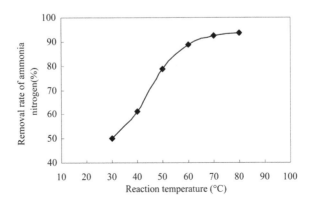

Figure 4. Effects of reaction temperature on the removal rate of ammonia nitrogen.

2.2.2 *Impacts of oxidizer amount on the removal rate of ammonia nitrogen*

When 10 g of Fe-Ce/5A molecular sieve was used as catalyst, 50 ml of water sample respectively reacted with 5, 10, 15, 20 and 30 ml of H_2O_2 for 60 min at 60°C. According to Figure 5, the removal rate of ammonia nitrogen was up to the maximum 90.8% when 15 ml of H_2O_2 was added to water sample. Then it went down slowly as the increase of H_2O_2 amount, showing that the increase of H_2O_2 amount could not improve the removal rate of ammonia nitrogen. It is because that excessive H_2O_2 could react with hydroxyl groups to generate $HO_2\cdot$ group which would react with H_2O_2 to produce H_2O and hydroxyl groups, that is, H_2O_2 was consumed by itself.

2.2.3 *Influences of catalyst amount on the removal rate of ammonia nitrogen*

When the amount of Fe-Ce/5A molecular sieve added was 2, 5, 10, 15 and 20 g respectively, 50 ml of water sample and 15 ml H_2O_2 reacted for 30 min at 60°C. The amount of catalyst added concerns treatment efficiency and cost. As shown in Figure 6, the removal rate of ammonia nitrogen went up with the increase of catalyst amount, and the maximum 90.8% appeared when 10 g of Fe-Ce/5A molecular sieve was added. Afterwards, the removal rate of ammonia nitrogen was nearly constant as the continuous increase of catalyst amount. It is because that the contact area of reactants with the catalyst enlarged with the increase of catalyst amount, but excess catalyst only increased cost instead of improving the removal rate. Thus, the best amount of Fe-Ce/5A molecular sieve added was 10 g.

2.3 *Kinetic analysis of ammonia nitrogen removal*

As the amount of Fe-Ce/5A molecular sieve added was 10 g, 50 ml of water sample and 15 ml H_2O_2 reacted at 60°C, and the changes of removal rate of ammonia nitrogen with reaction time were studied. From Figure 7, we could see that the removal rate of ammonia nitrogen increased rapidly with reaction time prolonging, and the reaction achieved a balance when

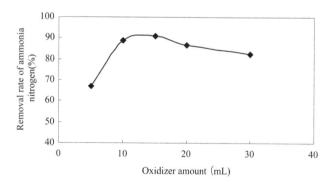

Figure 5.   Effects of H_2O_2 amount on the removal rate of ammonia nitrogen.

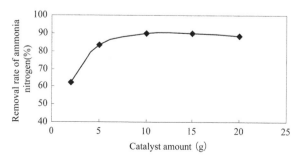

Figure 6.   Effects of catalyst amount on the removal rate of ammonia nitrogen.

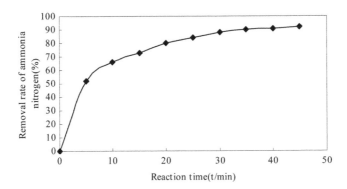

Figure 7.   Impacts of reaction time on the removal rate of ammonia nitrogen.

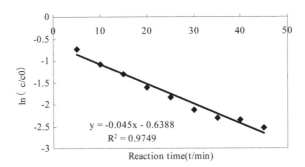

Figure 8.   First-order kinetic fitting of ammonia nitrogen and reaction time.

the reaction carried out 30 min. Hydroxyl groups reacted with ammonia nitrogen quickly and generated constantly during the reaction progress, so steady assumption could be carried out when other conditions were constant. According to absolute rate theory, it is assumed that the oxidation reaction process of ammonia nitrogen can be regarded as pseudo first order reaction, and its kinetic equation was $dc/dt = Kc$, where c is ammonia nitrogen concentration (mol/L); t is reaction time (min); K is reaction rate constant (min^{-1}). The changes of ammonia nitrogen concentration with reaction time was fitted by means of nonlinear least square method. As shown in Figure 8, reaction rate constant K was 0.045 min^{-1}, and correlation coefficient R was 0.987. In a word, the removal reaction of ammonia nitrogen conformed with pseudo first order kinetic equation.

3   CONCLUSIONS

In this study, active iron catalysts with 5A molecular sieve as the carrier were prepared firstly, and then they were used in the treatment of ammonia nitrogen in landfill leachate pretreated by MBR by using CWPO, so as to study the effects of preparation process of the catalyst, assistants and reaction conditions on the removal rate of ammonia nitrogen.

First, when steeping fluid concentration was 2 mol/L, active iron catalysts with 5A molecular sieve as the carrier had good catalytic activity after being roasted for 3 h at 400°C, and the removal rate of ammonia nitrogen was high. The addition of suitable assistants to the active iron catalyst could improve its activity significantly, and cerium was superior to lanthanum.

Second, when 10 g of Fe-Ce/5A molecular sieve was used as catalyst, and landfill leachate pretreated by MBR reacted with 15 ml of H_2O_2 for 30 min at 60°C, the removal rate of ammonia nitrogen was up to 90.8%, that is, ammonia nitrogen concentration decreased from 253 to 23 mg/L, reaching the national emission standard 25 mg/L (GB16889-2008).

Third, the kinetic analysis of ammonia nitrogen removal shows that the removal reaction of ammonia nitrogen conformed with pseudo first order kinetic equation.

Fourth, our preliminary results indicate that low concentrations of ammonia nitrogen could be removed better by using CWPO, and it is feasible to use this method to deeply treat landfill leachate pretreated by MBR.

ACKNOWLEDGMENTS

Supported by the Project of Agricultural Key Programs for Science and Technology Development of Ningbo (2011C11006) and Science and Technology Plan Project of Ningbo City (Application technology on controlling landfill odor).

REFERENCES

Arena, F. Trunfio, G. & Negro, J. 2008. Optimization of the MnCeOx system for the catalytic wet oxidation of phenol with oxygen (CWAO). *Applied Catalysis B: Environmental* 85(1–2):40–47.

Bamwenda, G. R. & Arakawa, H. 2000. Cerium dioxide as a photocatalyst for water decomposition to O_2 in the presence of Ce^{4+} and Fe^{3+} species. *Journal of Molecular Catalysis A: Chemical* 161(1–2):105–113.

Castrillon, L., Fernandex-nava, Y. & Ulmanu, M. 2010. Physical-chemical and biological treatment of MSW landfill leachate. *Waste Management* 30:228–235.

Cezar, C., Carmen, T. & Matei M. 2003. Catalytic wet peroxide oxidation of phenol over Fe-exchanged pillared beidellite. *Water Research* 37:1154–1160.

Guelou, E., Barrault, J., Fournier J., et al. 2003. Active iron species in the catalytic wet peroxide oxidation of phenol over pillared clays containing iron. *Applied Catalysis B* 44(1):1–8.

Halil, H., Sezahat, A.U., Ubeyde, I., et al. 2009. Stripping/flocculation/membrane bioreactor/reverse osmosis treatment of municipal landfill leachate. *Journal of Hazardous Materials* 171(1–3):309–317.

Li, Y.L., Yuan, H.Y. & Wang, S.P. 2011. Pilot-scale study on advanced treatment of MBR effluent using O_3-BAC combined process. *Chinese Journal of Environmental Engineering* 5(6):1237–1240.

Luo, P. & Fan, Y.Q. 2009. Preparation of CuO/γ-Al_2O_3 catalyst and its application in wet catalytic oxidation. *Chinese Journal of Environmental Engineering* 3(5):782–786.

Masende, Z.P.G. & Kuster, B.F.M. 2003. Ptasinski KJ, et al. Platinum catalytic wet oxidation of phenol in a stirred slurry reactor. *App. Cata* 41(3):247–267.

Niu, N., Dong, X.W., Dong, X.L., et al. 2011. Treatment of methyl orange by catalytic wet peroxide oxidation. *Journal of Dalian Polytechnic University* 30(1):62–64.

Rossignol, S., Madier, Y. & Duprez, D. 1999. Preparation of zirconia-ceria materials by soft chemistry. *Catalysis Today* 50(2):261–270.

Tang, F.X., Cao, G.P. & Liu, J.L. 2012. Status quo of landfill leachate treatment and its development of treatment technique. *Journal of Hebei United University* 34(1):116–120.

Wei, L., Tao, H. & Qi, X.Z. 2010. Treatment of stabilized land-fill leachate by the combined process of coagulation/flocculation and powder activated carbon adsorption. *Desalination* 264(1–2):56–62.

Wu, D., Shang, H.L. & Ning, G.X. 2009. Study on preparation of advanced stage landfill leachate by O_3-H_2O_2 and TiO_2/GAC. *Technology of Water Treatment* 35(11): 65–68.

Wu, Z.M., Wei, C.H. & Wu, C.F. 2004. Treatment of simulated wastewater containing acid red B by wet peroxide oxidation. *Acta Scientiae Circumstantiae* 24(5):809–814.

Wu, L.S., Li, Y. & Gong, S.Z. 2012. Practical study of modification of fly ash and its application to the advanced treatment of landfill leachate. *Chinese Journal of Environmental Engineering* 6(2):529–534.

Xie, L., Hu, Y.Y. & Yang, R.C. 2002. Treatment of high concentrations of methyl orange by catalytic wet peroxide oxidation and its mechanism. *Environmental Engineering* 20(2):72–74.

Xu, Z.Y., Zeng, G.M., Yang, Z.H., et al. 2010. Biological treatment of landfill leachate with the integration of partial nitrification, anaerobic ammonium oxidation and heterotrophic denitrification. *Bioresource Technology* 30:79–86.

Zhang, L.M. & Hong, R. 2011. Reaction mechanism and degradation course of azo dyes by catalytic wet peroxide oxidation (CWPO). *Chinese Journal of Environmental Engineering* 5(9):2032–2038.

Hydraulic Engineering – Xie (Ed.)
© 2013 Taylor & Francis Group, London, ISBN 978-1-138-00043-8

Nutrient load during flood events of a typical river flowing into the Chao Lake, China

Yin Chu, You-Hua Ma & Shan-Shan Zheng
School of Resources and Environment, Anhui Agricultural University, Hefei, China

Christian Salles & Marie-George Tournoud
Université Montpellier 2, HydroSciences Montpellier (UMR 5569), France

ABSTRACT: Two summer events (Event06 and Event08) of Fengle River, which flows into the Chao Lake, were intensively sampled at Taoxi section. The water samples were analyzed for Total Nitrogen (TN), Ammonia, Nitrate, Total Phosphorus (TP) and Dissolved Phosphorus (DP). Hourly discharge data were also available at the same section. Concentration and instantaneous loads varied during floods demonstrating a general rising and recessing pollutograph and peak loads usually arrived before peak discharge. Six methods were applied to calculate event total load showing unobvious variation. Totally 192.8t-TN and 5.0t-TP in Event06 and 73.5t-TN and 3.7t-TP in Event08 were exported to the section and maybe until the Chao Lake. About 85.3% (for Event06) and 93.1% (for Event08) of TN total load were ammonia and nitrate loads, which were similar. Particulate phosphorus dominated TP event total loads, among which 38.8% and 43.5% respectively for the two events were DP loads.

1 INTRODUCTION

Rivers flow important pollutant to their receiving waters, i.e. lake, reservoir or coastal water. To control pollution of these waters it is important to know the amount and variation of pollutants transported from these rivers so that specific measures can be applied to reduce loads from tributary sources. River pollutant fluxes demonstrate high temporal variation especially in floods (Blanco et al. 2010). Routine monthly sampling protocols, which have been applied by the administration departments of environment and hydrology, can help understand interannual or seasonal variation (Meng et al. 2007, Chu et al. 2009) but cannot offer important information of pollutant export during flood period (Oeurng et al. 2011). During flood, specific sampling protocol is needed to study the dynamics of pollutant export and to calculate fluxes.

Load estimation methods fall into two categories: real data and empirical methods (Salles et al. 2008). Real data methods are the base to understand pollutant dynamics, to estimate loads and apply further modeling work. Pollutant flux from rivers generally can be estimated from the product of concurrent discharge and concentration measurements at river outlets. Continuous or near-continuous discharge measurement has been technically and economically practicable. But it is usually less available for the measurement of concentration due to the cost or technical difficulties. Different methods have been applied to estimate tributary loading depending on data availability and demonstrating different complexity, accuracy and bias (Littlewood 1992, Salles et al. 2008). Studies were essentially focused on annual load estimates and much less attention had been paid to load calculation at flood scale.

Chao Lake is the fifth largest fresh lake in China, which is located in the middle of Anhui province. Eutrophication has been the major environmental and ecological problem for the lake which receives water flow and pollutant load from about 10 rivers. Here specific sampling protocol had been conducted at a section of one of the rivers. The objectives of this study were to understand instantaneous load dynamics during flood and to estimate total event loads with different evaluated methods.

2 MATERIAL AND METHODS

2.1 Fengle river and the catchment

Fengle river is one of the main tributaries of Chao lake. (Fig. 1). The measuring and sampling section is located in Taoxi town. The catchment area is 1500 km² and main stream length is about 50 km. The catchment elevation is between 6–463 m. Land use includes agriculture (about 45%), forest (39%), town and roads (10%) and water area (ponds and river, 6%). No large cities or industry factories are located in this catchment. Average rainfall is about 1070 mm calculated from total annual precipitation of the past 20 years, which demonstrated high interannual and innerannual variation. Most precipitation happens in summer and spring. Heavy storms and floods are dominant in summer season. A typical storm can last several days. Consecutive flood might last longer time.

2.2 Sampling and analysis

The sampling site was located at Taoxi section which is about 500 m downstream part of Fengle Bridge. Grab samples were taken in the middle part of the section by using the cable across the section. During the rising period samples were taken hourly and then after the peak the sampling interval was prolonged to 4-hour, 6-hour and 12-hour at the recessing period.

Monitoring species included Total Nitrogen (TN), ammonia (NH4-N), Nitrate (NO3-N), Total Phosphorus (TP) and Dissolved Phosphorus (DP). Alkaline potassium persulfate digestion-UV spectrophotometric method for TN (GB11894-1989), Nseeler's reagent colorimetric method for NH4-N (GB7479-1987) and UV spectrophotometric method for NO3-N (HJ/T346-2007) were applied. DP was analyzed on filtered samples using the molybdate blue method and TP was determined on unfiltered samples by potassium peroxodisulphate digestion before analysis with ammonium molybdate (GB11893-1989).

2.3 Rainfall and runoff data

The Taoxi station of Hydrology has been applying measurement of rainfall at several rain gauges all over the catchment and water level at Taoxi section. Hourly rainfall and water level

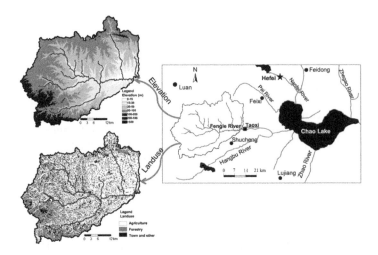

Figure 1. Location, elevation and landuse of Fengle catchment.

Table 1. Event total load estimation methods (Salles et al. 2008).

Formula		Formula	
$TL = \sum_{i=1}^{n} q_i c_i t_i$	(1)	$TL = V \cdot median \, (c_i)$	(4)
$TL = V \sum_{i=1}^{n} \dfrac{c_i}{n}$	(2)	$TL = \sum_{i=1}^{n} \overline{q_i c_i} t_i$	(5)
$TL = V \dfrac{\sum_{i=1}^{n} q_i c_i}{\sum_{i=1}^{n} q_i}$	(3)	$TL = \sum_{i=1}^{n} q_i c_{interp} t_i$	(6)

Note: c_i is the ith sample concentration, c_{interp} is the linear interpolation of the concentration, q_i is the instantaneous discharge at time of sampling, t_i is the corresponding time interval (i.e. half the interval of time from the preceding to the following sample), n represents the number of samples, V is the flood water volume estimated using the hourly discharge.

data were obtained from the following website: http://yc.wswj.net/new_yc/yc_web_main.htm. Discharge was calculated from water level data through stage-discharge relationships.

2.4 Load calculation

Instantaneous load is the product of concentration and concurrent discharge. Event total load was calculated with 6 evaluated methods reported by Salles et al. (2008). The methods fall into two groups: the first group included method 2, 3 and 4 in which the flux was the product of flood water volume and a represented concentration during the whole flood period. The volume was based on hourly data, so it was the same for the 3 methods. The difference was the way to estimate represented concentration. In the second group, the flux was the sum of represented load during the whole flood period (method 1, 5 and 6).

3 RESULTS AND DISCUSSION

3.1 The two monitored events

Both of the two monitored events were in the summer of 2010 (Fig. 2): one was in June (Event06) the other was in August (Event08). Total annual rainfall of 2010 was 1635 mm demonstrating a humid year. Total event rainfall of Event06 was 101.5 mm which lasted about 25 h. Peak flow arrived Taoxi section 21 hours after peak rain. Rising limb of the hydrograph was steeper that the recessing limb. The total rainfall of Event08 was 70 mm which was precipitated in several storms. The hydrograph demonstrated a gentler rising limb than Event06.

3.2 Concentration and instantaneous load variation

The statistics of nutrient concentration and discharge during the two monitored flood events were presented in Table 2. The minimum, average and maximum of all the species of Event06 were higher than Event08. That was because agricultural activities were dominant in Fengle catchment. June was the planting season of rice and a lot of manure had been applied before transplanting the seedling. The following fertilizations usually take place in June and July. Event08 happened at the end of August when the sources of nutrient from fields had been much reduced. Besides, pollutant export might also be linked with flood dynamics.

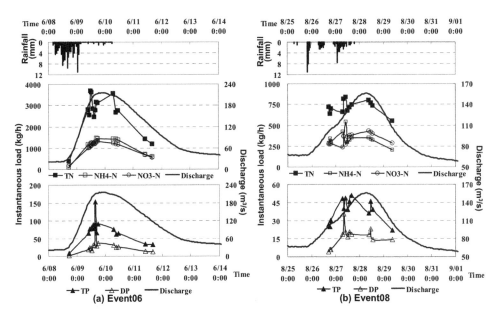

Figure 2.   Hyetograph, hydrographs and pollutograph of the two monitored events.

Table 2.   Statistics of nutrient concentration and discharge during the two monitored flood events.

Event	TN (mg/L)	NH4-N (mg/L)	NO3-N (mg/L)	TP (mg/L)	DP (mg/L)	Discharge (m³/s)
Event06						
Min	2.80	0.86	1.29	0.07	0.02	22.6
Max	5.56	1.94	1.72	0.20	0.12	216.6
Average	4.05	1.68	1.56	0.11	0.04	131.3
Event08						
Min	1.35	0.54	0.55	0.06	0.01	65.4
Max	1.98	1.18	0.83	0.11	0.08	156.4
Average	1.62	0.75	0.74	0.08	0.04	109.4

Event06 was more intense than Event08 and so had stronger washing off and transportation capacity. Dilution effect of high peak flow was less important for Event06.

According to the routine monthly data during 2010, the minimum, average and maximum of NH4-N and TP were 0.16, 1.46, 0.38 mg/L and 0.01, 0.21, 0.06 mg/L, respectively, which were lower than that of the two floods except for the maximum concentrations of TP for both events and the maximum concentration of NH4-N for Event08.

Since the instantaneous load is the product of concurrent concentration and discharge, its dynamics is the combination of variation of the two variables. Figure 2 showed that the loads of all species generally demonstrated rising and recessing pollutograph form and peak loads usually arrived before peak discharge.

3.3   Event total load

Event total loads of the five species of both events were calculated with the six evaluated methods. The results were presented in Table 3.

The differences between calculated event fluxes with the 6 methods were not obvious with the coefficients of variation being between 3% and 7% for the 5 species of the two events. Since the discharge data were continuous and complete, the calculation results depended

Table 3. Event total loads of 5 species for the two monitored events with 6 methods.

Event	Method	TN (t)	NH4-N (t)	NO3-N (t)	TP (t)	DP (t)
Event06	1	198.4	90.5	82.0	5.00	1.96
	2	196.8	84.3	78.4	5.27	2.06
	3	201.7	87.7	79.7	5.48	2.16
	4	192.2	86.5	79.7	5.11	1.86
	5	184.1	83.3	76.1	4.62	1.82
	6	183.4	82.8	75.9	4.60	1.81
	Average	192.8	85.8	78.1	5.00	1.90
	Uncertainty*	9.15	3.85	3.05	0.45	0.15
Event08	1	79.1	34.0	37.8	3.77	1.63
	2	74.4	35.5	34.2	3.90	1.68
	3	73.5	35.0	34.3	3.92	1.71
	4	69.8	32.8	35.1	3.89	1.76
	5	72.2	31.4	34.5	3.50	1.50
	6	72.2	31.4	34.6	3.49	1.50
	Average	73.5	33.4	35.1	3.70	1.60
	Uncertainty*	4.60	2.05	1.80	0.40	0.15

*Uncertainty was estimated as half of the difference between the minimum and maximum of the calculated fluxes with the 6 methods.

on the concentration sampling intensity, concentration variation and calculation methods (Littlewood 1992). The values of discharge and concentration are not at the same scale and discharge contributes more to calculated loads than concentration. Nevertheless, variation of concentration may produce some difference in estimated loads, e.g. for TP and DP. The differences were less important than that of flood events of south France coastal rivers (Salles et al. 2008).

Flood total loads calculated by the three represented concentration methods (i.e. method 2, 3 and 4) were generally higher than method 5 and 6, except for TN load of Event08. The two events were not evenly sampled and intensified sampling period of high flow period with high concentration will slightly over-estimate the calculated total load. As for method 5 and 6, the low concentration or low instantaneous load at the beginning and at the recessing part were reinforced and so demonstrated lower calculated total load for all the species of both events. Method 1 which uses an instantaneous load to represent load of the calculation period between sampling moments demonstrated higher variation than other methods.

Event total load of each species was represented as the average of values calculated with the 6 methods (Table 2). Totally 192.8t-TN and 5.0t-TP in Event06 and 73.5t-TN and 3.7t-TP in Event08 were exported to the section and maybe until the Chao Lake. About 85.3% (for Event06) and 93.1% (for Event08) of the TN total load were ammonia and nitrate loads, which were similar. Particulate phosphorus dominated the TP event loads, among which 38.8% and 43.5% respectively for the two events were DP loads.

4 CONCLUSIONS

The study showed that concentration and instantaneous loads varied during floods demonstrating a general rising and recessing pollutograph and peak loads usually arrived before peak discharge. Flood transported more pollutants in terms of concentration and loads during flood period than non flood period. Ammonia and nitrate loads dominated TN event loads while particulate phosphorus dominated TP event loads.

Event total loads calculated with the 6 evaluated methods demonstrated unobvious variation. To obtained a stable and reasonable estimation of event total load a sampling protocol with at least one at the beginning, one just before pick and one at the late recessing and method 2, 3, 5 and 6 were proposed when continuous discharge data were available.

ACKNOWLEDGMENTS

This study was supported by Project of Non-profit Research Foundation for Agriculture from the Ministry of Agriculture, P.R. China (No. 201003014), the key project from Department of Education of Anhui province, P.R. China (KJ2010A111) and Program Cai Yuanpei 2011–2013.

REFERENCES

Blanco AC, Nadaoka K, et al. 2010. Dynamic evolution of nutrient discharge under stormflow and baseflow conditions in a coastal agricultural watershed in Ishigaki Island, Okinawa, Japan. Hydrological Processes, 24:2601–2616.

Chu Y, Xia SX. 2009. Study on the characteristics of pollutant export from Fengle River. Journal of Anhui Agricultural University, 36(3):476–482.

Littlewood IG. 1992. Estimating contaminant loads in rivers: a review. Institute of Hydrology, Crowmarsh Gifford, Wallingford, Oxfordshire, OX10 8BB, UK, Report No. 117.

Meng W, Yu T, et al. 2007. Variation and influence factors of nitrogen and phosphorus transportation by the Yellow River. Acta Scientiae Circumstantiae, 27(12):2046–2051.

Oeurng C, Sauvage S, Coynel A, et al. 2011. Fluvial transport of suspended sediment and organic carbon during flood events in a large agricultural catchment in southwest France. Hydrological Processes, 23:2365–2378.

Salles C, Tournoud MG, Chu Y. 2008. Estimating nutrient and sediment flood loads in a small Mediterranean river. Hydrological Processes, 22:242–253.

Hydraulic Engineering – Xie (Ed.)
© 2013 Taylor & Francis Group, London, ISBN 978-1-138-00043-8

Study on permafrost distribution in Qinghai-Tibet highway based on ASTER data

Wang Kun, Chen Lichun & Wei Bin
Jilin Communications Polytechnic, Changchun, China

Jiang Qigang
College of Geo-exploration Science and Technology, Jilin University, Changchun, China

ABSTRACT: High spatial resolution remote sensing images have great potential in the simulation of permafrost distribution. In this paper, based on ASTER data, selecting five parameters of land surface temperature, elevation, equivalent latitude, vegetation index and land surface humidity index to build multivariate analysis model, and simulating the permafrost distribution in Qinghai-Tibet highway. Compared the results with the simulation results based on MODIS data; it shows that the results based on ASTER data intended better with permafrost distribution map. Melting district exist in Simulation results, which is the same with the results of MODIS data, meeting the permafrost actual distribution law. No matter which data simulation results appear melting district, although the distribution area is different, the position appears broadly similar, which show the effect is ubiquitous in the permafrost distribution, it need to be given proper attention in the follow-up studies.

1 INTRODUCTION

It is recognized that high spatial resolution remote sensing image has great potential in permafrost distribution simulation, which will play an increasingly important role in the field of remote sensing of permafrost. But so far, it has not seen the full advantage of the permafrost mapping work carried out by the space borne high-resolution sensors (Cao et al. 2006). For the widely distributed range of permafrost study area in Qinghai-Tibet Plateau, the cost of using high-resolution remote sensing images on permafrost research will be high. And too many scenes of image, time is difficult to obtain consistent; therefore high-resolution data does not have the potential to promote in a wide range. However, it can be used in some areas, such as along the Qinghai-Tibet highway or boundary area of permafrost regions with seasonal permafrost zone.

The ASTER remote sensing image is selected as the main data source to study the local permafrost distribution, which is equipped with earth observation satellite EOS-Terra in December 1999, and launched a high-resolution sensor with 14-band. It includes 3 visible and near-infrared bands with 15 m spatial resolution, 6 shortwave infrared bands with 30 m spatial resolution, as well as 5 thermal infrared bands with 90 m spatial resolution, and has the same track stereoscopic viewing (black-white stereo pair) (Zhu et al. 2003, Li et al. 2004). Due to the unique high-latitude, surface characteristics and climatic environment, cloud cover is very much in Qinghai-Tibet Plateau. Therefore the image covered the specified area is less able to meet little cloud cover and close time. This paper selected 6 scene ASTER images in April 13, 2002 which covered parts of Qinghai-Tibet highway (Fig. 1). They were purchased from Japanese Earth Remote Sensing Data Analysis Centre (http://www.ersdac.or.jp).

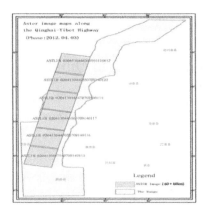

Figure 1.   Distribution range of ASTER data.

2   PREPARATION FOR MODEL DATA

It is unrealistic to consider all factors affecting the permafrost distribution in the model. Therefore we need to select several major factors affecting the permafrost distribution in the process of modeling. Comprehensive consideration the main factors impact on the permafrost distribution of Qinghai-Tibet highway and environmental information can be provided by RS for permafrost studies, selecting surface temperature, elevation, equivalent latitude, soil moisture, Normalized Difference Vegetation Index (NDVI) as model factors in this paper.

2.1   *Surface temperature*

Due to the difficulty of obtaining atmospheric parameters, inversion algorithms of land surface temperature for ASTER summed up are ASTER TES algorithm, multi-channel algorithm and radioactive transfer algorithm (Li et al. 2001, Liu et al. 2003, Mao & Tang et al. 2005, 2006, Wang et al. 2005).This paper is selecting ASTER TES algorithm, programming it in the MATLAB environment to get the land surface temperature inversion results of the study area (Fig. 2).

2.2   *Elevation*

ASTER is a sensor with stereo viewing capability, which consist 3 sub-systems namely the Visible and Near-Infrared Subsystem (VNIR), Shortwave Infrared Subsystem (SWIR) and Thermal Infrared Subsystem (TIR). Except vertical downward imaging, the third band of VNIR subsystem also has a rear-view imaging, the resulting stereoscopic image pair is the basis for extracting DEM (Lv et al. 2008). In this paper, using ASTER DEM software to extract DEM, got the DEM distribution of the study area (Fig. 3).

2.3   *Equivalent latitude*

Because of the high spatial resolution, ASTER data can provide richer information of surface features. So some local topographic factors on permafrost can be well represented. Based on algorithm of equivalent latitude, using ASTER DEM data obtained the equivalent latitude of the study area (Fig. 4).

2.4   *Vegetation index*

Calculating the Normalized Difference Vegetation Index (NDVI) using ASTER data, the formula is:

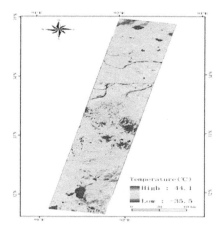

Figure 2.   Land surface temperature inversion results of the study area.

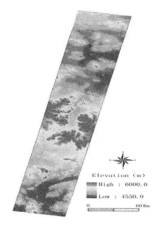

Figure 3.   DEM of the study area.

$$NDVI = \frac{ASTER3 - ASTER2}{ASTER3 + ASTER2} \qquad (1)$$

The data recorded in the ASTER sensor is image brightness (DN) not the reflectance value, so it needs to be converted into apparent reflectance, and then using formula (1) to calculate the NDVI of the study area (Fig. 5).

2.5   *Surface humidity index*

Selecting band 3 and band 5 of the ASTER data to construct normalized water index NDWI, this was used to represent the surface humidity. Calculated as follows (Qin et al. 2008):

$$NDWI = (\rho_3 - \rho_5)/(\rho_3 + \rho_5) \qquad (2)$$

In the formula, ρ_3 represents the reflectance of band 3, ρ_5 represents the reflectance of band 5. Resample band 5 to 15 m in the calculation to match it with band 3. NDWI distribution within the study area is shown in Figure 6.

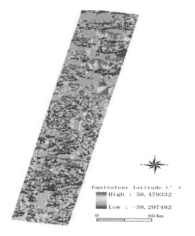

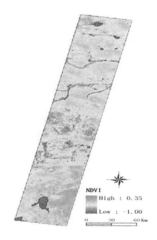

Figure 4. Equivalent latitude of the study area.

Figure 5. NDVI distribution of study area.

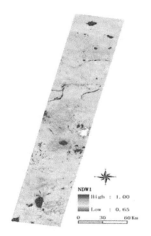

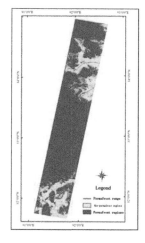

Figure 6. NDWI distribution of study area.

Figure 7. Simulation results of ASTER data.

Table 1. Correlation analysis of surface variables and permafrost existence.

			P	H	Ts	NDWI	Equlat	NDVI
Spearman's rho	P	Correlation coefficient	1.000	.465**	−.228*	−.258*	−.266**	−.336**
		Sig. (2 tailed)	.	.000	.000	.000	.000	.000

**indicates when the significance level is 0.01, the sample points was significantly related. *indicates when the significance level is 0.05, the sample points was significantly related. 2-tailed indicates bilateral inspection. P is a categorical variable of permafrost existence, H is elevation, Ts is the land surface inversion temperature, NDWI is the normalized water index, NDVI is the normalized vegetation index, Equlat is equivalent latitude.

3 MODEL CALCULATION

3.1 *Correlation analysis*

Using sample data in the study area to have a spearman rank correlation analysis (Table 1). Seen from table 1, when the significance level is 0.01, elevation, vegetation index and equivalent latitude are significantly related with permafrost distribution. And vegetation index and elevation have the most significant impact. When the significance level is 0.05, land surface temperature and normalized water index are significantly related with permafrost distribution, and vegetation index and elevation have the most significant impact.

3.2 Model building

Using Binary Logistic command to have logistic analysis in SPSS 16.0, set probability cut-off point of the model predicting as 0.5, the model will predict using the cut-off value mentioned above. After setting, established a logistic regression equation:

$$P = 1/1 + \exp(-0.02x_1 - 0.122x_2 + 44.199x_3 - 3.765x_4 + 0.032x_5 + 86.492) \tag{3}$$

where P is the probability of permafrost exist, x_1, x_2, x_3, x_4, x_5 respectively represent the elevation (H), land surface inversion temperature (Ts), Vegetation Index (NDVI), Normalized Water Index (NDWI) and Equivalent Latitude (Equlat).

3.3 Simulation results

Building model in the ERDAS Modeler and calculating the permafrost existence probability based on the formula (3), get the permafrost existence probability distribution of the study area. In order to compare with the multivariate analysis simulation results of MODIS data, selecting $P > 0.8$ to simulate the permafrost distribution of the study area, and get the simulation results of the corresponding probability (Fig. 7).

3.4 Comparative analysis of the simulation results

Subset the simulation results of multivariate analysis model of MODIS data using the range of ASTER data (Fig. 8). Compared the simulation results with the permafrost map, melting districts exited both in the simulation results of MODIS data and ASTER data, but the scope of the melting districts significantly expanded in the corresponding location simulation results of ASTER data.

Permafrost in the south island regions, simulation results of ASTER data fit very well with permafrost map. The space resolution of Remote sensing data improved, the more rich topographic information it can reflect, especially in some areas of island permafrost distribution is more favorable.

Compared the simulation results of ASTER data with permafrost geographic base map, it can be seen that the melting districts occurred mainly in the range from Tongtian river

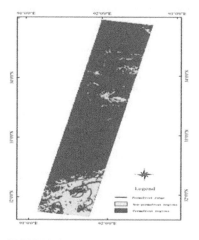

Figure 8. Simulation results of MODIS data.

to Tuotuo river, banded distribution along the river. Therefore, the melting districts in the simulation results of ASTER data are reasonable, in line with the actual distribution law of the permafrost.

4 CONCLUSION

1. Realized the permafrost distribution modeling using ASTER data in the range of Qinghai-Tibet road. Compared and analyzed the simulation results, the accuracy of ASTER data simulation results improved. Compared with the simulation results of MODIS data, the results of ASTER data fit better with the permafrost map. It is due to the high spatial resolution of ASTER data, which can reflect the local terrain and surface variables distribution more detailed.
2. The simulation results exited melting districts, overlay the simulation results and permafrost map, the melting districts mainly distributed in the range from Tongtian river to Tuotuo river, banded distribution along the river. Therefore, the melting districts in the simulation results of ASTER data are reasonable, in line with the actual distribution law of the permafrost.
3. Regardless of the data and simulation methods, simulation results existed melting districts. Although the distribution area is different, the position is similar, which reflected the impact is widespread in the simulation of permafrost, it should be given proper attention in the follow-up studies.

REFERENCES

Cao, M.S., Li, X., Chen, X.Z. et. al., 2006. Cryosphere Remote Sensing. Beijing:Science Press: 1–10.
Li, H.T, Tian, Q.J. 2004. An Introduction to ASTER Data and ASTER Mission. Remote Sensing Information, (3): 53–55.
Li, X.W., Wang, J.F., Wang, J.D, et al. 2001. Multi-angle thermal infrared remote sensing Beijing: Science Press:84–127.
Liu, Z.W., Dang, A.R., 2003. A Retrieval Model of Land Surface Temperature With ASTER Data and Its Application Study, Progress In Geography(5): 507–513.
Lv, Y.F. 2008. A Study of DEM and Digital Geomorphologic Mapping Based on ASTER Image. Fujian Normal University: 16–23.
Mao, K.B, Qin, Z.H, Xun B. 2005. Method for Land Surface Temperature Retrieval from ASTER Data. Journal of Institute of Surveying and Mapping, (1): 40–42.
Mao, K.B., Shi, J.C., Qin, Z.H. 2006. A Four-Channel Algorithm for Retrieving Land Surface Temperature and Emissivity from ASTER Data. Journal of Remote Sensing, (4): 593–599.
Mao, K.B., Tang, H.J. 2006. A Split-window Algorithm for Retrieving Land-Surface Temperature from ASTER Data. Remote Sensing Information (5): 7–11.
Qin, P., Chen, J.F. 2008. A Comparison between NDVI and SAVI for Vegetation Spatial Information Retrieval Based on ASTER Images: A Case Study of Huadu District, Guangzhou. Tropical Geography, 28(5): 419–422.
Tang, S.H., Zhu, Q.J., Su L.H. 2005. Thermal infrared tes al gorithm based on corrected alpha difference spectrum. Journal Infrared Millimeter and Waves (4): 286–290.
Tang, S.H., Li, X.W., Wang, J.D., et al. 2006. Science in China (Series D:Earth Sciences (7): 663–671.
Wang, F.M., Tian, Q.J., Guo, J.H. 2005. Retrieveing land surface temperature of mountain areas in southern china based on aster data. Remote Sensing For Land & Resources (1): 30–33.
Zhu, L.H., Qin, Q.M., Chen, S.J. 2003. The reading of aster data from file and the application of aster data. Remote Sensing For Land & Resources, 56(2): 59–63.

Hydraulic Engineering – Xie (Ed.)
© 2013 Taylor & Francis Group, London, ISBN 978-1-138-00043-8

Study on enhanced treatment of low temperature and low turbidity micro-polluted water by ozone pre-oxidation

Yuan Jian, Li Man, Wen Guojiao & Yan Weiwei
*School of Environmental and Municipal Engineering, Tianjin Institute of Urban Construction (TIUC),
Tianjin, China*
Tianjin Key Laboratory of Aquatic Science and Technology, TIUC, Tianjin, China

ABSTRACT: Because of characteristic of the viscosity, small size of water quality particle, less pollutant content and hard to be degraded, low temperature and low turbidity micro-polluted water has continuously been the problem of drinking water treatment. With the water quality index of turbidity, UV_{254}, COD_{Mn} and THMFP, the experiment has a study on the effect of enhanced direct filtration by ozone pre-oxidation in low temperature and low turbidity micro-polluted water treatment. The result shows, ozone dosage is 1.5 mg/L, pre-oxidation time is 10 min, effluent turbidity can drop to 0.3 NTU. The best of COD_{Mn} removal rate can reach to 59% and the best of COD_{Mn} removal rate can reach to 57%. The ozonation has a better removal efficiency of trihalomethane precursors, and the best removal rate can reach to 79%.

1 INTRODUCTION

Impurities are mainly dispersed as fine colloid disperse system in the low temperature and low turbidity water (Huo, 1998). Colloidal particle has a strong dynamic stability and condensed stability because of its small particle size, less content, uniform distribution and electronegativity. The specific performance are the large viscosity of the water, slow flocculation reaction and small and not easy sedimentation generated flocculation body (alum flowers) (Wang, 2009). So it has become a difficulty in the processing of water purification technology. And the water quality and water temperature of surface water in northern area of our country changes a lot with the influence of geographic conditions and seasonal climates. There is about 4–5 mouths are winter time in a year and the water temperature maintains in the 10°C below. And there are three month time the water temperature even less than 5°C. And the water turbidity also reduces to 30 NTU below. And all these factors form typical low temperature and low turbidity water (Liu, 2005). On the other hand, the growing of the pollutants discharged into water especially those trace amounts of poisonous and harmful organic pollutants that is difficult to break down and easy to be enriched in organisms and have potential carcinogenic effects which will do great harm to human health with the rapid development of industry and economy in recent years makes the city water supply face the problem of slightly polluted water when it is in the in low temperature and low turbidity period. The pollutants in slightly polluted source water are mainly organic pollution, though its concentration is low, character is complex. At present most urban waterworks still use chlorine disinfection, the combination of chlorine and organic can form chlorine disinfection by-products such as THMS which are more harmfulness to human body, and they make the water quality declining. And with the continuously improvement of people's life quality and determination methods, people's requirement of water quality become more strict than before and the corresponding water quality standard also continuously improves. Therefore, the purification treatment of low temperature and low turbidity micro-polluted raw water has become a very important and urgent new topic (Li, 2010). Ozone is a oxidant which has strong oxidation capacity, some researches show that, Ozone can make many organic pollutants in water oxidation decomposition, and have the effect of antiseptic, decoloring pollutants, excepting algae and improving the effect of flocculation. This study combines the research situation of low temperature and low turbidity water and micro-polluted water, add the preoxidation process which use ozone as oxidizer on the basis of the study of direct filtration, and making a deep study of it though changing the ozone dosing quantity and pre-oxidation time, in order to improve water quality and reduce the cost waterworks operation.

2 EXPERIMENTS

2.1 *Experimental water*

Composition of the water used in this study.

Water quality index	Temperature/ (°C)	Turbidity/ (NTU)	COD_{Mn}/ (mg/L)	UV_{254}/ (cm^{-1})	NH_4^+-N (mg/L)	pH
Range of rariation	3–7	3.53–7.20	2.67–5.89	0.08–0.13	0.08–0.13	7.3–8.1

2.2 *Experimental equipment*

1. Ozonator: Type CF—YG40, using built-in High concentration oxygen plant as the oxygen supply source.
2. Ozone contact oxidation apparatus: total length of glass cylinder is 1200 mm, inside diameter is 70 mm, diffuser disc is produced from the glass sand.
3. Filter candle: be made from material of plexiglass, height is 2000 mm, inner diameter is 50 mm, using anthracite/quartz sand as double filter material, diameter of quartz sand under layers is 0.8–1.0 mm, thickness is 400 mm, diameter of anthracite on the top is 1.0–2.0 mm, thickness is 500 mm.
4. Tube mixer: total length is 500 mm, inside diameter is 50 mm.
5. Flocculating tank: be made from material of plexiglass, consists of 4 compartments, each case's size is 200 × 200 mm.

2.3 *Experimental analysis methods*

pH: using glass electrode method, type of the instrument is Orion868-2.
Turbidity: type of the instrument is HACH2100AN.
Permanganate index: using acid potassium permanganate method.
UV254: using ultraviolet spectrophotometry method.

UV254 is refers to the ultraviolet absorbance of water sample in the wavelength of 254 nm, it is a representation of Organism which have conjugated double bond and Benzene ring structure, and it can be used as the substitution parameter of TOC and THMS precursors (Xu, 2000, A.D, 1995).

Ammonia nitrogen: using Nessler's reagent photometric method.
Total generation potential of THMS: using P&T GC-MS method.

2.4 *Experimental process*

The prophase micro-flocculation direct filtration experiment shows that the best micro flocculation time is 5 minutes, filter speed is 10 m/h, removal efficiency of turbidity can reach to 93.5%, but the removal rate of UV_{254} and COD_{Mn} is only 15% and 17%. This study adds the ozone experiment on the basis of micro-flocculation direct filtration experiment and keeps the preoxidation time 10 minutes, and first changes the ozone dosing quantity from 0.5 mg/L, 1.0 mg/L, 1.5 mg/L to 2.0 mg/L (using iodine volume

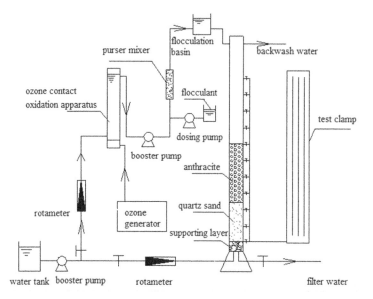

Figure 1. Experimental equipment diagram.

method measure ozone concentration (MOHC, 2007)) then changes the preoxidation time from 5 minutes, 10 minutes, 15 minutes to 20 minutes on the basis of the optimal dosing quantity. Last has a analysis of the change of the water index such as turbidity UV_{254}, COD_{Mn} and THMFP.

3 RESULTS AND DISCUSSION

3.1 *Effluent quality with ozone dosage change*

As can be seen from Figure 2, when the pre-oxidation time is constant, and adding a small amount of ozone, 0.5 mg/L, then turbidity of the effluent can be reduced to 0.3 NTU, then when the ozone dosage is increased, the turbidity of the water is almost Remain constant. It is because ozone pre-oxidation can enhance the removal efficiency of the NTU of low temperature and low turbidity water (Zhao, 2009). The coagulation aid effect can be explained in surface characteristics of particulate matter in water. When there is Natural Organic Matter (NOM) exist in the water, NOM will be fully adsorbed on the surface of the colloid and tiny particles and form envelope, and it leads to the chemical characteristics of colloid and tiny particles can't be showed, and it is not benefit to Congealment. But adding low doses of ozone, can make NOM molecular weight reduce, and it can remove the envelope of colloid and tiny particles, and make chemical characteristics of colloid and tiny particles can be showed, and it is benefit to Congealment (Liu, 2010).

As can be seen from Figure 3, when the pre-oxidation time is constant, ozone pre-oxidation has a better removal efficiency of UV254, In the condition of ozone dosing quantity is 0.5 mg/L, the removal rate of UV254 can reach to 44.6%, the UV254 value decreases and the removal rate increases with the increase of ozone dosing quantity, and when ozone dosing quantity is 2 mg/L, the removal rate reached to 59%. This is because the high removal rate of UV254 of ozone pre-oxidation relates to the mechanism of ozone in water. In the neutral or nearly neutral water, direct oxidation and indirect oxidation of ozone play the same important role. The major role of the direct oxidation is organics that contains unsaturated bonds, and Indirect oxidation has oxidation performance than the direct oxidation, and it is not selective, and it is also able to oxidize organic pollutants containing double bonds and benzene ring structure, and these substances is the main representational material of UV254,

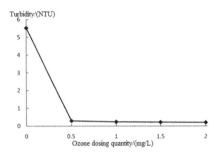

Figure 2. The variation of effluent turbidity with does of ozone.

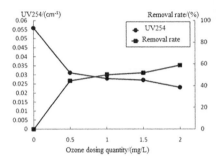

Figure 3. The variation of UV_{254} removal rate effluent UV_{254} with does of ozone.

so ozone pre-oxidation can effectively reduce the content of organic matter of UV254 in water (J.H, 1983).

As can be seen from Figure 4, in the process of ozone pre-oxidation, the removal rate of COD_{Mn} increased with the increase of ozone dosage at first, however, when the ozone dosage reach to 1.5 mg/L, removal rate of COD_{Mn} stabilized at about 57%, then there is almost no change with the increase of ozone dosage.

3.2 *Effluent quality with pre-oxidation time change*

It can be seen from Figure 5, compared to Figure 2, both In the case of a very short time of pre-oxidation and ozone dosage is very small, the turbidity of the water significantly reduces, but then increases ozone dosage there will have little effect on effluent turbidity, on the contrary, extends pre-oxidation time effluent turbidity will have a certain degree of change. The lowest effluent turbidity even lower than 0.1 NTU, but then water turbidity is very small, so the effect is not very obvious.

As can be seen from Figure 7, the ozonation on COD_{Mn} occurs mainly between 5 min–15 min, during this time the removal rate of COD_{Mn} increases when the contact time increases, but when the contact time is more than 15 min removal rate of COD_{Mn} is always stable in 57%.

3.3 *Ozonation of water trihalomethane formation potential*

As can be seen from Figure 8, ozone dosage and pre-oxidation time have a greater impact on the removal of THMS precursors. When pre-oxidation time change between 10 minutes and 15 minutes, water THMFP decreased with the increase of ozone dosage, pre-oxidation time is less than 10 minutes or higher than 15 minutes, it did not show a certain regularity. When pre-oxidation time is 5 minutes, the descending order of THMFP's removal rate on different dosage is 1.5 mg/L > 2.0 mg/L > 0.5 mg/L > 1.0 mg/L, and when the pre-oxidation time is 5 minutes, the dosage of ozone is both 0.5 mg/L and 1.5 mg/L, the removal rate of THMFP shows maximum removal, continue to increase the pre-oxidation time, the removal rate

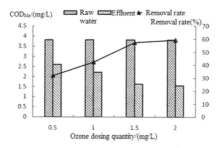

Figure 4.   The variation of COD$_{Mn}$ removal rate with with does of ozone.

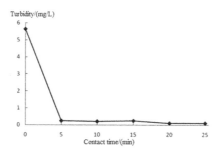

Figure 5.   The variation of effluent turbidity preoxidation time.

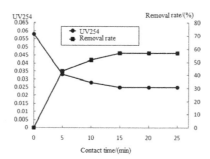

Figure 6.   The effect of preoxidation time on UV254.

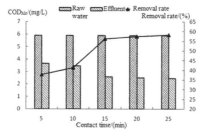

Figure 7.   The effect of preoxidation time on COD$_{Mn}$.

drops greatly. Therefore, the removal of trihalomethane precursors by ozone pre-oxidation, need to based on the quality of raw water and use experimental method to determine the optimal dosage and pre-oxidation time.

   Through the above analysis, it shows that the ozonation removal of trihalomethane precursors did not show obvious regularity. But on the whole, controls the pre-oxidation time in

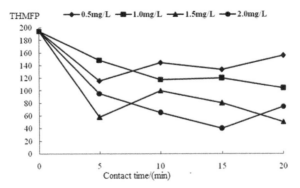

Figure 8. The variation of THMFP with preoxidation time.

10–15 min, there will be a better effect of the removal of trihalomethane precursors, and in this rage removal rate of trihalomethane precursor can reach to 79.4%.

4 CONCLUSION

1. The ozonation strengthen direct filtration treatment can enhance the removal efficiency of low temperature and low turbidity micro-polluted water, and dosing 0.5 mg/L ozone, the turbidity can dropped to 0.3 NTU, Ozone dosage is 1.5 mg/L, pre-oxidation time is 20 min, the effluent turbidity can be reduced to 0.1 NTU. The ozonation has a better removal efficiency of UV254 substances, and the best removal rate can reach to 59%. And the removal efficiency of COD_{Mn} is relatively obvious, the best removal rate can reach to 57%.
2. The ozonation has a better removal efficiency of trihalomethane precursors, and the best removal rate can reach to 79%. Under the same ozone dosage, THMFP does not change regularly with the reaction time changes, and under the same time, THMFP does not change regularly with the ozone dosage changes.

REFERENCES

Bao jiu Xu. 2000. Theory Of Water Treatment. Beijing: China Architecture & Building Press.
Eaton A.D. 1995. Measuring. UV-Absorbing Organics A Standard Method. J. AWWA, 87(2):86–90.
Hai long Liu, Zheng jian Li, Hong Zhao, et al. 2010. Pre-ozonation enhanced coagulation and its safety in treatment for water channeled from Yellow River. Environmental Pollution & Control, 32 (4):58–61.
Hoigne J, Bader H. Rate 1983. Constants of reactions of ozone with organic and inorganic compounds in water I. Non-dissociating Organic Compounds [J]. Water Research, 17(2):173–183.
Hong yuan Liu, Yan Zhang. 2005. Enhanced Technology And Project Cases In Water Treatment. Bei jing: Chemical Industry Press.
Liang Zhao, Xing Li, Yan ling Yang. 2009. Review of ozone preoxidation technology in drinking water production. Water Technology, 4(3):6–10.
Ming xin Huo, Xin yuan Liu. 1998. Analysis of quality characteristics of low temperature and low turbidty water. China Water & Wastewater, 14(6):33–34.
Ministry of Health of the People's Republic of China. 2007. China National Standardization Management Committee. GB/T5750.11-2006, Drinking Water Sanitation Standardized Testing Method (Disinfectant index). Beijing: China Standard Press.
Yang Li, Xiao rui Duan, Jian Yuan, et al. 2010. Research progresses in low temperature and low turbidity micro-polluted water treatment. Technology & Application, 5:52–55.
Yong Wang, Shuo Yang, Xin Wen. 2009. On influential factors on processing technique of low temperature and low turbidity water and its analysis. Shanxi Architecture, 35(22):181–182.

Hydraulic Engineering – Xie (Ed.)
© 2013 Taylor & Francis Group, London, ISBN 978-1-138-00043-8

Fuzzy assessment of oil spill impact and consequence for offshore petroleum industries

Yan Lu
China Offshore Environment Services Co., Ltd., Tianjin, P.R. China
Ocean University of China, Qingdao, P.R. China

Wenpu Wei
China National Offshore Oil Co., Beijing, P.R. China

Jia Wang
Ocean University of China, Qingdao, P.R. China

Yong Yang & Wei An
China Offshore Environment Services Co., Ltd., Tianjin, P.R. China

ABSTRACT: With the increasing exploration and development of offshore petroleum project in China, the potential risk of oil spill gradually attract the public attentions. In order to effectively respond to oil spill incident, and limit the spilling influence and to bring it under control, it is necessary to establish a consequence assessment model for offshore oil spill incident. In this paper, a consequence assessment was developed using fuzzy comprehensive model. The consequence assessment indices of spills include not only the spilling volume, property of spilled oil and spilling location, also the elements of spilling detection and impact controlling. Each of the evaluated factors were suggested a relative evaluation criterion and assigned the impact ranking. According to the consequence assessment, it is easily to determine the spilling consequence level (very low, low, moderate, high or very high) and make a decision whether the response resource allocated to the site could be enough.

1 INTRODUCTION

With the increasing exploration and development of offshore petroleum project in China, the potential risk of oil spill gradually attract the public attentions. According to the statistics from Ministry of Transport of P.R. China, over two thousand of oil spill incidents were happened through 1973 to 2006; total amount of spillage was accumulated up to 37000 tons (Shao C. & Zhang Y. 2009), causing the serious influence to marine environment and local economy. Especially since the oil spill incident of Penglai 19-3 of China in June 2011. a serious public discussion was emerged that how to promptly determine the impact or consequence of oil spill to marine environment and ecology in an objective manner. To effectively respond to oil spill incident, and limit the influence or consequence of the incident and to bring it under control, it is necessary to establish a consequence assessment model for offshore oil spill incident.

Evaluation of oil spill impact (consequence) is an element to assess the oil spill risk; the major assessment factors for consequencein the previous work (French McCay. 2003a, b, 2009; NRC, 2002; Wirtz K.W. & Baumberger N. 2007) were the environmental impact, such as spillage, toxicity and the fate of spilling oil. While the consequence-control factors which consist of the limitation measure of oil lease, respond time for oil spill, emergency respond resource, etc., should not be ignored according to API document (2000) of risk inspection for offshore petroleum industries. In this paper, the potential impact indices of spills on environmental influence combined with consequence controlling were assessed using fuzzy comprehensive model, with some modifications based on experts' opinion. Each of the evaluated factors in our work were developed a relative evaluation criterion and assigned the impact ranking.

2 METHODOLOGY

The present fuzzy assessment model is an extension of AHP method suggested by Saaty (Lin S.& Zhang Z., 1992). As a decision-making model, the fuzzy comprehensive evaluation is expected to lower the uncertainties in the data by using experts' experience and try to transfer the lexical knowledge to numerical values, which are easily integrated in an evaluation process. The detailed procedure of applying fuzzy assessment approach is described in the following paragraphs.

2.1 Indexes of evaluation and the criterions

The index system is a Fuzzy subset composed of the evaluation factors with the number of n. As follows:

$$U = \{U_1, U_2, \dots U_n\} \tag{1}$$

In the evaluation process, the evaluation set include all the appropriate levels or scores of evaluation criteria. To assist decision maker to identify the consequence degree, scored the consequence level ranging from 1 to 7 are predefined, which imposes a numeric measure on the performance of alternatives, as follows:

$$V = \{V_1, V_2, \dots V_m\} = \{1, 2, 3.5, 5, 7\} \tag{2}$$

where, m = the number of grade divided, and $m = 5$ in this paper; V_m represent the consequence level of Very Low (VL), Low (L), Moderate (M), High (H), and Very High (VH), respectively from V_1 to V_5.

2.2 Determine the weight and fuzzy membership matrix

In fuzzy integrated evaluation, the weight of the index are determined regard to their contrition to the consequence level of oil spill incident. The weight distribution can be showed as:

$$A = \{A_1, A_2, \dots A_n\} \tag{3}$$

The evaluation results to factor u_i composed single-factor fuzzy evaluation subset $R_i = (R_{i1}, R_{i2}, \dots R_{im})$; the fuzzy mathematics matrix $R = (r_{ij})_{n \times m}$, as follows:

$$R = \begin{bmatrix} r_{11} & r_{12} & \cdots & r_{1m} \\ r_{21} & r_{22} & \cdots & r_{2m} \\ \vdots & \vdots & \cdots & \vdots \\ r_{n1} & r_{n2} & \cdots & r_{nm} \end{bmatrix} \tag{4}$$

2.3 Integrated evaluation

Combined the predefined first order fuzzy membership matrix (R_i) and the weight vector (A_i), the consult of initial evaluation for criterion i can be calculated through the following formula:

$$B_i = A_i \cdot R_i = (w_1, w_2 \dots, w_m) \cdot \begin{bmatrix} r_{11} & r_{12} & \cdots & r_{1m} \\ r_{21} & r_{22} & \cdots & r_{2m} \\ \vdots & \vdots & \cdots & \vdots \\ r_{n1} & r_{n2} & \cdots & r_{nm} \end{bmatrix} (i = 1, 2 \dots n) \tag{5}$$

And the total assessment matrix **B** is obtained as equation 6; the final result is determined by the weighted average method, based on the fuzzy comprehensive evaluation index b_j in this paper.

$$\mathbf{B} = \mathbf{A} \cdot \mathbf{R}; \quad R = \begin{bmatrix} B_1 \\ B_2 \\ \vdots \\ B_n \end{bmatrix} = \begin{bmatrix} A_1 & \cdot & R_1 \\ A_2 & \cdot & R_2 \\ \vdots & & \\ A_n & \cdot & R_n \end{bmatrix}; \quad v = \frac{\Sigma b_j v_j}{\Sigma b_j} \tag{6}$$

where, b_j = the value obtained by fuzzy comprehensive evaluation; v_j = corresponding evaluation criterion value.

3 CONSEQUENCE ASSESSMENT INDEX SYSTEM OF OIL SPILL

3.1 *Assessment index system*

Risk assessment of oil spill consequence for offshore petroleum industries should take some basic elements into account; the evaluated elements not only include the spilling volume, propertyand location of spilling oil, also comprise the elements to evaluate whether or not promptly control and limit impact of oil leaking (Kent Muhlbauer W., 2001). Thus, the consequence assessment index system for offshore oil spill in this paper is a multi-factor system (as Fig. 1), which intergrades the evaluation of hazard (impact) of spilling oil (U_1) and impact controlling (U_2). The hazard of oil is closely correlated with the spillage amounts (U_{11}), environmental sensitivity (U_{12}) as well as property of oil (U_{13}). The element of impact controlling is used to discuss the efficiency of response against a spill incident for the offshore petroleum industry. As known, a good response and emergency capability for offshore spilling can limit the extents polluted by oil spilling and reduce the economic loss in operative manner. Detection of spills (U_{21}) together with the emergency capability (U_{22}) are suggested to be the important factors for impact controlling; the former is assessed by the reliability of detection or forecasting devices for oil spilling (for example, whether there is an automatic detection for spills applied in the offshore oil field), and the latter is measured by the factors of oil-spilling limitation devices (U_{221}) and response and treatment capability for spills (U_{222}). In our work, we select six sub-indexes to intergrately evaluate the factor U_{222}, including the emergency communication, emergency train, oil spill contingency plan, resource allocation time, maintained condition of the oil spill resource as well as recovery service of offshore oil spill.

Based on the statistics of offshore oil-spilling incident in recent decades and the opinions from the petroleum experts, the weight could be predefined. The relative weights of index presented in this paper could be adjusted in accordance with specific condition.

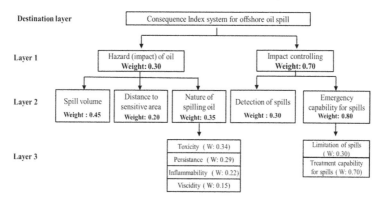

Figure 1.   Consequence evaluation system for offshore oil spill.

3.2 *Membership degree matrix of evaluation factors*

Because of different types of evaluation factors are very complicated, the grading standards of most factors are not defined in this paper. We take the membership matrix of spilling volume (U_{11}) and detection of spills (U_{21}) for instance, in order to illustrate how to grade and quantify the evaluation factors.

According to the standards from State Ocean Agency of P.R. China, the oil spill rank on the basis of the spilling volume can be divided into three level. Level one is low-influence to marine, and the spilling amount is suggested to be; ton, level II represent mid-influence to marine, and the corresponding amount is 10 ton; the last level means high-influence, and the corresponding value should up to be 100 ton. Thus, the membership function of spilling volume is established based on the relative standards of SOA, and developed referring to expert opinion.

According to Figure 2, the membership function of spilling volume can be determined as follows:

For consequence criterion V_1:

$$V_1(t) = \begin{cases} 1, & t \leq 0.5 \\ 1-(t-0.5)/0.5, & 0.5 < t \leq 1 \\ 0, & t > 1 \end{cases} \tag{7}$$

For consequence criterion V_2:

$$V_2(t) = \begin{cases} 0, & t \leq 0.5 \\ (t-0.5)/0.5, & 0.5 < t \leq 1 \\ 1-(t-1)/9, & 1 < t \leq 10 \\ 0, & t > 10 \end{cases} \tag{8}$$

For consequence criterion V_3:

$$V_3(t) = \begin{cases} 0, & t \leq 1 \\ (t-1)/9, & 1 < t \leq 10 \\ 1-(t-10)/30, & 10 < t \leq 40 \\ 0, & t > 40 \end{cases} \tag{9}$$

For consequence criterion V_4:

$$V_4(t) = \begin{cases} 0, & t \leq 10 \\ (t-10)/30, & 10 < t \leq 40 \\ 1-(t-40)/60, & 40 < t \leq 60 \\ 0, & t > 60 \end{cases} \tag{10}$$

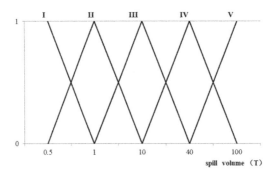

Figure 2. Membership function of spilling volume.

334

Table 1. Membership matrix of the criterion for detection of spills.

Type of spilling detection device	V1	V2	V3	V4	V5
Active detection device	0.6	0.4	0	0	0
(based on the change of pressure or flow rate of transporting oil)					
Passive detection device	0	0.2	0.4	0.4	0
(detected when there is oil leaking outside pipeline or wellhead)					
No detection device or device losing efficacy	0	0	0	0.2	0.8

For consequence criterion V_5:

$$V_5(t) = \begin{cases} 1, & t \leq 40 \\ (t-40)/60, & 40 < t \leq 100 \\ 0, & t > 100 \end{cases} \tag{11}$$

According API guideline, the type of spilling detection devices have an important role to fast response for oil spill incident. As recommended, the active detection system based on the suddenly change of pressure or flow rate of transporting oil can respond to the signal of leaking promptly. The criterion of detection of spills and the corresponding membership matrix are described in Table 1.

4 EXAMPLE

In this paper, we take example with an oil spill incident happened in Bohai waters and discuss the reliability of the developed frame structure of consequence assessment for offshore oil spill.

The spills was happened in south western of Bohai offshore waters because of operational failure of drilling mudinjection, and the spilling location was approximately 27.4 km from the nearest sensitive area along the wind direction. The water temperature there was 12 °C and wind speed was 7.4 km/h. The total amount of oil spilling, as predicted, was 3 ton. The leaked oil was heavy crude oil; alkane and cycloalkane were the main ingredient. The viscosity of oil was over 10000 mPa·s. Although detected the phenomenon of leaking by operator, oil spilling was controlled and limited promptly. The response of the petroleum company was strictly in compliance with the oil spill contingency plan and contact the service department of spilled oil recovery and treatment. The resources and devices could be allocated to the spilling site about 2.5 h.

In fuzzy comprehensive model, the last layer factors should be calculated with their weight and membership matrix. According to the evaluated factor system established in Figure 1, the results of the third layer evaluation factors using fuzzy comprehensive model are shown below:

For the property of oil:

$$B_{13} = A_{13} \cdot R_{13} = (0.34, 0.29, 0.22, 0.15) \cdot \begin{bmatrix} 0.2 & 0.6 & 0.2 & 0 & 0 \\ 0 & 0.2 & 0.6 & 0.2 & 0 \\ 0 & 0.2 & 0.6 & 0.2 & 0 \\ 0 & 0.2 & 0.6 & 0.2 & 0 \end{bmatrix}$$

$$= (0.068, 0.336, 0.464, 0.132, 0)$$

For the emergency capability of spills:

$$B_{22} = A_{22} \cdot R_{22} = (0.30, 0.70) \cdot \begin{bmatrix} 0.2 & 0.7 & 0.1 & 0 & 0 \\ 0.6 & 0.3 & 0.1 & 0 & 0 \end{bmatrix} = (0.48, 0.42, 0.1, 0, 0)$$

The value of the consequence for the oil spill incident were determined by above method (as Table 2), and the overall consequence level can be divided into 5 ranks according to evaluation criterion (see Section 2.1).

Table 2. Consequence level of oil spill incident.

Evaluated factors	Initial evaluation	Overall evaluation	Consequence level
Hazard (impact) of oil	(0.02, 0.50, 0.39, 0.09, 0)	(0.24, 0.35, 0.17, 0.07, 0.17)	$V = 3.05$ (moderate)
Impact controlling	(0.34, 0.29, 0.07, 0.06, 0.24)		

Under the process of fuzzy comprehensive assessment, the consequence result for the oil spill incident was determined as moderate level. If the value of consequence level is more than 3.5, it suggest the oil spill incident has a high impact to environment and society. Meanwhile the additional emergency resource should be allocated to the spill site in order to reduce or control the influence by the spilled oil.

5 CONCLUSION

The developed consequence assessment of offshore oil spill with application of fuzzy comprehensive assessment model can provide reliable information to control the influence by an oil-spilling incident. In this paper, the consequence assessment indices of spills include not only the spilling volume, property of spilled oil and spilling location, also the elements of spilling detection and impact controlling. With some modifications based on experts' opinion, each of the evaluated factors in our work were developed a relative evaluation criterion and assigned the impact ranking. According to the consequence assessment, it is easily to determine the spilling impact level (very low, low, moderate, high or very high) to local marine environment and society and make a decision whether the response resource allocated to the site could be enough. Thus the consequence assessment approach described in this paper is proposed to prediction of spilling influence by offshore petroleum industries.

ACKNOWLEDGE

Funding for this study was provided by the Marine Public Welfare Research Project of China (No. 201205012). We greatly appreciate the valuable comments by Z. Zhang who has sufficient experience of offshore petroleum development.

REFERENCES

American Petroleum Institute (API). 2000. Risk-based inspection base resource document. *API publication-581.*

French McCay D. 2003a. Development and application of damage assessment modeling: example for North Cape Oil Spill. *Marine Pollution Bulletin.* 47(9–12):341–359.

French McCay D. 2003b. Oil spill impact modeling: development and validation. *Environmental toxicology and chemistry*, 23(10): 2441–2456.

French McCay D. & Beegle Krause C.J., et al. 2009. Oil spill risk assessment—relative impact indices by oil type and location. 32nd AMOP technical seminar on environmental contamination and response, 655–681. Emergencies Science Division, Ottawa, Canada.

Kent Muhlbauer W. 2001. Pipeline risk management manual. Gulf Publishing Company.

Lin S. & Zhang. Z. 1992. Fuzzy risk analysis of harbor engineering investment by hierarchy system approach. *China Ocean Engineering.* 6(1):87–94.

National Research Council (NRC). 2002. Oil in the sea III: inputs, fates and effects. *National Academy Press.* Washington, USA.

Shao C. & Zhang Y., et al. 2009. Application in oil-spill risk in harbors and coastal areas using fuzzy integrated evaluation model. *MCDM 2009*, 681–688. Berlin: Springer-Verlag.

Wirtz K.W. & Baumberger N., et al. 2007. Oil spill impact minimization under uncertainty: evaluating contingency simulations of the Prestige accident. *Ecological economics.* 61:417–428.

Hydraulic Engineering – Xie (Ed.)
© 2013 Taylor & Francis Group, London, ISBN 978-1-138-00043-8

Impact analysis and simulation of South China Sea SSTA towards monsoon atmospheric LFO

Ni Wenqi
General Staff Meteorological and Hydrological Space Weather Terminus, Beijing, China

Jiang Guorong
University of Science and Technology of the People's Liberation Army Institute of Meteorology, Nanjing, China

Liu Ruihui & Wang Yibai
General Staff Meteorological and Hydrological Space Weather Terminus, Beijing, China

ABSTRACT: Using Lanczos bandpass filtering, the EOF empirical orthogonal decomposition and wavelet analysis method analyzes the 1982–2010 NCEP zonal wind field data, synthesis of East Asia atmospheric Low-Frequency Oscillation (LFO) characteristics of cycle, strength and propagation in positive and negative South China Sea (SCS) SST anomalies years in the summer. Secondly, using regional climate model RegCM3, making increasing and reducing the SCS SST sensitive experiments to explore the impact of SCS SST anomalies towards monsoon LFO characteristics. The results show that: the unusually high SCS sst has the trend of making the time of Southward spreading of Summer Monsoon LFO changing to the northward spreading ahead of time, and shortening the cycle. Both the positive and negative SCS sst numerical experiments make the time of Southward spreading of Summer Monsoon LFO changing to the northward spreading ahead of time, the strong value of LFO oscillation moves south to the lower latitude by 5–7°. The abnormal negative experiment can play a very significant role in stretching atmosphere LFO cycle, strengthening the 60–65 day cycle oscillation anomaly and weakening 25–30 day and 18-day cycle oscillation abnormally.

1 INTRODUCTION

The time scale between 10 days and 100 days of atmospheric motion is called atmospheric Low-Frequency Oscillation (LFO). Intraseasonal Oscillation (ISO) which is also known as the 30–60 day and 10–20 days quasi-biweekly oscillation are most significant atmospheric low-frequency oscillations. These two types of oscillation are the most outstanding performance in the monsoon region (Li. 1991). Study shows that the evolution of the monsoon system (monsoon onset, active, interrupt and retreat) has significant low-frequency oscillation characteristics, LFO is considered to be an important component of monsoon variability. Therefore, improving the awareness of the LFO is of great significance for improving short-term forecast (Lin. 2008.).

The air-sea interaction not only affects the intensity of the tropical atmosphere ISO, but also affects the frequency and intensity of the high-latitude atmosphere ISO. Horel (Horel. 1981) pointed out that the obvious equatorial SST anomalies can not only affect the tropical atmospheric circulation and climate, but also the high latitudes. He Jinhai (2006) believes that global warming makes a series of changes towards the tropical low-frequency oscillation intensity, distribution, propagation and cycle, certifys the impact of SST atmosphere towards ISO from another angle (He. 2006). Lu Weisong, Qiu Mingyu pointed out: Forcing changes has modulation to middle and high latitudes modulation. Liu Peng, Lu Weisong pointed out: in the beginning of the El Nino and El Nino maturity period, both the central and eastern equatorial Pacific and equatorial Indian Ocean sea surface temperature, atmospheric low-frequency oscillation cycle has shortened in the process of gradually warming. Qiuming Yu, Lu Weisong, Wang Shangrong pointed out: Enhancing (weakening) the strength of the Indian Ocean SSTA unipolar thermal forcing and the base flow have the trend of shorting (extending) the atmospheric low-frequency oscillation cycle. Therefore further study is necessary for the impact of SCS sst towards the low frequency oscillation. This article first synthetic analyses atmospheric LFO spreading, cycle and the strength characteristics during the summer in June–August after the SCS sst positive and negative anomaly period from April to July during 1982–2010, and uses regional climate model RegCM3 testing the impact of SCS sst towards the characteristics of the monsoon LFO.

2 SYTHETIC ANALYSIS OF THE CHARACTERISTICS OF THE ATMOSPHERIC LFO IN SOUTH CHINA SEA TEMPERATURE ANOMALIES YEARS

From 1982–2010 totally 29 years of NCEP data, choosing the years of typical positive and negative anomalies in April to July of SCS SSTA, sythetic analyzes monsoon atmospheric LFO characteristics in these years respectively during June–August. Reveals meridional propagation characteristics of monsoon atmospheric LFO after the lancoz filtering, EOF empirical orthogonal function decomposition and wavelet analysis, exports the first feature of LFO characteristics in SCS sst positive and negative anomalies years in order to reveal the impact of SCS SSTA towards the monsoon atmospheric LFO characteristics.

Data from Tables 1 and 2 obviously shows the sum of the first three principal components have basically been able to characterize the original variables, the first characteristic variance is 48.40% and 38.14% of the total variance. As you can see in the periodogram of first feature, atmospheric low frequency zonal wind in monsoon region presents several major cycles. During positive anomalies years, zonal wind (Fig. 1A) before July generally performs for 25 d and 55 d cycle oscillation, but after July performs for 30 d and 70 d cycle oscillation. The oscillation of 70 d cycle is weaker than SST Negative anomalies years, mainly in the 30 d; During SST negative anomalies years (Figure 1B) generally performs for 30 d cycle oscillation before July, after the summer (June–August) the monsoon atmospheric low-frequency per-

Table 1. Total variance of S and fitting rate of first three models of meridional distribution of the low-frequency zonal wind in positive anomalies years.

ρ_1	ρ_2	ρ_3	$\sum\limits_{h=1}^{3}\rho_h$
0.4840	0.2621	0.0892	0.8353

Table 2. Total variance of S and fitting rate of first three models of meridional distribution of the low-frequency zonal wind in negative anomalies years.

ρ_1	ρ_2	ρ_3	$\sum\limits_{h=1}^{3}\rho_h$
0.3814	0.3068	0.1069	0.7951

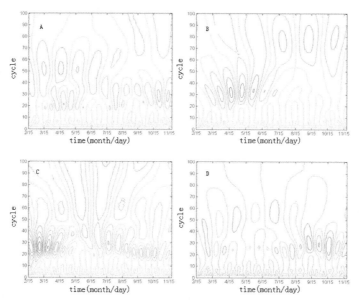

Figure 1.   The first feature of the cycle analysis of the low-frequency zonal wind u (A and B), low-frequency meridional wind v (C and D) at positive (A and C), negative (B and D) SCS SSTA anomalies years (/day).

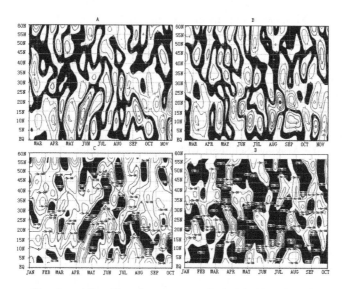

Figure 2.   Time profile of meridional low-frequency zonal wind (m/s) and vorticity (m/s²) in positive (A), negative SCS SSTA anomalies years (B).

forms for 70–90 d cycle, other cycles are weak. Meridional wind's first characteristic variance shows the main cycle is 30 d from June to August during the SCS SST positive anomaly years (Figure 1C), but in negative anomaly years (Figure 1D) the main cycle is 55 d, the second cycle is 25 d, the main cycle is stretched compared with positive anomalies years. So visibly positive anomaly makes the atmospheric meridional wind LFO cycles in monsoon region shortened. This result consists with Liu Peng, Lu Wersong's result: when no metter the central and eastern equatorial Pacific sst or the equatorial Indian Ocean sst is in gradually warming period, atmospheric low-frequency oscillation cycle all has shortened trend.

Figure 2A and B are the evolve over time of meridional profile of low-frequency zonal wind in the negative or positive anomaly years. When monsoon happens, the propagation of low-frequency zonal wind changes from southward spreading to northward spreading instead. In positive years, this change is earlier, and spreading to higher latitude. From vorticity field (Figure 2C), the time of southward speading transfering to northward spreading is earlier,and spreading to higher latitude too. In this way, SCS SSTA creates the climate anomaly of middle and high latitude area. In negative years, the transformation of low-frequency east wind to west wind in SCS is later than positive years, the same as the transformation from southward spreading to northward spreading. Different from positive years, low-frequency energy is concentrated in south of 25 N and spreads north to about 30 N and then stops. The northward progation of vorticity field (Figure 2D) is not obvious.

3 NUMERICAL SIMULATION ANALYSIS OF THE IMPACT OF ABNORMAL SCS SST TOWARDS EAST ASIAN SUMMER MONSOON ATMOSPHERE LFO CHARACTERISTICS

Through one-way nested technology, the regional climate model (RegCM3) uses the simulation outputs of the global atmospheric model or the measured data as boundary conditions and initial field to improve simulation accuracy. Relative to the global circulation models, regional climate model has higher resolution, better characterizes the terrain and vegetation, has a variety of advanced parameterization schemes. So it is a powerful tool to study the change of regional climate (Zhao. 1998.).

He Jinhai, Ding Yihui (He. 2001) pointed out in the book "Definition of date when South China Sea Summer Monsoon happens and Monsoon Index" that 1988 is a year in which monsoon broke normally, many cases occurred in 1988 can represent the characteristics of the climate state (Song. 1996). So this article uses NCEP atmospheric reanalysis data in 1988 and weekly average sst data from OISST as the modal input, and runs the regional climate model RegCM3.

Simulation area is East Asia area and the Western Pacific sea (90°–130°E, 0°–45°N), the center point is at 25°N, 110°E, the grid number of north-south is 120, east-west is 100, the horizontal resolution of 60 km. Vertical direction is divided into 18 layers non-uniform mode, with overhead pressure 10 hpa. Cumulus convective parameterization scheme selects Grell scheme based on the Arakawa-Schubert with closure assumption. The planetary boundary layer scheme uses Hotslag non-local planetary boundary layer scheme. Radiation process uses the latest CCM3 radiative transfer program. Land surface the process uses BATS1E, including one layer of vegetation, a layer of snow and three ayers of soil. Side of border of modal uses index relaxation time-varying-border scheme, the border buffer chooses 10 laps. The integrating period is from December 1 in 1987 to December 31 in 1988, December in 1987 isn't analyced as a spin-up time. In sensitivity experiments 2°C is added and decreased to the South China Sea area (100°–120°E, 0°–20°N), and chooses April to July as the anomalies maintenance phase. By analyzing differences between sensitive experiment and control experiment, dicuses the specific impact of SCS sst anomalies on the East Asian summer monsoon LFO characteristics.

From the model results, both positive and negative experiments make the transforming time of direction of propagation from southward to northward advances to the end of April or early May. The extremal values of both experiments are all concentrated in low latitudes, strong values exists by 30°N. Strong value of oscillation moves southward by 5–7°. The main difference between the two experiments is range of low frequency easterly wind is bigger than control experiment at the low latitudes in the negative experiments before the end of April or early May.

The same as control experiment, the main cycle of positive experiment (Figure 3A) is 25–30 d, the second cycle is 60–65 d, and the third is about 18 days. Both 18 d and 25 d cycle osillation are stronger than control experiment, this is to say the 25–30 d cycle is shortened to the 18-day oscillation. 60 days oscillation is weaker than the control experiment

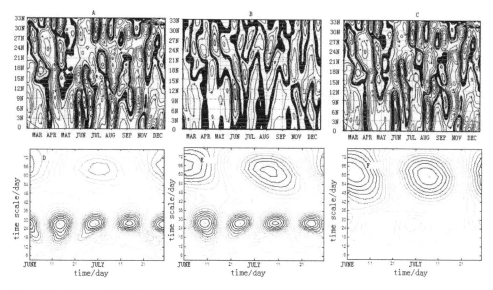

Figure 3.   Time profile of meridional low-frequency zonal wind (m/s) in increasing (A), control (B), decreasing (C) experiment and the first characteristic cycle (D–F) respectively (/day).

by a certain degree. Therefore we can say, the SCS sst abnormal warming has the trend of shortening the monsoon atmospheric LFO 30 day cycle to 18 days. 60–65 d cycle oscillation is weaker. Figure 3B is cycle distribution of the 850 hPa low frequency zonal wind in negative sst experiment. Obviously 60–65 d cycle is stronger than the control experiment, and becomes the main cycle instead of 25–30 d; 25–30 day cycle oscillation weakens a lot, 18 d cycle oscillation almost doesn't exist. It reveals the SCS sst negative experiment can play a very significant role in stretching atmosphere LFO cycle, strengthening the 60–65 d cycle oscillation anomaly and weakening 25–30 day and 18-day cycle oscillation abnormally. This phenomenon is not difficult to explain, because the atmosphere obtains energy from the sea at SST anomalies warm phase, atmospheric fluctuations oscillation speed up and cycle shortens; anomalous cold sst makes atmosphere loses energy, atmospheric wave oscillation slows down, cycle is extended.

4   CONCLUSION

1. The synthesis analysis of the East Asian monsoon atmospheric low-frequency movement cycle reveals unusually high SCS sst has the trend of transfering southward spreading to northward spreading ahead of time and shortening the cycle, and spreading to higher latitude. In this way, positive SCS sst creates the climate anomaly of middle and high latitude area.
2. Both positive and negative experiments make the transform time of direction of propagation from southward to northward advances to the end of April or early May. Oscillation strong value moves southward by 5–7°. The SCS sst abnormal warming has the trend of shortening the the monsoon atmospheric LFO 30 day cycle to 18 days. 60 to 65 days cycle oscillation is weaker. In negative experiment, obviously 60–65 d cycle is stronger than the control experiment, and becomes the main cycle instead of 25–30 d; 25–30 day cycle oscillation weakened a lot, 18 d cycle oscillation almost doesn't exist. It reveals the SCS SSTA negative experiment can play a very significant role in stretching atmosphere LFO cycle, strengthening the 60–65 day cycle oscillation anomaly and weakening 25–30 day and 18-day cycle oscillation abnormally.

REFERENCES

Chen, L.X. 1988. Westward propagation low-friquency oscillation and its teleconnections in the eastern hemisphere. *Acta, Meteor. Sinica*, 2, 300–312.

Chang, C.P. 1977. Viscous internal gravity waves and low-frequency oscillation in the tropics. *J.Atmos. Sci*: 34,901–910.

Dickinson, R.E. & Errico R.M. & Bates, G.T. 1989. A regional climate model for the western United States. *Clim Change*: 15, 383–422.

Giorgi F. 1995. Perspectives for regional earth system modeling. *Global Planet Change*: 10,23–42.

Giorgi F. & Marinucci M.R. 1996. Improvements in the simulation of surface climatology over the European region with a nested modeling system. *Geophys.Reophys.Res. Lett.* 23,273–276.

Giorgi F. & Marinucci M.R. &Visconti G. 1990. Use of a limited-area model nested in a general circulation model for regional climate simulation over Europe. *J Geophys Res*. 95:D11,18413–18432.

He J.H. & Dong M. Jiang G.R. 2006. The impact of global warming towards the tropical atmospheric intraseasonal oscillation characteristics and its numerical simulation study. *Meteorology and Disaster Reduction Research*: 29(1): 17–21.

Horel J.D. & Wallace J.M. 1981. Planetary-scale atmospheric phenomena associated with the Southern Oscillation [J]. *Mon Wea Rev*, 109: 813–829.

He J.H. & Ding Y.H. & Xu H.M. 2001. SCS summer monsoon setting date and Monsoon Index. *Beijing*: *China Meteorological Press*.

Krishinamurti, T.N. & Subrahmanyam D. 1982. The 30–50 day mode at 850 mb during MONEX. *J.Atmos.Sc*: 39,2088–2095.

Li C.Y. 1991. Atmospheric low-frequency oscillations [M]. *Beijing: China Meteorological Press*.

Lin A.L. & Liang J.Y. & Gu D.J. 2008. The Impact of Tropical Intraseasonal Oscillation towards East Asian monsoon region and Research progress of different time scale Changes. *Journal of Tropical Meteorology*, 24 (1):11–19.

Lau K. & Yang S. 1996. Seasonal variation, abrupt transition and intraseasonal variability associated with the Asian summer monsoon in the GLAGCM. *J Climate*, 9(5):965–985.

Lau N.C. & Lau K.M. 1986. The structure and propagation of 40–50 day oscillations appearing in a GFDL general circulation model. *J.Atmos. Sci*. 43,2023–2047.

Li C.Y. & Jia X.L. & Dong M. 2006. Numerical simulation comparative study of atmospheric intraseasonal oscillation. *Meteorological Science*, 64(4):412–419.

Madden R.A. & Julian P.R. 1972. Description of global scale circulation cells in the tropics with 40–50 day period. *J.Atmos.Sci*, 29,1109–1123.

Madden R.A. & Julian P.R. 1971. Detection of a 40–50 day oscillation in the zonal wind in the tropical Pacific. *J.Atmos.Sci*, 28,702–708.

Murakami M. 1984. 30–40 day global atmospheric changes during the northern summer 1979. *GARP Special Report*, No. 44,113–116.

Pielke R.A. et al. 1992. A comprehensive meteorological modeling system-RAMS. Meteorol.Atmos. Phys:49, 69–91.

Qian Y. & Giorgi F. 1999. Interactive coupling of regional climate and sulfate aerosol models over eastern Asia. *J.Geophys.Res.*104,6501–6514.

Qiu M.Y. & Wang S.R. 2006. Numerical experiments of the impact of Indian Ocean SSTA towards low-frequency oscillation in high latitudes. *Journal of Tropical Meteorology*, 22(6):605–611.

Song Y.K. & Chen L.X. 1996. The numerical simulation of the low-frequency oscillation in summer. *New research progress of Asian monsoon—cooperative research papers on Sino-Japanese Asian monsoon mechanism. Beijing*: China Meteorological Press.

Seth A. & F. Giorgi. The effect of domain choice of summer precipitation simulation and sensitivity in a regional climate model. *J Clim*, 11:2698–2712.

Wang P.X. & Wu H.B. & Xu J.J. 1994. Complex empirical orthogonal function analysis and intuitive display. *Nanjing Institute of Meteorology*, 17,4:448–454.

Xin F. & Xiao Z.N. & Li Z.C. 2007. The relationship of 1997 South China flood season precipitation anomalies and atmospheric low-frequency oscillations. *Meteorological*: 33.

Zhao Z.C. & Ra Y. 1998. Study progress of regional climate model in 1990s. *Meteorological Science*, 56(2):113–119.

Zhen S.L. & Deng Z.W. 1999. Wavelet climate diagnostic techniques. *Meteorological Press*: 33.

Hydraulic Engineering – Xie (Ed.)
© *2013 Taylor & Francis Group, London, ISBN 978-1-138-00043-8*

Species composition and distribution of ichthyoplankton in the Huanghe River estuary

Yanyan Yang, Zhenbo Lv, Fan Li, Zhongquan Wang & Chunxiao Sun
Shandong Marine and Fishery Research Institute, Shandong Provincial Key Laboratory of Restoration for Marine Ecology, Yantai, China

Qiang Xu
Yantai Marine Safety Administration, Yantai, China

ABSTRACT: To gain a better understanding the inference of the Water and Sediment Discharge Regulation (WSDR) project on the species composition and distribution of ichthyoplankton in the estuary and adjacent area, we investigated surveys at 13 stations in June and July. The results show that, 1280 eggs, 5866 larvae are collected, the dominant fish eggs are Dotted gizzard shad (*Konosiru spunctatus*), silver sillago (*Sillag osihama*), Joyner's tongue-sole (*Cynoglossus joyneri*); the dominant larvae are Japanese anchovy (*Engraulis japonicas*), Tapertail anchovy (*Coilia mystus*) and the gobiidae fishes. The numbers of species (S), abundance (N), Margalef richness index (D), Shannon-Wiener diversity index (H') and Pielou's evenness index (J') were calculated. The result showed that the values of S, N, H', D are low, but the value of J' is high in the middle of WSDR.

1 INTRODUCTION

In marine ecosystems, ichthyoplankton are important prey, larvae are also the predators, both with one of the important link in the marine food chain (Ruijing Wan & Yanwei Jiang 1989; Ruijing Wan et al. 2008; Akihiko. K & Shigeki. S 1983; Kazumasa. H et al. 1987). Because of the nutrition salt and organic detritus from the Huanghe water system, the primary productivity and food diversity are higher in the Huanghe River estuary, where is the spawning ground, feeding ground and fattening ground (Jingyao Deng & Xianshi Jin 2000; Xinhua Zhu et al. 2001). Species composition and distribution of ichthyoplankton in the Huanghe River estuary depends on the fishery resources of the Yellow sea and Bohai sea, which also affects the complement of fishery resources in this area. Since the 1980s, many ichthyologists paid attention to the influence to breeding of fishes by hydrological geographic features in this area (Xinhua Zhu et al. 2001; Whitfield A.K 1989; Weinstein. M.P et al. 1980). In 1970s, 1980s and 1990s, they studied the structure of biological communities, primary productivity and biodiversity (Zinan Zhang et al 1990; Ruihua Lv & Mingyuan Zhu 1992; Yumu Jiao 1998; Peng Wang 1999), also, they studied fishery resources in the Huanghe River estuary (Yumu Jiao & Jiayi Tian 1999; Jiayi Tian 2000; Fan Li & Xiurong Zhang 2001). There are no public reports about the species composition and distribution of ichthyoplankton in the Huanghe River estuary during the period of WSDR.

This study on the basis of trawl data in 2011 in the Huanghe River estuary, we analyzed the dynamic changes of the species composition and distribution of ichthyoplankton, in order to provide the basic data, which will help to scientifically asses the ecological impact by WSDR.

2 MATERIALS AND METHODS

2.1 *Survey stations and voyage distribution*

Samples were collected at 13 stations between 119°05′–119°27′E, 37°44.5′–38°00′N in the Huanghe estuary. (Fig. 1).

2.2 *Sample techniques*

At these stations, samples were all conducted by the specification 'specification for oceanographic survey-Part 6: Marine biological survey' (GBT 12763.6–2007). Ichthyoplankton samples were collected horizontally with large plankton net (mouth diameter 80 cm, 280 cm in length) at 10 min duration at a speed of 2.0 kn on the sea surface. Samples saved with 5% formalin seawater solution, then made qualitative and quantitative survey by dissecting microscope after sorted.

2.3 *Data analyze*

We processed data by statistical software SPSS17.0, we investigated environmental factors contained depth, surface salinity (SSS), surface water temperature (SST) and so on, which relations to community diversity and quantitative distribution of ichthyoplankton caught by surface trawl (Jie Han et al.2004). We drew quantity plane distribution figure of ichthyoplankton by software ArcGIS9.0.

2.4 *Ecological dominance*

Ecological dominance is used to determine the community importance in species by using Index of relative importance of Pinkas (IRI) (Pinkas. L et al. 1971). IRI = N × F (N means accounting the percentage of one certain kind of ichthyoplankton in total; F means the percentage accounting N appeared stations in total), the value of IRI is greater than or equal 1000, which kinds are dominant species, the value is between 100~1000, which kinds are important species.

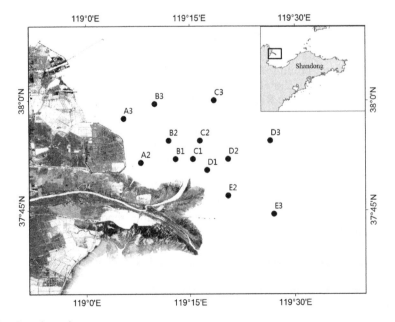

Figure 1.   Sample stations.

2.5 Biodiversity

Species richness was estimated by a Margalef richness index (*D*) (Margalef. R 1958). Species diversity by a Shannon-Wiener diversity index (*H*) (Krebs 1989). and Pielou's evenness index (*J'*) (Pielou 1966) were used for evenness.

3 RESULTS AND ANALYSIS

3.1 Species composition

We collected 1280 eggs, 5866 larvae in three times. After appraisal, there were 15 kinds altogether (7 eggs, 12 larvae), in which, 13 kinds identified to species, belonging to 13 genera, 11 families and 7 orders. 1 taxa of gobiidae fishes as only family level, 1 taxa of eggs were unidentified (Table 1). For suitable temperature, warm water species were dominant, for suitable habitat continental shelf pelagic-neritic fish and demersal fish were dominant (Table 2).

The dominant fish eggs were Dotted gizzard shad (*Konosiru spunctatus*), silver sillago (*Sillag osihama*), Joyner's tongue-sole (*Cynoglossus joyneri*), the dominant larvae were Japanese anchovy (*Engraulis japonicas*), Tapertail anchovy (*Coilia mystus*) and the gobiidae fishes. The dominant species of inchthyoplankton changed from the species of warm temperate and continental shelf pelagic-neritic fish to that of warm temperate and continental shelf demersal fish (Table 3).

F, floating eggs; A, adherent eggs; O, ovoviviparous; CD, continental shelf demersal fish; CBD, continental shelf benthopelagic fish; CPN, continental shelf pelagic-neritic fish; CRA, continental shelf reef-associated fish WT, warm temperate species; WW, warm water species; CT, cold temperate species; OEP, oceanicpelagic fish;

Table 1. The species of fish eggs and larvae in the Huanghe River estuary.

Species	Appreaing time			Egg type	Developmental stage			Eco-type	
	BW	MW	AW		Eggs	Larvae	Juvenile	Suitable temperature	Habitat type
Konosirus punctatus	+		+	F	+	+		WT	CPN
Engraulis japonicus	+	+	+	F	+	+		WT	CPN
Thryssa kammalensis			+	F		+		WW	CPN
Strongylura anastomella		+	+	A		+		WT	OEP
Liza haematocheila	+	+		F		+	+	WT	CPN
Lateolabrax japonicus	+	+		F		+	+	WT	CRA
Sillago sihama	+	+	+	F	+			WW	CRA
Eupleurogrammus muticus			+	F	+	+		WW	CBD
Trichiurus lepturus	+			F	+			WT	CBD
Synechogobius ommaturus	+			A		+		WT	CD
Gobiidae sp.	+	+	+	A		+			
Cynoglossus joyneri	+		+	F	+			WT	CD
Syngnathus acus	+			O		+		WT	CD
Hippocampus coronatus			+	O		+		CT	CD
ND		+	+	F	+				

345

Table 2. The species composition of fish eggs and larvae in the Huanghe River estuary and its adjacent area.

Eco-type	Number of species		
	BW	MW	AW
WT	8	4	4
WW	1	1	3
CT	0	0	1
CBD	1	0	0
CD	3	0	2
CPN	3	2	3
CRA	2	2	1
OEP	0	1	1

Table 3. Dominant species composition of ichthyoplankton in the Huanghe River estuary and its adjacent area.

Project	Survey time	Species	IRI
Egg	BW	*Konosirus punctatu*	1081.15
		Sillago sihama	333.58
		Engraulis japonicus	187.64
	MW	*Sillago sihama*	1427.83
	AW	*Cynoglossus joyeri*	1844.09
		Sillago sihama	427.54
Larvae	BW	*Engraulis japonicus*	2245.32
		Liza haematocheilus	249.48
	MW	*Engraulis japonicus*	3924.38
		Liza haematocheilus	260.76
	AW	Gobiidae sp.	3759.77

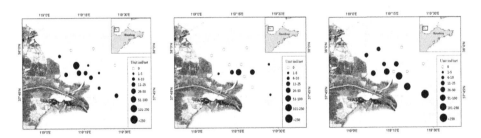

Figure 2. Horizontal distribution of eggs in the Huanghe estuary River and its adjacent area.

3.2 Quantity distribution

Before WSDR, the number of eggs collected were 270 in 10 stations, the occurrence frequencies of eggs were 76.9%, average density of the eggs was 20.8. Middle of WSDR, the number of eggs collected were 114 in 6 stations, the occurrence frequencies of eggs were 46.2%, average density of the eggs was 8.8. After WSDR, the number of eggs collected were 896 in 9 stations, the occurrence frequencies of eggs were 69.3%, average density of the eggs was 68.9. (Figs. 2 and 3).

3.3 Biodiversity

Table 4 shows that the values of S, N, H', D are low, but the value of J' is high in the middle of WSDR.

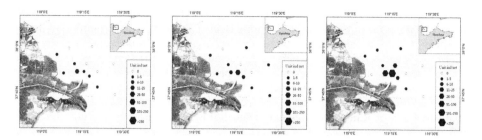

Figure 3. Horizontal distribution of larvae in the Huanghe River estuary and its adjacent area.

Table 4. Contrast of number of species, individual and diversity of ichthyoplankton in the Huanghe River estuary and its adjacent area.

Time	S	N	D	J'	H'
BW	11	307	0.578	0.361	0.697
MW	8	173	0.411	0.565	0.646
AW	10	6773	0.556	0.364	0.707

Table 5. The pearson correlation analysis of the quantities of ichthyoplankton in the Huanghe River estuary and its adjacent area.

Project	Number of eggs	Number of larvae
Number of eggs	1.000	−0.075
Number of larvae	−0.075	1.000
Salinity	0.199	−0.156
Transparency	−0.129	−0.252
Temperature	0.292	0.113
Chlorophyll	−0.097	0.870**
Dissolved oxygen	0.238	−0.125
Active phosphate	0.005	−0.068

**indicates $P < 0.01$ very significant correlation, *indicates $P < 0.05$ significant correlation, other indicates no correlation relationship.

3.4 *Ecological environment factors impact analysis*

Table 5 shows that the number of eggs have no significant correlation ($P > 0.05$) to environment factors, the number of larvae have very significant positive correlation to the chlorophyll a ($R = 0.870$, $P < 0.01$) and have no significant correlation to other environment factors ($P > 0.05$).

4 DISCUSSION

4.1 *The survey about species of eggs and larvae composition and quantity distribution diversification*

We collected 16 species of eggs and larvae, the majority of spawning broodstock was warm temperate in three species warm temperate, warm water and cold temperate; in term of species composition of eggs and larvae, we compared with other results such as 1982 (June, September), 1984 (May), 1989 (June, September) and 2007 (May, July) in this sea area, it shows that the number of species decreased, but approached 2007 (Shan dong province science and technology committee, China Science and Technology Press; Xiaodong Bian et al. 2010). Other, the same to 2007 (Xiaodong Bian et al. 2010), we also didn't find the eggs and larvae of economical species such as *Pampus argenteus, Paralichthys olivaceus, Nibea albiflora* and *Johnius belangerii*, which were spawning during the same period. There were very many dead eggs during 3 times survey, the percentage of dead eggs was 59.3%, human factors such as the main stream construction, fishing and pollution, which damaged the environment and lead to the disappear of part species, this phenomenon also appeared in interannual variability of Raide Aveire estuary (Bauchot M.L & Prasa 1987). Under the double interference of fishing and environmental conditions changing, the species of fish resource and fish community changed greatly, thus, the supplement species changed greatly too.

In term of quantity distribution, research has shown that as the runoff of Huanghe decreasing, the distribution area of eggs and larvae in Huanghe estuaryin concentrating to the estuaries Gate (Peng Wang 1999). In this study, we got the same conclusion, but the density of eggs and larvae were the highest after water and sediment discharge regulation in mid-July, the study in 1985 showed that the highest eggs density was in late June, the highest eggs density was in late July. The conclusion can't be determined whether the spawning period of eggs has changed in Huanghe River estuary, it's need next research to prove.

4.2 Biodiversity

The values of *S, N, H', D* are low, but the value of *J'* is high in the middle of WSDR. the species diversity index in community is the balanced response to specie number and individual number among the species, this shows that the specie structure of eggs and larvae was relatively unstable affected by the sediment into the sea and changing runoff in the middle of WSDR. Connel (Connel. J.H 1978) gave the intermediate disturbance theory, that was the intermediate disturbance can maintain the higher diversity, but excessive disturbance was on the contrary. Sanders (Sander. H.L 1968) also pointed out that low specie diversity was relevant to increased disturbance and the changing non-biological environmental conditions.

4.3 Dominant species

The study of Bian Xiaodong in 2007 showed that, the dominant species of eggs and larvae was *Konosirus punctatu* in May, the dominant species of eggs and larvae was *Cynoglossus joyeri* in July. The study of LiuShuang in 2009 showed that, in May, the dominant fish eggs were *Konosirus punctatu* and *snapper*; the dominant larvae were *Japanese sardinella, Larimichthys Polyactis* and *flathead mullet*; in August, the dominant fish eggs was *cymoglossus robustus*; the dominant larvae were *Pseudosciaena polyactis* and *cymoglossus robustus*. In this study, before the water and sediment discharge regulation (June), the dominant species of eggs and larvae were *Konosiruspunctatus* belonging to Clupeidae and *Engraulis japonicus* belonging to Engraulidae, after the water and sediment discharge regulation (July) were *Cynoglossus joyneri* belonging to Cynoglossidae and fish belonging to Gobiidae, the species from warm temperate, continental shelf pelagic-neritic fish changed to warm temperate, continental shelf demersal fish, and formatted the dynamic pattern that different species of eggs and larvae correspond with specific environment and seasonal changes.

ACKNOWLEDGEMENTS

The authors would like to thank everyone involved in sampling surveys and analysis, and the reviewers for their helpful comments and suggestions. This research was supported by the National Marine Public Welfare Research Project (200905019) and fund of 'Taishan scholar of aquatic nutrition and feed'.

REFERENCES

Akihiko. K & Shigeki. S. 1983. Diurnal changes in vertical distribution of Anchovy eggs and larvae in the Western Wakasa Bay. *Bull Jap Sac Sci Fish*, 50(8):1285–1292.

Bauchot, M.L & Prasa. 1987. *Guiade Los Pecesde Marde Espafiayde Eurpa.* Bracelona: Ediciones Omega.

Connel, J.H. 1978. Diversity in tropical rainforests and coralreefs. *Science*. 199:1302–1310.

Fan Li, Xiurong Zhang. 2001. Impact of variation of water and sediment fluxes on sustainable use of marine environment and resources in the Huanghe River estuary and adjacent sea I. Reason for the cut-off water flow and large decrease of runoff and developing trend of them. *Studia marina sinica,* 43:51–59.

General Administration of quality supervision, Inspection and quarantine of the People's Republic of China, Standardization administration of the Repulic of China. 2007. *Specifications for oceanographic survey-part 6: Marine biological survey.* Beijing: StandardsPress of China, 56–62.

Jiayi Tian. 2000. The diversity of phytoplankton in the sea areas nearby Huanghe delta. *Marine Environmental Science,* 19(2):38–42.

Jie Han, Zhinan Zhang, Zishan Yu. 2004. Macrobethic community structure in the southern and central Bohai sea, China. *Acta ecologica sinica,* 24(3):531–537.

Jingyao Deng, Xianshi Jin. 2000. Study on fishery biodiversity and its conservation in Laizhou Bay and Yellow River estuary. *Zoological Research,* 21(1):76–82.

Kazumasa, H., Tsuneo, G. & Mitsuyuki, H. 1997. Diet composition and prey size of larval anchovy Engraulis japonicas, in Toyama Bay, southern *Japan Sea. Aquatic Ecology*, 47:67–78.

Krebs, C.J. 1989. *Ecological Methodology.* New York: Harper Collins Publishers.

Margalef, R. 1958. Information theory in ecology. *General System,* 3:36–71.

Pielou, E.C. 1966. The use of information theory in the study of ecological succession. *Journal of Theoretical Biology*10: 370–383.

Ping Wang, Yan Jiao, Yiping Ren, Chongjun Zhong, Hao Yu. 1999. The biodiversity features survey of caught fish in spring of Laizhou Bay and Huanghe estuaryin. *Transactions of Oceanology and Limnology,* 1:40–44.

Pinkas, L, Oliphamt, M.S. & Ierson, I.L.K. 1971. Food habits of albacore, bluefin tuna, and bonito in Californian waters. ClifDep *Fish Game Fish Bull,* 152:1–105.

Ruihua Lv, Mingyuan Zhu. 1992. The primary productivity of the inshore waters in Shandong. *Journal of Ocean ography of Huanghai & Bohai seas,* 10(1):42–47.

Ruijing Wan, Gao Wei, Shan Sun, Xianyong Zhao. 2008. The spawning ecology of Engraulis japonicus in south Shandong peninsula spawning grounds I. The quantity distribution of Engraulis japonicus eggs and larvae. *Acta Zoologica Sinica,* 54(5):785–797.

Ruijing Wan, Yanwei Jiang. 1989. The survey and ecological Research of bony fish eggs and larvae in Yellow sea. *Marine Fisheries Research Institute,* 19(1):60–73.

Sander, H.L. 1968. Marine benthic diversity: a comparative estudy. *American Naturalist,* 102:243–282.

Shan dong province science and technology committee. *The comprehensive survey reports about coastal zone and tideland resources in Shandong province.* 1991. The comprehensive survey report of survey area in Huanghe estuaryin. Beijing: China Science and Technology Press.

Weinstein, M.P, Weiss, S.L, Hodson, R.G & Gerry, L.R. 1980. Retention of three taxa of postlarval fishes in an intensively flushed tidal estuary, Cape Fear river, *North Carolina. Fish Bull US,* 78:419–435.

Whitfield, A.K. 1989. Ichthyoplankton interchange in the mouth region of a southern African estuary. *Mar Ecol Prog Ser,* 54(1–2):25–33.

Xiaodong Bian, Xiumei Zhang, Tianxiang Gao, Ruijing Wan, Peidong Zhang. 2010. The species composition and quantity distribution of eggs and larvae during spring and summer 2007 in Huanghe estuaryin. *Journal of Fishery Sciences of China,* 17(4):815–826.

Xinhua Zhu, FengmiaoDong Liu, Weiwei Xian. 2001. Spationtemporal pattern and dominant component of fish community in the Yellow River estuary and its adjacent waters. *Studia Marina Sinica,* 43:141–151.

Yumu Jiao, Xinhua Zhang, Huixin Li. 1998. Influence on fish diversity in the sea area off the Huanghe River estuary by the cutoff of water supply. *Transactions of Oceanology and Limnology,* 4:48–53.

Yumu Jiao, Jiayi Tian. 1999. Zooplankton diversity around the Huanghe River Delta. *Marine environmental science,* 18(14):33–38.

Zhinan Zhang, Lihong Tu, Zishan Yu. 1990. Preliminary study on the macrobenthos in the Huanghe River estuary and its adjacent waters. *Journal of Ocean University of Qingdao,* 20(2):45–52.

Hydraulic Engineering – Xie (Ed.)
© 2013 Taylor & Francis Group, London, ISBN 978-1-138-00043-8

Research on calculation methods of SO_2 emission factor in the industrial sector in China

Ming-Hao Liu
North China Electric Power University, Hebei, China

Zhi-Quan Wu
China National Water Resources & Electric Power Materials & Equipment Co., Beijing, China
North China Electric Power University, Hebei, China

Zhong-He Han
North China Electric Power University, Hebei, China

ABSTRACT: Based on the hypothesis that the average sulfur content rate of energy used in the same area remains unchanged or change can be ignored in the short term, the SO_2 emission factor in each industrial sector and the average sulfur content rate of coal used in each province in China are calculated by the regression of the statistical data of sulfur-containing energy usage and SO_2 emission, and a correction method for calculating the SO_2 emission factor in each province is put forward. Finally, the SO_2 emission factor of each industrial sector in Beijing is calculated by the correction method and the accuracy of the SO_2 emission factor modified is verified with the energy statistical data in each sector in Beijing.

1 INTRODUCTION

SO_2 is one of the main pollutants in the atmosphere and an important index to measure atmospheric pollution level. Monitoring and controlling SO_2 emission has become an important task in the environmental protection. The SO_2 emission is mainly comes from the combustion of sulfur-containing energy and some production process in the nature. For most industrial sectors, the SO_2 emission is decided by the combustion quantity and the average sulfur content rate of the energy. And for Mining and Quarrying, Coking, Chemical industry, Smelting and Pressing of Metals, etc, there is a part of SO_2 emission comes from production process. At present, there are some scholars have calculated the SO_2 emission factor of various sulfur-containing energy and unit production through the experimental research and statistical analysis, etc[1-4]. These emission factors are mainly focused on a specific area or a kind of energy, while there are few literatures about calculating the SO_2 emission factor of the sulfur-containing energy in sector. If SO_2 emission is calculated by the same SO_2 emission factor in different sectors, there will be a certain deviation in some sector. This article analyzed the SO_2 emission factor of the sulfur-containing energy in each industrial sector with the statistical data of sulfur-containing energy usage and SO_2 emission in China, and put forward a simple correction method for calculating the SO_2 emission factor in each industrial sector in Chinese provinces with the average sulfur content rate of the energy used in each province and China.

2 SO_2 EMISSION FACTOR IN THE INDUSTRIAL SECTOR IN CHINA

2.1 *The sulfur-containing energy*

At present, the energy used mainly includes fossil fuels (coal, oil, and natural gas) and non-fossil energy (hydropower, nuclear power, wind power, solar energy and biomass energy, etc).

Among them, fossil energy and biomass energy are the mainly sulfur-containing energy. Because of biomass energy is used seldom, the SO_2 emission caused by biomass energy can be ignored. The principal ingredient of gas is methane, and it almost does not contain sulfur. Therefore, the sulfur-containing energy is mainly refers to coal and oil.

2.2 Sector division

According to the SO_2 emission composition of each industrial sector in China in 2010, it can be found that Mining and Washing of Coal, Extraction of Petroleum and Natural Gas, etc discharge more than 5% of SO_2 in the production process, and the other industrial sectors are all less than 5%. Due to Production and Distribution of Electric Power and Heat Power discharges a lot of SO_2, and the energy used in the sector is mainly power coal, so it is taken as a single sector. And other sectors which discharge less than 5% of SO_2 in the production process are taken as a sector, called the other industrial sector in this article, in which SO_2 emission is mainly from the combustion of sulfur-containing energy. Considering the SO_2 emission of Manufacture of General and Special Purpose Machinery, Manufacture of Electrical Machinery and Equipment, Recycling and Disposal of Waste, Production and Distribution of Gas, Production and Distribution of Water is low, these sectors are integrated to the other industrial sector. The industrial sectors which discharge more than 5% of SO_2 in the production process are all taken as a single sector. Therefore, according to the composition of SO_2 emission in each industrial sector, the industry is divided into 10 sectors shown in Table 1.

2.3 Regression analysis

In order to analyzing the relationship between the sulfur-containing energy usage and the SO_2 emission, the linear regression analysis is conducted by collecting the statistical data of energy usage and SO_2 emission in each industrial sector in 2004–2010, the regression results are collected into Table 1. The statistical data of energy usage is from the annual China energy statistics yearbook, refers to the coal and petroleum used in each sector. Considering the usage of petroleum is relatively less, and the average sulfur content rate of petroleum is about 1/3 of coal used in China[5]. Therefore, this article integrates the usage of petroleum to coal by timing 0.35. The statistical data of SO_2 emission is from the annual China environment statistical yearbook, refers to the total amount of finally SO_2 emission and SO_2 emission reductions in each sector, in this way, it can eliminate the influence of SO_2 mitigation rate.

Table 1. The regression results of the SO_2 emission and the sulfur-containing energy usage in each sector.

Sector	Regression coefficient	t	Sig.	R square	F	Sig.
Industrial total	0.0128	46.016	0.000	0.997	2117	0.000
Mining and washing of coal	0.0054	24.289	0.000	0.990	590	0.000
Extraction of petroleum and natural gas	0.0101	11.643	0.000	0.958	136	0.000
Mining and processing of metal ores	0.0429	21.309	0.000	0.987	454	0.000
Mining and processing of nonmetal and other ores	0.0114	18.956	0.000	0.984	359	0.000
Processing of petroleum, coking, processing of nuclear fuel	0.0061	26.357	0.000	0.991	696	0.000
Chemical industry	0.0117	27.474	0.000	0.992	755	0.000
Manufacture of non-metallic mineral products	0.0096	16.072	0.000	0.977	258	0.000
Smelting and pressing of metals	0.0174	28.315	0.000	0.993	802	0.000
Production and distribution of electric power and heat power	0.0151	25.512	0.000	0.991	651	0.000
The other industrial sector	0.0116	40.028	0.000	0.996	1602	0.000

The results in Table 1 show that the regression models can all meet the 1% significant level of F test, the models have high reliability, and R square value are close to 1, fitting goodness of models are high. Regression coefficients can all meet the 1% significant level of t test, it indicates that the regression coefficients can describe the relationship between the sulfur-containing energy usage and the SO_2 emission effectively.

The regression coefficient of the other industrial sector is 0.01165, and the standard coal coefficient is 0.7143 in China[6]. According to the material balance calculation method of SO_2 emission (formula 1), the average sulfur content rate of standard coal in the other industrial sector can be calculated, the result is 1.02% which is close to the result calculated by Li Xin, the average sulfur content rate of standard coal in China he calculated is 1.1%[5]. In the same way, the average sulfur content rate of standard coal of Production and Distribution of Electric Power and Heat Power can be calculated, the result is 1.32% which is slightly higher than the result calculated by Li Xin, the average sulfur content rate of power coal in China he calculated is 1.15%[5]. For these sectors which discharge more than 5% of SO_2 in the production process, as SO_2 emission is affected by production process strongly, so the fluctuation of regression coefficients is relatively obvious. Where, the coefficient of Mining and Washing of Coal is the least, which is 0.0054, the energy used by the sector is mainly coal, in addition to part of the energy are used in combustion, and part of the coal are lost in the mining and washing process. The coefficient of Mining and Processing of Metal Ores is the largest, which reaches 0.429. The SO_2 discharge ratio of industry production process is as high as 74.86% in the sector in 2010, the SO_2 emission is mainly determined by production process.

$$SO_2 \text{ emission} = \text{fuel coal consumption} \times \text{sulfur content rate} \times 0.8 \times 2 \times (1\text{-desulfurization rate}) \tag{1}$$

where, desulfurization rate is refers to the proportion of SO_2 emission reduction in the total SO_2 emission produced by fuel coal combustion. In this article the desulfurization rate is 0.

3 SO_2 EMISSION FACTOR IN THE INDUSTRIAL SECTOR IN PROVINCE

Since there is some difference in the average sulfur content rate of the energy used in different provinces, in order to calculating the SO_2 emission factor in each sector in provinces, the SO_2 emission factor in each industrial sector in China needs to be modified with the average sulfur content rate of the energy used in each province and China.

3.1 Regression analysis

In order to calculating the average sulfur content rate of the energy used in each province and China, the linear regression analysis is conducted by collecting the statistical data of energy usage and SO_2 emission in each province in 2004–2010. The data source and preprocessing in this section are consistent with the section 1.3.

The regression results shows that the regression models can all meet the 1% significant level of F test, the models have high reliability, and R square value are close to 1, fitting goodness of the models are high. Regression coefficients can all meet the 1% significant level of t test, it indicates that the regression coefficients can refused to zero hypothesis significantly and can describe the relationship between the sulfur-containing energy usage and the SO_2 emission effectively. According to formula (1) and the standard coal coefficient, the average sulfur content rate of the coal used in each province can be calculated, and the results are collected into Table 2 and showed in Figure 1.

From Figure 1 and Table 2, it is known that the average sulfur content rate of the coal used in northeast China is the lowest and southwest China is the highest. The average sulfur content rate of the coal used in Gansu, Jiangxi, Guangxi, and Chongqing is relatively high, Jilin, Heilongjiang, Beijing, Hebei, Tianjin and Shanghai is relatively low.

Table 2. The average sulfur content rate of the coal used in each province.

Region	Average sulfur content rate (%)	Region	Average sulfur content rate (%)	Region	Average sulfur content rate (%)
China	1.04	Zhejiang	1.20	Hainan	1.06
Beijing	0.65	Anhui	1.41	Chongqing	2.51
Tianjin	0.86	Fujian	0.88	Sichuan	1.48
Hebei	0.83	Jiangxi	2.83	Guizhou	2.02
Shanxi	1.19	Shandong	0.99	Yunnan	1.84
Neimenggu	1.07	Henan	1.21	Shanxi	1.15
Liaoning	1.08	Hubei	1.11	Gansu	2.88
Jilin	0.50	Hunan	1.27	Qinghai	1.04
Heilongjiang	0.54	Guangdong	1.27	Ningxia	1.07
Shanghai	0.89	Guangxi	2.67	Xinjiang	0.93
Jiangsu	1.08				

Figure 1. The average sulfur content rate of the coal used in each province.
The regression analysis of Tibet, Hong Kong, Macao, Taiwan is not conducted for data missing, and the white parts indicate the regions without data in Figure 1.

3.2 Correction of SO₂ emission factor

The SO_2 emission factor in each industrial sector and the average sulfur content rate of the coal used in each province have been calculated in the section 1.3 and section 2.1. To calculate the SO_2 emission factor in each industrial sector in provinces, the SO_2 emission factor in each industrial sector in China is modified in line with formula (2).

$$a_{ij} = \frac{s_i}{s_0} a_{0j} \quad (i = 1, 2, ..., 30 \quad j = 1, 2, ..., 10) \tag{2}$$

where, a_{ij} and a_{0j} indicates the SO_2 emission factor in the j industrial sector in the i province and China, s_i and s_0 indicate the average sulfur content rate of the coal used in the i province and China.

3.3 Empirical analysis

From Table 2, it is known that the average sulfur content rate of coal used in Beijing is 0.65%, and China is 1.04%. The SO_2 emission factor in each industrial sector in Beijing is calculated by modifying the SO_2 emission factor in each industrial sector in China in line with formula (2), and the results are listed on Table 3.

According to the SO_2 emission factor in Table 3 and energy usage in each industrial sector in Beijing in 2010 and 2011, the industrial SO_2 emission included the part of reduced in Beijing is calculated, and the results are 188.1 thousand tons and 186.5 thousand tons in

Table 3. The SO$_2$ emission factor in each industrial sector in Beijing.

Sector	Emission factor	Sector	Emission factor
Mining and washing of coal	0.0034	Chemical industry	0.0073
Extraction of petroleum and natural gas	0.0063	Manufacture of non-metallic mineral products	0.0060
Mining and processing of metal ores	0.0268	Smelting and pressing of metals	0.0109
Mining and processing of nonmetal and other ores	0.0071	Production and distribution of electric power and heat power	0.0094
Processing of petroleum, coking, processing of nuclear fuel	0.0038	The other industrial sector	0.0073

2010 and 2011. There are difference of 0.59% and 0.81% with the statistical data released by Beijing bureau, which are 187 thousand tons and 185 thousand tons in 2010 and 2011. The results prove that the industrial SO$_2$ emission in Beijing can be calculated accurately with the SO$_2$ emission factor modified.

3.4 Calculation of SO$_2$ emission in industrial sector

In order to eliminating the influence of SO$_2$ mitigation rate on the emission factors, the SO$_2$ emission factor in each industrial sector is calculated with the statistical data of SO$_2$ emission included the part of reduced. Therefore, the SO$_2$ mitigation rate is need to be considered when calculate the total SO$_2$ emission in industrial sector with the SO$_2$ emission factor, the specific calculation formula is as formula (3).

$$S = \sum_{i=1}^{n} e_i a_i (1 - \eta_i) \tag{3}$$

where, S indicates the total SO$_2$ emission in industrial sector, e_i indicates the sulfur-containing energy used in the i industrial sector, a_i indicates the SO$_2$ emission factor in the i industrial sector, η_i indicates the SO$_2$ mitigation rate in the i industrial sector.

4 CONCLUSION

Based on this hypothesis that the average sulfur content rate of the energy used in the area maintains unchanged or the change can be ignored, the linear regression analysis is conducted by collecting the statistical data of sulfur-containing energy usage and SO$_2$ emission, and the following conclusions are obtained.

1. The SO$_2$ emission factor in each industrial sector in China is calculated. Where, the SO$_2$ emission factor in the other industrial sector is 0.0116. And the SO$_2$ emission factor in Production and Distribution of Electric Power and Heat Power is 0.0151. And the SO$_2$ emission factors in the sectors which discharge more than 5% of SO$_2$ in the production process fluctuate relatively obvious, the largest factor is 0.429, and the least factor is 0.0054.
2. The average sulfur content rate of the coal used in each province in China is calculated. Where, the average sulfur content rate of the coal used in Gansu, Jiangxi, Guangxi, and Chongqing is relatively high, Jilin, Heilongjiang, Beijing, Hebei, Tianjin and Shanghai is relatively low.
3. A simple correction method for calculating SO$_2$ emission factor is put forward.
4. The SO$_2$ emission factor in each industrial sector in Beijing is calculated by the correction method and an empirical analysis is conducted with the statistical data in Beijing.

ACKNOWLEDGEMENT

This study was supported by the "Study on the Beijing Energy Saving and Emission Reduction Support System Based on Energy Saving Potentials" of the special funds of Beijing Co-construction.

REFERENCES

B Zhao, JZ Ma. 2008. Development of an air pollutant emission inventory for Tianjin. *Acta Scientiae Circumstantiae* 28(2): 368–375.
CESY. 2011. *China Energy Statistical Yearbook 2011*. Beijing: China Statistics Press.
China Bureau of Statistics, Ministry of Environment Protection. 2011. *China statistical yearbook 2011*. Beijing: China Statistics Press.
David G, Gregory R, Carmichael, etc. 1999. Energy Consumption and Acid Deposition in Northeast Asia. *Royal Swedish Academy of Sciences* 28(2): 135–143.
Kato N, Akimoto H. 1992. Anthropogenic emissions of SO_2 and NOx in Asia: emission inventories. *Atmospheric Environment* 26 A(16): 2997–3017.
WU Zhiquan, Zhang Shiping, MA Meiqian, etc. 2011. Energy Demand Forecast and Pollution Emission Tendency in Beijing. *Energy Technology and Economics* 23(11): 41–45.
Xin Li, Ying Duan. 2002. Two Main Problems Exist in Controlling SO_2 Pollution. *Northern Environment* (03): 18.

Hydraulic Engineering – Xie (Ed.)
© 2013 Taylor & Francis Group, London, ISBN 978-1-138-00043-8

Analysis of the anomalous torrential rain caused by NESAT and the cold mass

Anning Gao & Ruibo Zhang

Guangxi Meteorological Observatory, Nanning, Guangxi, China

ABSTRACT: By using the traditional observation datasets, analysis was made to the events of the anomalous 3-days continuing torrential rain caused by the interaction between the typhoon NESAT after weakened and the cool mass. It shows that (1) NESAT weakened to be the tropical depression after making landfall over the northern part of Vietnam. The eastward cold mass from Qing -Zang Plateau penetrated the western area of south China in the low layer, and met with the south airflow on the east side of NESAT in the southern part of Guangxi, thus the mesoscale convergence line was formed, promoting and keeping the dynamic convergence and rising flow in the southern part of China. (2) The strong southeast stream on the east side of NESAT transferred large amount of moisture to the upper of Guangxi, providing sufficient vapor for the continuing torrential rain. (3) As NESAT weakened and dissipated, the vertical wind shear grew small over the southern Guangxi on the outer circulation of NESAT, which favored for the accumulation of thermal and vapor and the converging rising, as well as developing of the convective cloud cluster and the releasing of large amount of latent heat from water condensation, resulting in the important positive feedback of the amplitude of heavy rain.

1 INTRODUCTION

The 17th typhoon, NESAT is the most powerful one that making landfall on China in 2011. It was generated in the northwest pacific at 08:00 (BJT, Beijing time) 24 September, and enhanced to be the severe typhoon at 23:00 (BJT) 26 September, moving up south China sea at 14:00 (BJT) 27 September and making landfall on the coastal area of Wengtian, Wenchang county in Hainan province. The maximum wind speed was up to 42 m/s (force 14 wind) near the centre on which it made landfall, then move up the Qiongzhou Strait, and made landfall again along the coastal area of Jiaowei of Xuwen county in Guangdong province with typhoon intensity (35 m/s, force 12 wind) at 21:15 (BJT), and weakened to be a severe tropical storm over the northern part of Beibu Gulf at 05:00 (BJT) 30 September, and made landfall over the region of Guangning in the northern part of Vietnam at around 11:30 (BJT), then weakened to be the tropical depression at 20:00 (BJT), and after that the Chinese Central Observatory halted to number it as typhoon.

Torrential rain lasted for 3 days in Gunagxi area affected by NESAT. There were 6 towns in Guangxi with the rainfall over 500 mm from 20:00 (BJT) 29 September to 20:00 (BJT) 2 October, covering the area of Shiwanshan Park in Shangsi County (713.2 mm), Fucheng in Yinhai region (578.4 mm), Shikang in Hepu county (556.7 mm), Wenli of Lingshan County (521.3 mm), Luxu of Binyang County (504.5 mm). There were 170 towns with the fainfall from 250 to 499 mm, 467 towns from 100 to 250 mm and 289 towns from 50 to 100 mm respectively, according to the rain gauges data in Guangxi province (Fig. 1). Severe loss of agriculture, traffic and personnel property occurred resulted from the heavy rain caused by NESAT. The statics shows 41 towns in 9 cities suffered disaster, around 3,231,500 disaster toll, 7 death and 1 life lost, and the direct economic loss was up to 2.688 billion RMB.

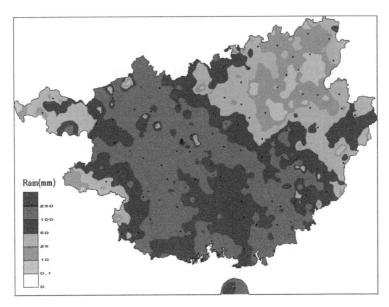

Figure 1. Distribution of rain from 20:00 (BJT) 29 Sep 2011 to 20:00 (BJT) 2 Oct 2011 by the affecting of Nesat in Guangxi.

Why the torrential rain can last for 3 days in Guangxi after NESAT made landfall and weakened to be tropical depression, with its center not moving to the inner of Guangxi? There seems to be close relation of interaction between NESAT and the cold air mass. Lots of attention upon the precipitation amplitude caused by the interaction between the northwards typhoon and mid latitude cold mass have been made by meteorologists, some of whom have done the research about the typhoon precipitation affected by cold mass. He Lifu *et al* (2009), pointed out that abnormal heavy rain results from the well dynamics and thermodynamics conditions by the convection of the enhancing northeast stream from the cold mass and the east stream from the ocean at the northern part of typhoon; Zhang Xingqiang *et al* (2005), reveal that close relations exist among the cold mass, rainfall amplitude of typhoon and increasing precipitation under the condition of promotion of rising movement, as the front arrives in the peripheral of typhoon, resulting in the releasing of baroclinic potential energy, in the aspect of the barotropic and baroclinic instability and the releasing of latent heat of condensation; Li Qingcai *et al* (1998), confirmed that there is a clear baroclinicity in the torrential rain area caused by the interaction between the western system and the typhoon, baroclinic energy might be another kind of energy to maintain the residual depression after typhoon made landfall. In different aspects, research aboved promote the acquaintance to the forming mechanics on the torrential rain amplitude by interaction between cold mass and tropical depression, and to improve the forecast level of torrential rain of typhoon. However, these achievements focus primarily on the typhoon that made landfall in the offing, or the low pressure got to the inland, few cases were studied on the interaction between the weaken tropical depression far from inland and the cold mass, that lead to the abnormal continuing torrential rain. In this paper, analysis is made on the abnormal torrential rain resulting from interaction between cold mass and the tropical depression after the 17th typhoon, NESAT made landfall on the northern part of Vietnam, to reveal the cause of formation of abnormal torrential rain caused by typhoon alike.

2 THE REAL PROCESS

The severe typhoon, NESAT continued to move northwest after it made landfall on Wenchang, Hainan at 14:30 (BJT) 29 September, then stayed in the Beibu Gulf for 14 hours.

It made landfall again on the coastal area of Guangxi in northern part of Vietnam at 11:30 (BJT) 30 September with its maximum wind up to wind 11 (30 m/s), the lowest pressure of 980 hpa near the center. It waned to be tropical depression at 20:00 (BJT) in the northern part of Vietnam. On the impact of NESAT, from 20:00 (BJT) 29 September to 20 (BJT) 30 September, torrential rain occurred in the southern part of Guangxi, within 24 hours, with the rainfall over 100 mm in 237 towns, over 250 mm in 8 towns. The maximum rainfall was 462.3 mm. What's worth mentioning is that the rain intensity didn't get weakened after NESAT waned to be tropical depression in the northern part of Vietnam. From 1 to 2 October, impacted by the remaining cloud of depression and the cold mass, long-lasting torrential rain weather happened in the south and the west of Guangxi, with the rainfall over 100 mm in 338 towns, over 250 mm in 16 towns. The maximum rainfall was 467.4 mm in Fucheng, Yinghai region, Beihai.

3   CHARACTERISTIC OF ENVIRONMENTAL FIELDS

3.1   *Background of large scale circulation for 500 hPa*

NESAT has a stable moving track, high intensity, fast moving speed and big scope of impact. Analysis for the 500 hpa large scale circulation shows that NESAT moved to the northwestward at a mean speed of 22 km/hr during 27 to 30 September, impacted by the enhancing west moving of subtropical high, and directed by the southeast flow in the south of NESAT. Since a translot moved south in Mongolia, the north branch of front position by south, suppressing the possibility of the ridge of subtropical high moving north, keeping the ridge of subtropical high stable at around 25°N, resulted in NESAT's stable track and fast moving. From 30 September to 1 October, a little trough split up from the north branch and moving eastward, guiding cold mass to impact south China vie Hetao area. Under the influence of cold advection back to the trough, ridge of tropical high drop southwards to 22°N around, and the west point of 588 line in synoptic chart was in the east path of South China, nearly paralleling to the depression center of NESAT after it made landfall and weakened in the northern part of Vietnam. Meanwhile, impacted by the southeast flow in the southwest side of tropical high, vapor from south China sea continually transport to Guangxi, providing a favor environment for the long-lasting precipitation (Figure 2).

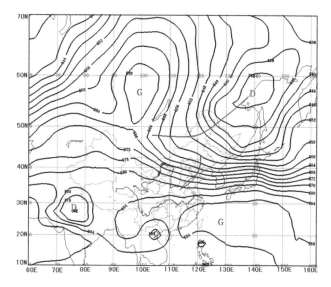

Figure 2.   Circulation at 500 hPa at 20:00 (BJT) 30 Sep 2011.

3.2   *Analysis on cold mass invasion*

Related research document (Shen W.X. *et al,* 2002) shows that rainfall amplitudes are frequently connected with cold mass invasion, and can reach maximum with easy as the front in the surrounding of typhoon. As the low pressure of the weakened NESAT was in the northern part of Vietnam, 500 hpa translot in Mongolia split up a small trough and moved eastwards to the East China. South China was on the bottom of the shallow trough, so that the cold mass that moving south crawlled slowly below 850 hpa and got gradually to the west part of South China.

At 20:00 (BJT) 28 September, in the charts of 850 hpa of variable-temperature, the cold mass was in the southern part of Sichuan and the northern of Guizhou, with a region of big value of −4—6°C, and the southern part of Guizhou and Guangxi were in the region of rising temperature. At 20:00 (BJT) 29 September, cold mass moved south and the region of minus temperature was over the southeastern part of Guizhou and the western part of South China, with the maximum value of −8°C, indicating that the cold mass in low layer intruded the typhoon circulation in diffusion. From 08:00 (BJT) to 20:00 (BJT) 30 September, region of minus variation of temperature continued to reach the west coast of South China, majority of cold mass had intruded into the typhoon circulation, as NESAT withered away quickly.

At 20:00 (BJT) 30 September, on the flow field of surface charts after NESAT died out (Figure 3a), cold mass penetrated into the western part of South China in the low layer, flow by north prevailed over northern and middle of Guangxi, joined with the flow by south in the east of NESAT, and formed a 300 km-wide, transmeridional middle-scale convergence line in around 23°N, meanwhile, on the flow field of 850hpa, a torrent by east of 12–20 m/s was formed, resulting from the invasion of cold mass and the clear enhancement of the flow by east from the south of the Yangtze River to the north of South China. That caused the enhancement and maintenance of dynamic convergence and rising movement, and resulted in the obvious enhancement of torrential rain process.

For furthrely revealing the role of cold mass on torrential rain amplitude, a cross section of temperature advection is chosen along northwest to southeast (105–112°E, 23–26°N) of torrential rain region (Figure 4a). It can be seem that, at 20:00 (BJT) 30 September, the cold advection reached Nanning around through the southwestern part of Guizhou (Anshun) via the northwestern part of Guangxi (Hechi), and extended from northwest to southeast, with the thickness up to the height of 850 hPa, and meanwhile, a very powerful warm advection was transported to Guangxi from the northwestwards. As the cold mass intruded, a cold matress was formed in the middle north of Guangxi, which enabled the warm moisture of NESAT to climb over. The structure of cold advection in low layer and warm advection in high layer enabled the atmospheric stratification to get stable, alter the mono behavior of

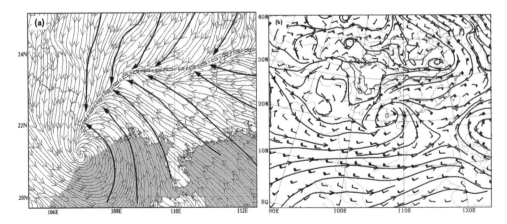

Figure 3.   Surface field (a) and upper field at 850 hPa (b) at 20:00 (BJT) 30 Sep 2011.

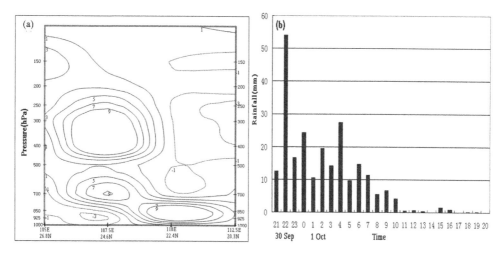

Figure 4. (a) Profile of temperature advection along 105–112°E at 20:00 (BJT) 30 Sep 2011 (b) Hourly rainfall at Binyang Station (No 59238) from 21:00 (BJT) 30 Sep 2011 to 20:00 (BJT) 1 Oct 2011.

warm precipitation in typhoon, and turn to be the mixture precipitation, or the long-lasting stratiformis precipitation, resulted in the occurrence of torrential rain. Hourly Rain distribution in Binyang, Guangxi, (Figure 4b) showed that total rainfall was up to 223 mm in the process from 21:00 (BJT) 30 September to 20:00 (BJT) 1 October, which broke the historical record of maximum rainfall in a day. Within 24 hours, there were nearly 10 hours in which the hourly rain intensity was over 10 mm, of which the maximum was 54 mm occurred at 20:00 (BJT) 30 September.

4 CHANGING CHARACTERISTIC OF VARIOUS PHYSICAL QUANTITIES

4.1 Vertical wind structure in the south of Guangxi

As NESAT went westwards over the surface of the Beibu Gulf, or in the south of 21°N, then weakened and died out after it made landfall in the northern part of Vietnam, its centre didn't reach the inland of Guangxi. The 3-day abnormal torrential rain weather in Guangxi, impacted on the circulation of NESAT, was connected with the plentiful moisture transported by the deep flow by south from South China Sea. Analysis for cross-section on the daily sounding shows that, at 20:00 (BJT) 29 September, the wind below 700 hPa over Nanning was northeastwards, then it turned to rotate; at 08:00 (BJT) 30 September, the wind at 850 hPa turned to be 30 m/s by east, and at 700–300 hPa turned accordantly to be by southeast with the maximum over 24 m/s which near 500 hPa,central value of 30 m/s; at 20:00 (BJT) 30 September, the wind over Nanning continued to rotate, at 925–250 hPa was accordantly by southeast, 16–24 m/s. The powerful jet stream transported moisture to over Guangxi, providing abundant vapor resource for the occurrence of torrential rain.

4.2 Flux and divergence of vapor

The abundant of vapor is one of the important condition for torrential rain, however, it seems to be more important for the convergence and the vertical transportation of vapor in the low layer. Large-scale sustaining torrential rain will not occur in the absence of vapor convergent and vertical transportation, even if the vapor is abundant. During the period when the peripheral circulation of NESAT influenced the west part of South China, vapor content over Guangxi was very abundant. Cross-section on vapor flux from the northwest to the southeast in the rainfall region shows that, at 20:00 (BJT) 30 September, maximum value of vapor flux occurred from the Leizhou Peninsula to the southern part of Guangxi, with the thickness of 925–700 hPa, maximum central value of 22 g/cm·hPa·s that in the southeast of Guangxi, meanwhile, cross-section on vapor flux divergence shows that, the low layer of 925–850 hPa over the whole western part of South China was the main region of vapor collection and convergence, with the flux divergence of −40—48 gcm²·hPa·s. The phenomenon like this of vapor collection and convergence lasted to 08:00 (BJT) 2 October. High intensity and rainfall in the south of Guangxi in 36 hours, and the clear amplitude of torrential rain, were related closely with the plentiful vapor transportation and convergence.

4.3 *Vertical velocity*

After NESAT waned in the northern part of Vietnam, the region of big value for vertical velocity in peripheral circulation of the depression center, was maintained in the south of Guangxi. On the cross-section for vertical velocity (Figure 5a), at 20:00 (BJT) 30 September, the whole troposphere over the western part of South China is almost in the region of rising movement, of which the maximum is located from Beihai to Nanning. Region of big value for vertical velocity of $−35 \times 10^{-3}$—55×10^{-3} hPa/s is in 850–300 hPa, of which the maximum center is near 600 hPa, with the value of $−60 \times 10^{-3}$ hPa/s, in Figure 5b, it is showed that, at 20:00 (BJT) 1 October, the whole troposphere over the western part of South China is still in the region of rising movement, though the convective height and intensity are somewhat weaken, vertical velocity over Beihai, Nanning, Hechi around, is more bigger relatively, which maintain the big value of $−20 \times 10^{-3}$—29×10^{-3} hPa/s at 850–400 hPa, maximum center at 700 hPa around, with the value of $−30 \times 10^{-3}$ hPa/s. The violent rising movement was the directive dynamics condition for the torrential rain amplitude from NESAT.

4.4 *Vertical wind shear (VWS)*

Related documents (Wang Y.D. 2009) show that, the magnitude of VWS impacts, to some extent, on the rain intensity caused by typhoon that made landfall. As the VWS is small, the rain field usually turns to concentrate with more strong rain intensity; while the VWS grows great, the rain field will be sporadic with weak rain intensity. In general, it is thought that, in tropical area, it will be favored for the happening and persisting of convective cluster

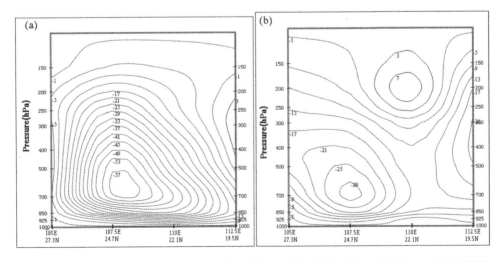

Figure 5.   Profile of vertical velocity along 105–112°E at 20:00 (BJT) 30 Sep 2011 (a) and 20:00 (BJT) 1 Oct 2011 (b).

of typhoon, as the VWS≤10 m/s. By using the sounding data over Nanning during the torrential rain period after NESAT made landfall and weakened in the northern part of Vietnam, the VWS of u200–u850 was calculated (Figure 6). It can be seen that, there is a reduced trend on the VWS from 08:00 (BJT) 29 September to 08:00 (BJT) 30 September, and it became more clear of the reduced trend after 20:00 (BJT) 30 September. In the previous 12 hours, VWS reduced from 11.6 m/s to 7 m/s, and 24 hours afterwards VWS over Nanning maintained a small value. At 20:00 (BJT) 1 October, VWS reduced to only 5.2 m/s. The small VWS and the CISK, will promote the accumulation of thermal and moisture,

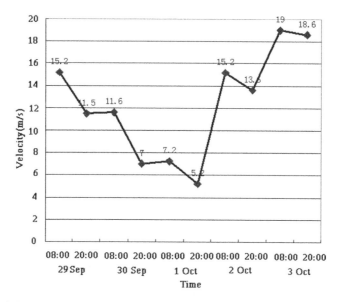

Figure 6. Evolution curse of vertical windshear between u200 and u850 over Nanning station from 30 Sep 2011 to 3 Oct 2011.

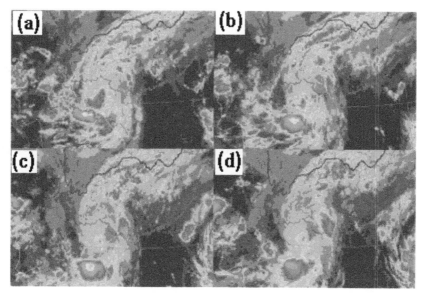

Figure 7. Infrared cloud images of FY2 at 20:00 (BJT) 30 Sep 2011 (a); 23:00 (BJT) 30 Sep 2011 (b); 02:00 (BJT) 1 Oct 2011 (c); 05:00 (BJT) 1 Oct 2011 (d).

enable the disturbing temperature and moisture to greatly exceed that of environment, make for moisture convergence and rising movement, facilitate to the initializing and developing of convective clusters, therefore, they are one of the origins that bring about the persistent and the amplitude of torrential rain. By analyzing the cloud images from satellites, it can be shown that, at 20:00 (BJT) 30 September, a mesoscale convective cluster developed (Figure 7a); 3 hrs later or at 23:00 (BJT), the mesoscale cluster developed violently, as well as its area extended quickly (Figure 7b); at 02:00 (BJT) 1 October, the mesoscale cluster developed to its mature stage in a tight structure and a clear fringe, with the characteristic of MCS (Figure 7c); at 05:00 (BJT) 1 October, the structure of mesoscale convective cluster began to loose with a fuzzy boundary (Figure 7d), indicating the mesoscale convective cluster turned to be the decline phase.

5 PRELIMINARY CONCLUSIONS

1. After NESAT weakened and died out in the north of Vietnam, cold mass penetrated into the west of South China in the low layer, with the joining in the south of Guangxi, of the flow by north of cold mass and the flow by south in the east of NESAT, thus forming a mesoscale convergence line, meanwhile, the enhancing flow by east at 850 hPa field on the north of South China, met in the south of Guangxi, with the flow by southeast in the east of NESAT, to keep and sustain the dynamic convergence and rising movement over the south of Guangxi, in favor of the amplitude of torrential rain.
2. The stable and deep flow by southeast in the east of the NESAT depression transported continually the moisture to Guangxi from South China Sea, providing plentiful water resource for the lasting torrential rain.
3. After NESAT died out, the VWS in the peripheral circulation in the south of Guangxi turned to be small, in favor of the accumulation of thermal and vapor and the rising movement, as well as the initialing and developing of the convective cluster, releasing large amount of latent heat of condensation, thus played an important role of positive feedback to the amplitude of torrential rain.

REFERENCES

He, L.F. *et al.* 2009. The role of cold mass and topography in the torrential rain process of typhoon Talim. The meteorological technology 37(4):385–391.
Li, Q.C. *et al.* 1998. A mechanism study on the sudden enhancement of torrencial rain for typhoon that making landfall. Atmospheric Science 22(2):199–206.
Shen, W.X. *et al.* 2002. A Numerical Investigation of Land Surface Water on Landfalling Hurricanes. Journal of the Atmosphere Science 59:789–802.
Wang, Y.D. 2009. Impact for the vertical wind shear on the precipitation structure of typhoon that making landfall. Papers of the 26th annual meeting of CMA.
Zhang, X.Q. *et al.* 2005. The baroclinic unstability of the torrential rain for the long distance typhoon. Nanjing Meteorological College Acta 28(1):78–85.

Hydraulic Engineering – Xie (Ed.)
© *2013 Taylor & Francis Group, London, ISBN 978-1-138-00043-8*

A new model of industrial symbiosis optimization

Lin Feng & Jing Han Di
School of Environment and Nature Resources, Renmin University of China, Beijing, China

ABSTRACT: A multi-objective fuzzy optimization model of industrial symbiosis is established. It takes the maximum of synthetic benefits as objective function, nonlinear relative membership degree as weight coefficient, and total pollutant emission controls etc. as restriction conditions.

1 INTRODUCTION

Industrial symbiosis is a sub-field of industrial ecology (Magnus K, 2008; Chertow, 2000). It concentrates on the integration of geographically proximate industries in networks of material and energy exchange (Lowe EA, 1997). As a good means to both generate economic benefits for the companies involved and reduce pollution emissions, industrial symbiosis is gaining popularity worldwide (Meneghetti and Nardin, 2012). However, despite countless attempts in the past, many symbiosis programs have been unsuccessful due to multifarious reasons (Baas and Boons, 2004). It has been pointed out that symbiosis success depends not only on industrial chain design but also on optimization of operation (Karlsson and Wolf, 2008). Recently, industrial symbiosis optimization has attracted many scholars. They discuss how to optimize symbiosis networks in aspects of life cycle assessment, energy analysis, material flow analysis, stability & complexity of symbiosis network and etc. (Boons et al., 2011; Chertow MR, 2000) However, existing researches are mainly inclined to focus on ecologic efficiency or environment impact of industrial symbiosis. Synthetic considerations about efficiency involving social, economic, environmental and resources use are few. To make up for the deficiency, a new method of industrial symbiosis optimization integrating fuzzy set theory is discussed in this paper.

Traditional operational method can solve well in behavior system with clear definition. However, it can't work efficiently in complex system with multi behavior features. But, fuzzy set theory provides an efficient route to establish a relevant model and optimization approach for such system. Adopting the fuzzy optimal selection theory among multi-objectives (Cheng, 1998), we develop a multi-objective fuzzy optimization model for industrial symbiosis network.

2 OPTIMIZATION MODEL OF INDUSTRIAL SYMBIOSIS NETWORKS

Industrial symbiosis network has obvious complexity. It pursues the maximum of synthetic benefits of economy, society, resources & environment. Meanwhile, it is restricted by multi-elements like techniques, labor & capital, management, resources condition and pollution control. Suppose there are n enterprises in a industrial symbiosis network, and each enterprise owns m indexes of assessment. Thus, the feature matrix of this sample can be described as:

$$A = \begin{bmatrix} a_{11} & a_{12} & \cdots & a_{1n} \\ a_{21} & a_{22} & \cdots & a_{2n} \\ \cdots & \cdots & \cdots & \cdots \\ a_{m1} & a_{m2} & \cdots & a_{mn} \end{bmatrix} \qquad (1)$$

In formula (1), a_{ij} is the characteristic value of index i in sample j ($i = 1,2, \ldots m; j = 1,2, \ldots n$)

There are two types of enterprise evaluation indexes—quantitative and none-quantitative indexes. Quantitative indexes can be sorted into profitability one and consumption one. Profitability index is the higher, the better. Consumption index is the lower, the better (Li, 2003). As a dimensionless method, we use the following equations $e_{ij} = a_{ij}/\max a_{ij}$ and $e_{ij} = \min a_{ij}/a_{ij}$ to evaluate the relative membership degree between profitability and consumption indexes. To none-quantitative indexes, we use relative pair-wise comparison method to evaluate their relative membership degree (Chan, 2001). Then, we can get the relative membership matrix composed by these indexes above:

$$
E = \begin{bmatrix} e_{11} & e_{12} & \cdots & e_{1n} \\ e_{21} & e_{22} & \cdots & e_{2n} \\ \cdots & \cdots & \cdots & \cdots \\ e_{m1} & e_{m2} & \cdots & e_{mn} \end{bmatrix} = (e_{ij}) \tag{2}
$$

The relative membership degree of m indexes in sample j is:

$$
e_j = \left(e_{1j}, e_{2j}, \ldots e_{mj} \right)^T \tag{3}
$$

In general, different indexes have different weights because these m indexes in the samples influence the optimization to different degrees. Although the membership degrees of optimal objection and importance differ on the concept aspect, these two membership degrees are closely connected. Usually, if the membership degree of optimal objection is greater, we will set it a heavier weight. Define the membership by the concept of weights according to fuzzy concentration and transpose the matrix E to get the matrix of relative membership degree, whose target is to show the importance to the fuzzy concept:

$$
F = E' = (f_{ji})_{n \times m} \tag{4}
$$

Then, the relative membership degree vector about n schemes to object i is

$$
f_i = \left(f_{1i}, f_{2i}, \ldots f_{ni} \right)^T \tag{5}
$$

To illustrate the differences between target i and ideally important & unimportant targets, we set relative membership $f(i)$ and $1 - f(i)$ as weighing factors and then get the generalized weighted distance. Establish an objective function to minimize the sum of weighted distance squares, then we get the relative membership degree of targets to importance:

$$
f(i) = 1 / \left\{ 1 + \left[\sum_{j=1}^{n} \left(1 - f_{ji}\right)^{\alpha} / \sum_{j=1}^{n} \left(f_{ji}\right)^{\alpha} \right]^{2/\alpha} \right\} \tag{6}
$$

Applying the two-stage model of fuzzy optimization, we get the relative membership degree w_j of enterprise j with maximum synthetic benefits:

$$
w_j = \frac{1}{1 + \left\{ \sum_{i=1}^{m} \left[f_i \left(e_{ij} - 1\right) \right]^{\alpha} / \sum_{i=1}^{m} \left(f_i e_{ij}\right)^{\alpha} \right\}^{2/\alpha}} \tag{7}
$$

The multi-objective fuzzy optimization model is established as follows. The object is maximizing the synthetic benefits of the industrial symbiosis, and the corresponding restrictions are resources consumptions and pollution controls.

$$\max Z = \max \sum_{j=1}^{n} w_j y_j$$

$$S.T. \begin{cases} \sum_{j=1}^{n} y_i S_j \leq Q \\ \sum_{j=1}^{n} y_i P_w^j \leq O_w \\ \sum_{j=1}^{n} y_j P_g^j \leq O_g \end{cases} \quad S.T. \begin{cases} \sum_{j=1}^{n} y_j P_s^j \leq O_s \\ Y_c \geq y_j \geq 0, (j=1, \dots n) \\ etc. \end{cases} \tag{8}$$

In formula (8), w_j is the relative optimal membership degree calculated within the fuzzy optimal model when the synthetic benefits of firm j are maximized. y_j is the optimized output of firm j. n is the total number of the firms in this model. S_j is the industrial water quota of unit production of firm j. Q is the total amount of fresh water supply. P_w^j is the waste water discharge coefficient of firm j. O_w is the total quantity control of waste water discharge. P_g^j is the air pollutant emission coefficient of firm j. O_g is the total quantity control of air pollutant emission. P_s^j is the solid waste coefficient of firm j. O_s is the total quantity control of solid waste. Y_c is the maximum capacity of the firms in the planned years. $etc.$ refers to other factors which should be considered according to the actual situation.

3 CASE STUDY

Shihezi city is located in the Manas river basin of China. The industrial symbiosis networks are shown in Figure 1. For more concentrated thinking, we only select the well developed chemical industrial chain network above, taking PVC projects (1.2 million ton per year) among the network as a case to probe into the application of multi-objective fuzzy optimization model.

3.1 *Determine the relative membership degree of synthetic optimal benefits*

There are five firms—calcium carbide plant (Plant A), power plant (Plant B), chemical plant (Plant C), plastic plant (Plant D) and cement plant (Plant E)- in this network. And the considered quantitative feature values of each crop are as Table 1.

According to the quantitative assessment methods to qualitative indexes, we suggest to determine social benefits of each firm, using the experts' experience to judge importance

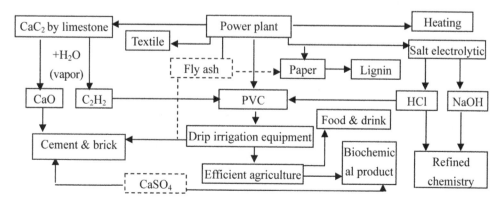

Figure 1.   Industrial symbiosis networks of Shihezi city in China.

Table 1. Considered quantitative feature values of each crop.

Index	Plant A	Plant B	Plant C	Plant D	Plant E
Economic benefits	11.62	10.88	7.86	7.63	30.22
Environmental benefits	279	250	231	265	221
Benefits of resources	84	55	101	121	178

Note: ①Economic benefits mean the average profits of each firm within recent 3 years × 100. ②Environmental benefits mean the sum of the disposal rate of solid waste and the removal rates of water and air pollutant × 100 during the latest 3 years of each firm. ③Benefits of resources mean the sum of consumption declining rates of fresh water, electricity and material per product × 100 during the latest 3 years of each firm. ④Data above are from cleaner production report of each industry (year 2010).

degree by pair-wise comparison. The main criterions for judgment are labor number per unit of output value, influence to local living standards, contributions to local industry development and importance in the whole industrial chain etc. Thus, combining the experts' options, the coherence priority matrix in pair-wise comparative importance for social benefit indexes of these five firms is acquired. Sort the sum of each row of the matrix in the descending order, then, we'll get the qualitative importance order of the factors set: power plant, plastic plant, calcium carbide plant, cement plant and chemical plant. Judging from the sorted pair-wise comparison and referring to the membership degrees between particle operator and ration mark, we get the membership degrees of none-quantitative indexes:

$$e_5 = (0.368, 1.0, 0.245, 0.636, 0.275)^T \qquad (9)$$

Thereupon, the relative membership matrix of index featured values of these five firms in the industrial symbiotic network is obtained. Applying the two-stage fuzzy optimization model ($\alpha = 2$), we get the relative membership degree w_j of firm j with max synthetic benefits:

$$w_5 = (0.7987, \quad 0.8198, \quad 0.7201, \quad 0.8308, \quad 0.8502) \qquad (10)$$

3.2 Definition of parameters in constraint equation

We only consider resources & environment constraint, because the economic and technological supports in this project are relatively strong. Determine the relative parameters in constraint equation (Table 2) on the basis of environment impact assessment reports and spot monitoring data by local bureau of environmental protection. Among these, solid waste is not listed due to zero discharge by utilization and disposition in the symbiosis network.

3.3 Result and analysis

Use software LINDO to solve this linear programming model. The optimization result for this network in 2012 is shown in Table 3.

1. The result illustrates that the optical outputs of each firm approach or equal to the maximal outputs in the planned years. On the one hand, it reflects that the symbiotic network is high efficient. On the other hand, it indicates that the expectant result of industrial symbiosis will be achieved only when each firm strictly takes the measures of the cleaner production and the utilization of three wastes in the environmental impact assessment report.
2. The calculated optimal output of Plant A is a little lower than the designed one, which shows that there is space of water saving, electricity saving and emission reduction

Table 2. Relative parameters in constraint equation.

Parameters	Plant A	Plant B	Plant C	Plant D	Plant E	Constraints	Quantities
Hg mg/t	–	–	–	0.17	–	Dusk	578t
COD kg/t	0.103	–	0.203	0.160	–	COD	700t
Dust kg/t	0.005	–	–	0.009	0.088	SO_2	29 433.6t
Water quota m^3/t	48.00	38.89	6.50	10.26	0.40	Water	99 744 000 m^3
Smoke kg/t	0.43	34.14	0.06	–	0.166	Smoke	5500t
SO_2 kg/t	0.140	682.18	–	–	0.058	Hg	55 kg
Max capacity t	2×10^6	1400	1.1×10^6	1.2×10^6	4.6×10^6		

Note: ①Units of the parameters of plant B are all MW and that of max capacity is MW. ②Fresh water quota of each firm is adopted by the data from "industrial water quota of Xinjiang Region". Fresh water quota of plant C is calculated by NaOH, and plant D is by PVC. ③Fresh water quota of plant A is unveiled estimated by 65% of that in Hebei Province.

Table 3. Optimization result for industrial symbiosis network, 2012.

	Plant A	Plant B	Plant C	Plant D	Plant E
Optimal outputs t	1 803 246	1400 MW	1.1×10^6	1.2×10^6	4.6×10^6
Synthetic benefits			6 978 320		

in this firm. Lean production management should be added into cleaner production mechanism of the symbiosis network. Then, by analyzing and auditing all the links like material, energy, operation and program, workmanship or operations with some squander will be improved, and the consumption & emission will be accordingly reduced.

3. If some changes happen to the industrial chain design, resources condition, output productivity and goal of pollutant control, the parameters in this model will also change, and the optimal result will be altered. If the designed output productivity and the optimal value of some unit (firm) greatly deviate, which means this unit contributes a little in this symbiotic network. Then, not only the rates of resources utilization and pollutant disposal in this unit should be enhanced, but also the design of whole industrial chain should be improved to increase complexity and stability of the symbiotic system, raising the overall efficiency of the network.

4 CONCLUSION

A fuzzy optimization model of industrial symbiosis networks is established in this paper, in which, maximum of synthetic benefits in economy, society, resources and environment is the object. Nonlinear relative membership degrees are taken as weighted coefficients. Resources limits, total emission controls of pollutants and so on are restriction conditions. It expressed the complexity of symbiosis system scientifically, and incarnated the connotations well that industrial symbiosis pursued win-win between economy and environment. So, it is theoretically more reasonable and practically useful in industrial symbiosis optimization. However, some deeper susceptibility analysis of the influence by the market variation, technique improvement and etc. are required later. We hope for further research with other scholars together.

ACKNOWLEDGMENTS

We gratefully acknowledge the financial support from Key Technologies R&D Program of China (2012BAC06B04), National High-Tech R&D Program of China (2012AA062607).

REFERENCES

Baas LW, Boons FA. 2004. An industrial ecology project in practice: exploring the boundaries of decision making levels in regional industrial systems. *Journal of Cleaner Production*. 12(8–10):1073–1085.

Boons F, Spekkink W, et al. 2011. The dynamics of industrial symbiosis: a proposal for a conceptual framework based upon a comprehensive literature review. *Journal of Cleaner Production*. 19(9–10):905–911.

Chan PT, Rad AB, Wang J. 2001. Indirect adaptive fuzzy sliding mode control. Part II: Parameter project and supervisory control. *Fuzzy Sets and Systems*. 119(1):156–161.

Cheng SY. 1998. *Engineering fuzzy set theory and application*. Beijing: National Defense Industry Press: 59–79.

Chertow MR. 2000. Industrial symbiosis: literature and taxonomy. *Annual Review of Energy and Environment*. 25:313–337.

Li L, Li YM. 2003. The design and stability analysis of adaptive system based on linear T–S fuzzy system. *Acta Automatica Sinica*. 29(6):1024–1026.

Lowe EA. 1997. Creating by-product resource exchanges: strategies for eco-industrial parks. *Journal of Cleaner Production*. 5(1–2):57–65.

Meneghetti A, Nardin G. 2012. Enabling industrial symbiosis by a facilities management optimization approach. *Journal of Cleaner Production*. 35:263–273.

Karlsson M, Wolf A. 2008. Using an optimization model to evaluate the economic benefits of industrial symbiosis in the forest industry. *Journal of Cleaner Production*. 16:1536–1544.

Author index

Printed and bound by CPI Group (UK) Ltd, Croydon, CR0 4YY

18/10/2024

01776251-0002